Arithmetic, Geometry, Cryptography and Coding Theory

CONTEMPORARY
MATHEMATICS

574

Arithmetic, Geometry, Cryptography and Coding Theory

13th Conference
Arithmetic, Geometry,
Cryptography and Coding Theory
CIRM, Marseille, France
March 14–18, 2011

Geocrypt 2011
Bastia, France
June 19–24, 2011

Yves Aubry
Christophe Ritzenthaler
Alexey Zykin
Editors

American Mathematical Society
Providence, Rhode Island

2000 *Mathematics Subject Classification.* Primary 11G10, 11G20, 11M38, 11R42, 11T06, 11T71, 14G10, 14G15, 14G50, 14Q05.

Library of Congress Cataloging-in-Publication Data

International Conference "Arithmetic, Geometry, Cryptography and Coding Theory" (13th : 2011 : Marseille, France)

Arithmetic, geometry, cryptography and coding theory : 13th Conference on Arithmetic, Geometry, Cryptography and Coding Theory, March 14-18, 2011, CIRM, Marseille, France : Geocrypt 2011, June 19-24, 2011, Bastia, France / Yves Aubry, Christophe Ritzenthaler, Alexey Zykin, editors.

p. cm. — (Contemporary Mathematics ; v. 574)

Includes bibliographical references.

ISBN 978-0-8218-7572-8 (alk. paper)

1. Abelian varieties–Congresses. 2. Dimension theory (Algebra)–Congresses. I. Aubry, Yves, 1965– II. Ritzenthaler, Christophe, 1976– III. Zykin, Alexey, 1984– IV. Geocrypt Conference (2011 : Bastia, France) V. Title.

QA564.I576 2011
510′–dc23 2012012152

∞ The paper used in this book is acid-free and falls within the guidelines
established to ensure permanence and durability.
Visit the AMS home page at `http://www.ams.org/`

10 9 8 7 6 5 4 3 2 1   17 16 15 14 13 12

Contents

Preface

The 13-th AGC^{2}T conference (Arithmetic, Geometry, Cryptography and Coding Theory) took place at CIRM (Centre International de Rencontres Mathématiques) in Marseille, France, on March 14–18, 2011. This international conference has been a major event in the area of arithmetic geometry and its applications for more than 25 years and more than 80 participants attended it this year. We thank all of them for creating a stimulating research environment. The topics of the talks extended from algebraic number theory to diophantine geometry, curves and abelian varieties over finite fields and applications to codes, boolean functions or cryptography. We especially thank the speakers Bruno Anglès, Stéphane Ballet, Burcu Baran, Régis Blache, Ivan Boyer, Nils Bruin, Alain Couvreur, Frédéric Edoukou, Arnaldo Garcia, Sudhir Ghorpade, Safia Haloui, Marc Hindry, Masaaki Homma, Everett Howe, Kamal Khuri-Makdisi, David Kohel, Aristides Kontogeorgis, Philippe Langevin, Gregor Leander, Elodie Leducq, Petr Lisonek, Kit-Ho Mak, Alina Ostafe, Ferruh Ozbudak, François Rodier, Karl Rökaeus, Cécilia Salgado, Jean-Pierre Serre, Vijaykumar Singh, Benjamin Smith, Henning Stichtenoth, Bianca Viray, Gabor Wiese and Yuri Zarhin for their lectures.

For the second time, the AGC^{2}T conference was twinned with Geocrypt (conference on Geometry and Cryptography) which took place in la Marana near Bastia, Corsica, France, on June 19–24, 2011. This conference focuses on algebraic geometry issues raised by cryptography. We would like to thank Christophe Arene, Xavier Caruso, Luca de Feo, Pierre Dèbes, Oumar Diao, Everett Howe, David Kohel, Reynald Lercier, Pascal Molin, Damien Robert, David Roe, Éric Schost, Benjamin Smith, Peter Stevenhagen, Marco Streng, Emmanuel Thomé and Osmanbey Uzunkol for their talks. We also thank the ANR CHIC (Hyperelliptic Curves Isogenies, Point Counting) for its financial support.

The editors would like to thank the staff of CIRM (Olivia Barbarroux, Muriel Milton and Laure Stefanini) and that of the Institut de Mathématiques de Luminy (Aurélia Lozingot and Corinne Roux) for their remarkable professionalism.

Finally, special thanks to the Saint Patrick's day which gave us an opportunity of a memorable party.

Contemporary Mathematics
Volume **574**, 2012
http://dx.doi.org/10.1090/conm/574/11429

Construction of a $\mathbb{k}$-complete addition law on Jacobians of hyperelliptic curves of genus two

Christophe Arène and Romain Cosset

ABSTRACT. In this paper we explain how to construct $\mathbb{F}_q$-complete addition laws on the Jacobian of an hyperelliptic curve of genus 2 classically embedded in $\mathbb{P}^{15}$, and provide the first explicit example of such addition laws in the genus 2 case. This is a generalization to abelian surfaces of the arithmetic completeness of addition laws on elliptic models firstly revealed with Edwards curves in 2007.

1. Introduction

Cryptographic protocols using abelian varieties, specifically elliptic curves and abelian surfaces, have been studied intensively for at least the past two decades. They are based on the discrete logarithm problem and pairings for which the computation of the addition of two points is central. In particular, one pays attention to two aspects. Obviously, the number of operations needed to compute the equations must be as small as possible. It appears that their domain of definition has also to be taken into account. Indeed, with the development of embedded cryptosystems, the theoretical resistance of the discrete logarithm problem is no longer sufficient to ensure the protocol security, we also have to deal with physical attacks. For instance the implementation of the usual formulæ on the Weierstraß model of an elliptic curve is vulnerable against side-channel attacks due to the use of different formulæ for a generic addition or a doubling. We refer to [**LM05**] for a possible alternative on genus 2 curve cryptosystems.

In this paper we only consider this second problem. Lange and Ruppert [**LR85**] first considered this problem geometrically. They worked on the existence of complete sets of addition laws, *i.e.* for all P, Q in $A\left(\overline{\mathbb{k}}\right)$ there is an addition law defined at (P,Q). See the definition of completeness in Definition 1.4 . Then Bosma and Lenstra [**BL95**] proved for an elliptic curve given in Weierstraß form that such a set has always cardinality greater than one, a fact generalized in [**AKR11**] to any abelian variety with a projective embedding. This motivates to restrict the additon laws to the points defined on a non algebraically closed base field, specifically a finite field. An addition law is said to be $\mathbb{k}$-complete if it is defined over $(A\times A)(\mathbb{k})$.

2010 *Mathematics Subject Classification.* Primary 68-04; Secondary 11-04.

Key words and phrases. Theta functions, Jacobian, genus 2 curve, addition law, completeness, embedding, line bundle, finite field.

The authors acknowledge the financial support by grant ANR-09-BLAN-0020-01 from the French ANR and the AXA Research Fund for the Ph.D. grant of the first author.

Explicit examples of $\Bbbk$-complete addition laws in genus 1 included Edwards curves [**Edw07, BL07**], twisted Hessian curves [**BKL09, FJ10**], even special Weierstraß models [**AKR11**]. See also [**Koh11**] for a large study of the structure of the space of addition laws on elliptic curves and the completeness of addition laws acted on by a torsion subgroup. But no explicit addition laws were known for higher genera. Our aim is to find an explicit $\Bbbk$-complete addition law on the Jacobian of genus 2 hyperelliptic curves.

In the first section we introduce the theta coordinates on the Jacobian of an hyperelliptic curve and explain the link with the classical Mumford coordinates. We then sketch the theory of addition laws. In section 2 we explain how to construct in practice an $\mathbb{F}_q$-complete addition law.

1.1. Theta functions of level 4, link with genus 2 curves. In this subsection, we are interested in arithmetic aspects of the Jacobian of a genus 2 curve. We work over $\mathbb{C}$ to simplify the introduction and the use of theta functions. But the results remain true over a finite field of characteristic different from two, as seen in Remark 1.3. For the classical theory of theta functions, the reader is referred to [**Mum83, Mum84**]. Let Ω be an element of the Siegel half-space:

$$\left\{\Omega \in \mathrm{Mat}_{2\times 2}(\mathbb{C}),\ \ {}^t\Omega = \Omega,\ \Im(\Omega) > 0\right\},$$

the classical Riemann theta function is defined by

$$\vartheta\,(z,\Omega) = \sum_{n\in\mathbb{Z}^2} \exp\left(i\pi\, {}^t n\Omega n + 2i\pi\, {}^t n\, z\right).$$

For all elements a et b of $\mathbb{Q}^2$, the theta function with characteristics a, b is defined by

$$\begin{aligned}(1)\qquad \vartheta\left[\begin{smallmatrix} a \\ b \end{smallmatrix}\right](z,\Omega) &= \exp\left(i\pi\, {}^t a\Omega a + 2i\pi\, {}^t a\ (z+b)\right)\vartheta\,(z+\Omega a + b, \Omega)\\ &= \sum_{n\in\mathbb{Z}^2} \exp\left(i\pi\, {}^t(n+a)\Omega(n+a) + 2i\pi\, {}^t(n+a)\,(z+b)\right).\end{aligned}$$

The characteristics are considered modulo $\mathbb{Z}^2$ since for all α, β in $\mathbb{Z}^2$ we have

$$\vartheta\left[\begin{smallmatrix} a+\alpha \\ b+\beta \end{smallmatrix}\right](z,\Omega) = \exp\left(2i\pi\, {}^t a\beta\right)\vartheta\left[\begin{smallmatrix} a \\ b \end{smallmatrix}\right](z,\Omega).$$

We will consider theta functions of level 4 which means that the characteristics live in $\frac{1}{2}\mathbb{Z}^2/\mathbb{Z}^2$.

A classical result of Lefschetz states that the theta functions of level 4 give an embedding of $\mathbb{C}^2/\Omega\mathbb{Z}^2 + \mathbb{Z}^2$ into $\mathbb{P}^{15}(\mathbb{C})$. For a proof see [**Mum70**, p. 29]. We denote by Θ the *theta divisor* defined by the image of the map from $\mathcal{C}$ to $\mathrm{Jac}(\mathcal{C})$ which sends a point $P \in \mathcal{C}$ to $(P) - (\infty)$. There exists an element $\mathcal{K}$ of $\mathrm{Jac}(\mathcal{C})$, called the *Riemann's constant*, such that the divisor induced on $\mathrm{Jac}(\mathcal{C})$ by the function ϑ is $\Theta_{\mathcal{K}}$, the translation of Θ by $\mathcal{K}$. See for example [**BL04**, Theorem 11.2.4]. The Formula (1) shows that the divisors associated to the theta functions of level 4 are the translates of $\Theta_{\mathcal{K}}$ by a 2-torsion point.

For the sake of readability we use the following notations:

Notation 1.1. [**Gau07**, Section 7.1] We index the sixteen theta functions of level 4 as follow:

$$\vartheta_1(z)=\vartheta\begin{bmatrix}{}^t(0\ 0)\\ {}^t(0\ 0)\end{bmatrix}(4z,\Omega),\qquad \vartheta_2(z)=\vartheta\begin{bmatrix}{}^t(0\ 0)\\ {}^t(\frac12\ \frac12)\end{bmatrix}(4z,\Omega),$$

$$\vartheta_3(z)=\vartheta\begin{bmatrix}{}^t(0\ 0)\\ {}^t(\frac12\ 0)\end{bmatrix}(4z,\Omega),\qquad \vartheta_4(z)=\vartheta\begin{bmatrix}{}^t(0\ 0)\\ {}^t(0\ \frac12)\end{bmatrix}(4z,\Omega),$$

$$\vartheta_5(z)=\vartheta\begin{bmatrix}{}^t(\frac12\ 0)\\ {}^t(0\ 0)\end{bmatrix}(4z,\Omega),\qquad \vartheta_6(z)=\vartheta\begin{bmatrix}{}^t(\frac12\ 0)\\ {}^t(0\ \frac12)\end{bmatrix}(4z,\Omega),$$

$$\vartheta_7(z)=\vartheta\begin{bmatrix}{}^t(0\ \frac12)\\ {}^t(0\ 0)\end{bmatrix}(4z,\Omega),\qquad \vartheta_8(z)=\vartheta\begin{bmatrix}{}^t(\frac12\ \frac12)\\ {}^t(0\ 0)\end{bmatrix}(4z,\Omega),$$

$$\vartheta_9(z)=\vartheta\begin{bmatrix}{}^t(0\ \frac12)\\ {}^t(\frac12\ 0)\end{bmatrix}(4z,\Omega),\qquad \vartheta_{10}(z)=\vartheta\begin{bmatrix}{}^t(\frac12\ \frac12)\\ {}^t(\frac12\ \frac12)\end{bmatrix}(4z,\Omega),$$

$$\vartheta_{11}(z)=\vartheta\begin{bmatrix}{}^t(0\ \frac12)\\ {}^t(0\ \frac12)\end{bmatrix}(4z,\Omega),\qquad \vartheta_{12}(z)=\vartheta\begin{bmatrix}{}^t(0\ \frac12)\\ {}^t(\frac12\ \frac12)\end{bmatrix}(4z,\Omega),$$

$$\vartheta_{13}(z)=\vartheta\begin{bmatrix}{}^t(\frac12\ 0)\\ {}^t(\frac12\ 0)\end{bmatrix}(4z,\Omega),\qquad \vartheta_{14}(z)=\vartheta\begin{bmatrix}{}^t(\frac12\ \frac12)\\ {}^t(\frac12\ 0)\end{bmatrix}(4z,\Omega),$$

$$\vartheta_{15}(z)=\vartheta\begin{bmatrix}{}^t(\frac12\ 0)\\ {}^t(\frac12\ \frac12)\end{bmatrix}(4z,\Omega),\qquad \vartheta_{16}(z)=\vartheta\begin{bmatrix}{}^t(\frac12\ \frac12)\\ {}^t(0\ \frac12)\end{bmatrix}(4z,\Omega).$$

Remark that the first ten theta functions are the even ones and the last six are the odd ones. For simplicity, we drop the Ω. The evaluation at 0 of these functions are called theta constants. We write them ϑ_i instead of $\vartheta_i(0)$.

Consider an hyperelliptic curve $\mathcal{C}$ of genus 2. Associated to this curve is its period matrix Ω which is an element of the Siegel half-space. The Abel-Jacobi map is an analytic isomorphism between $\mathrm{Jac}(\mathcal{C})$ and $\mathbb{C}^2/\Omega\mathbb{Z}^2+\mathbb{Z}^2$.

The Thomae formulæ, see [**Tho70**] or [**Mum84**, III.8], link the 4th power of the theta constants with the parameters of the curve. Up to isomorphisms, we can recover the theta constants by taking well chosen roots [**CR11**]. Assume that the curve is in Rosenhain form:

$$\mathcal{C}:\quad y^2=f(x)=x(x-1)(x-\lambda)(x-\mu)(x-\nu),$$

then the ordering $\{0,1,\lambda,\mu,\nu\}$ leads to the following relations:

$$\left(\frac{\vartheta_5}{\vartheta_1}\right)^4=\frac{\mu}{\lambda\nu},\qquad \left(\frac{\vartheta_7}{\vartheta_1}\right)^4=\frac{\mu(\nu-1)(\lambda-\mu)}{\nu(\mu-1)(\lambda-\nu)},$$

$$\left(\frac{\vartheta_3}{\vartheta_1}\right)^4=\frac{\mu(\nu-1)(\lambda-1)}{\lambda\nu(\mu-1)},\qquad \left(\frac{\vartheta_4}{\vartheta_1}\right)^4=\frac{\mu(\lambda-1)(\nu-\mu)}{\lambda(\mu-1)(\nu-\lambda)}.$$

We can take a square root of the preceeding quotients in an arbitrary way. The other squares of theta constants of level 4 are given by the formulæ:

$$\vartheta_6^2=\frac{1}{\nu}\frac{\vartheta_1^2\vartheta_4^2}{\vartheta_5^2},\qquad \vartheta_8^2=\frac{1}{\lambda}\frac{\vartheta_1^2\vartheta_7^2}{\vartheta_5^2},$$

$$\vartheta_2^2=(\nu-1)\frac{\vartheta_5^2\vartheta_6^2}{\vartheta_3^2},\qquad \vartheta_9^2=(\lambda-1)\frac{\vartheta_5^2\vartheta_8^2}{\vartheta_3^2},$$

$$\vartheta_{10}^2=\frac{\vartheta_1^2\vartheta_2^2-\vartheta_3^2\vartheta_4^2}{\vartheta_8^2},$$

where arbitrary square roots can be taken. Note that we may have to take a field extension to take these roots.

We need to have an explicit algebraic morphism between $\mathrm{Jac}(\mathcal{C})$ and the image in $\mathbb{P}^{15}(\mathbb{C})$ of the embedding by the theta functions of level 4. These formulæ can be found in [**CR11**] for the genus 2 case and in [**Cos11**] for the general case. Let $\{a_1, \ldots, a_5\}$ be the ordered roots of f and let

$$\eta_1 := {}^t\left[\tfrac{1}{2}, 0; 0, 0\right], \qquad \eta_2 := {}^t\left[\tfrac{1}{2}, 0; \tfrac{1}{2}, 0\right], \qquad \eta_3 := {}^t\left[0, \tfrac{1}{2}; \tfrac{1}{2}, 0\right],$$
$$\eta_4 := {}^t\left[0, \tfrac{1}{2}; \tfrac{1}{2}, \tfrac{1}{2}\right], \qquad \eta_5 := {}^t\left[0, 0; \tfrac{1}{2}, \tfrac{1}{2}\right], \qquad \eta_\infty := {}^t[0, 0; 0, 0].$$

For a subset S in $\{1, \ldots, 5, \infty\}$, we set

$$\eta_S = \sum_{i \in S} \eta_i,$$

and we define η'_S and η''_S to be the first and second part of η_S. This notation comes from the fact that if we denote by ∞ the point at infinity of $\mathcal{C}$ and by A_i the Weierstraß point with affine coordinate equal to $(a_i, 0)$ for $i = 1, \ldots, 5$ and $A_\infty = \infty$, then the divisor $\sum_{i \in S}(A_i) - \#S(\infty)$ is mapped to $\Omega\eta'_S + \eta''_S$ by the Abel-Jacobi map. In particular, if we denote by $\mathcal{U}$ the set $\{1, 3, 5\}$, then $\mathcal{K}$ is the point $\Omega\eta'_{\mathcal{U}} + \eta''_{\mathcal{U}}$ and Θ is the divisor of $\vartheta[\eta_{\mathcal{U}}] = \vartheta_{16}$ [**Cos11**].

Let $\circ$ denote the symmetric difference of two sets. All theta functions of level 4 can be written as $\vartheta\left[\eta_{\mathcal{U} \circ V}\right]$ and a subset V of $\{1, \ldots, 5\}$ of odd cardinality. For each such subset, van Wamelen [**vW98**] defines the function $t_V(z) := f_V \vartheta\left[\eta_{\mathcal{U} \circ V}\right](z)$, where f_V is a constant which is $f_V = \vartheta[0] / \vartheta\left[\eta_{\mathcal{U} \circ V}\right]$ for the even functions (i.e. $\#V = 3$) and which is, for the others,

$$f_1 = \frac{-1}{\sqrt{a_2 - a_1}} \frac{\vartheta_1\vartheta_5\vartheta_6\vartheta_8}{\vartheta_2\vartheta_3\vartheta_9\vartheta_{10}}, \qquad f_2 = \frac{-1}{\sqrt{a_2 - a_1}} \frac{\vartheta_5\vartheta_6\vartheta_8}{\vartheta_4\vartheta_7\vartheta_{10}},$$
$$f_3 = \frac{-1}{\sqrt{a_2 - a_1}} \frac{\vartheta_1\vartheta_6}{\vartheta_2\vartheta_4}, \qquad f_4 = \frac{1}{\sqrt{a_2 - a_1}} \frac{\vartheta_5}{\vartheta_3},$$
$$f_5 = \frac{-1}{\sqrt{a_2 - a_1}} \frac{\vartheta_1\vartheta_8}{\vartheta_7\vartheta_9}, \qquad f_\emptyset = f_{\{1,2,3,4,5\}} = \frac{-1}{\sqrt{a_2 - a_1}^3} \frac{\vartheta_5^2\vartheta_6^2\vartheta_8^2}{\vartheta_2\vartheta_3\vartheta_4\vartheta_7\vartheta_9\vartheta_{10}}.$$

The following theorem is a summary of results from van Wamelen.

THEOREM 1.2. *Let $D = (P_1) + (P_2) - 2(\infty)$ be divisor, not on the support of the divisor of a theta function, which corresponds to a vector $z \in \mathbb{C}^2/(\Omega\mathbb{Z}^2 + \mathbb{Z}^2)$. Let (x_i, y_i) be the coordinates of the point P_i, $i = 1, 2$. Write (u, v) for the Mumford's polynomials of D. For $k \in \{1, \ldots, 5\}$, and l, m two distinct elements of $\{1, \ldots, 5\} \setminus \{k\}$ we have*

$$u(a_k) = \frac{t_k^2(z)}{t_\emptyset^2(z)}, \qquad v(a_k) = \frac{Y_{k,m} - Y_{k,l}}{a_l - a_m},$$

$$Y_{l,m} := \frac{y_1(x_2 - a_l)(x_2 - a_m) - y_2(x_1 - a_l)(x_1 - a_m)}{x_2 - x_1} = c_{1,2} \frac{t_l(z) t_m(z) t_{\{l,m\}}(z)}{t_\emptyset^3(z)},$$

$$Y := y_1 y_2 = \prod_{l=1}^{5} \frac{t_l(z)}{t_\emptyset(z)},$$

where $c_{1,2}$ is just a sign ± 1 which can be chosen arbitrarily.

By evaluating u at the roots of f, we obtain formulæ for computing all the $\vartheta_i(z)^2/\vartheta_{16}(z)^2$ with $1 \le i \le 16$. To get the theta functions of level 4, we will use the doubling formulæ [**Gau07**]:

$$4\vartheta\left[\begin{smallmatrix}a\\b\end{smallmatrix}\right](2z)\,\vartheta\left[\begin{smallmatrix}a\\b\end{smallmatrix}\right]\vartheta\left[\begin{smallmatrix}0\\0\end{smallmatrix}\right]^2 = \sum_{\alpha,\beta\in\frac{1}{2}\mathbb{Z}^2/\mathbb{Z}^2} \exp\left(-4i\pi{}^t a\beta\right)\vartheta\left[\begin{smallmatrix}a+\alpha\\b+\beta\end{smallmatrix}\right](z)^2\,\vartheta\left[\begin{smallmatrix}\alpha\\\beta\end{smallmatrix}\right](z)^2,$$

$$4\vartheta\left[\begin{smallmatrix}a\\b\end{smallmatrix}\right](2z)\,\vartheta\left[\begin{smallmatrix}a\\0\end{smallmatrix}\right]\vartheta\left[\begin{smallmatrix}0\\b\end{smallmatrix}\right]\vartheta\left[\begin{smallmatrix}0\\0\end{smallmatrix}\right] = \sum_{\alpha,\beta\in\frac{1}{2}\mathbb{Z}^2/\mathbb{Z}^2} \exp\left(-4i\pi{}^t a\beta\right)\vartheta\left[\begin{smallmatrix}a+\alpha\\b+\beta\end{smallmatrix}\right](z)\,\vartheta\left[\begin{smallmatrix}a+\alpha\\\beta\end{smallmatrix}\right](z)\,\vartheta\left[\begin{smallmatrix}\alpha\\b+\beta\end{smallmatrix}\right](z)\,\vartheta\left[\begin{smallmatrix}\alpha\\\beta\end{smallmatrix}\right](z).$$

The first formula allows to recover the even theta functions. For the odd theta functions, we will use the second formula. The products on the right side can be expressed in terms of the constants f_V and the functions $Y_{l,m}$, Y and $u(a_i)$. Since we need to divide by some $u(a_i)$, we make the hypothesis that the divisor is not of 2-torsion. For instance, the second formula gives

$$\begin{aligned}
\vartheta_{16}(2z)\vartheta_1\vartheta_4\vartheta_8 &= \vartheta_1(z)\vartheta_4(z)\vartheta_8(z)\vartheta_{16}(z) - \vartheta_9(z)\vartheta_{12}(z)\vartheta_{13}(z)\vartheta_{15}(z)\\
&\quad + \vartheta_5(z)\vartheta_6(z)\vartheta_7(z)\vartheta_{11}(z) - \vartheta_2(z)\vartheta_3(z)\vartheta_{10}(z)\vartheta_{14}(z),\\
\vartheta_{16}(2z)\vartheta_1\vartheta_4\vartheta_8 &= \frac{t_{2,4}(z)t_{2,3}(z)t_{3,4}(z)t_\emptyset(z)}{f_{2,4}f_{2,3}f_{3,4}f_\emptyset} + \frac{t_{1,5}(z)t_2(z)t_4(z)t_3(z)}{f_{1,5}f_2f_4f_3}\\
&\quad + \frac{t_{3,5}(z)t_{4,5}(z)t_{2,5}(z)t_1(z)}{f_{3,5}f_{4,5}f_{2,5}f_1} + \frac{t_{1,3}(z)t_{1,4}(z)t_{1,2}(z)t_5(z)}{f_{1,3}f_{1,4}f_{1,2}f_5},\\
\frac{\vartheta_{16}(2z)\vartheta_1\vartheta_4\vartheta_8}{t_\emptyset^4(z)} &= \frac{Y_{2,4}Y_{2,3}Y_{3,4}}{u(a_2)u(a_3)u(a_4)}\frac{1}{f_{2,4}f_{2,3}f_{3,4}f_\emptyset} + \frac{Y_{1,5}Y}{u(a_1)u(a_5)}\frac{1}{f_{1,5}f_2f_3f_4}\\
&\quad + \frac{Y_{2,5}Y_{3,5}Y_{4,5}Y}{u(a_2)u(a_3)u(a_4)u(a_5)^2}\frac{1}{f_{2,5}f_{3,5}f_{4,5}f_1}\\
&\quad + \frac{Y_{1,2}Y_{1,3}Y_{1,4}Y}{u(a_1)^2u(a_2)u(a_3)u(a_4)}\frac{1}{f_{1,2}f_{1,3}f_{1,4}f_5}.
\end{aligned}$$

Remark 1.3. Although we have defined our theta function over $\mathbb{C}$, our results apply to other fields of characteristic different from 2. To prove this over a finite field, whiwh is the relevant case in cryptography, we can use Lefschetz's principle and reduction to prove all the results for ordinary varieties. In general we can always use the algebraic theory of theta functions [**Mum66, Mum67a, Mum67b**].

1.2. Addition laws. Now, we focus on the notion of addition law. Let $\Bbbk$ be a field and $A/\Bbbk$ be an abelian variety of dimension g. We assume that an embedding of A in some projective space $\mathbb{P}^r$ is fixed and given by a very ample line bundle $\mathcal{L} = \mathcal{L}(D)$ for D an effective divisor. We denote by $\iota : A \hookrightarrow \mathbb{P}^r$ the corresponding morphism and also assume in the sequel that this embedding is projectively normal. Recall that by definition this means that for every $n \ge 1$ the restriction map $H^0(\mathbb{P}^r, \mathcal{O}_{\mathbb{P}^r}(n)) \to H^0(A, \mathcal{L}^n)$ is surjective. This is the case in the classical settings where $\mathcal{L} = \mathcal{L}_0^{n_0}$ with $\mathcal{L}_0$ an ample line bundle and $n_0 \ge 3$ [**BL04**, p. 187].

Let I_1 (resp. I_2) be the homogeneous ideal in $\Bbbk[X_0, \dots, X_r]$ (resp. in $\Bbbk[Y_0, \dots, Y_r]$) defined by A. The *group morphism*

$$\mu : A \times A \to A, (x, y) \mapsto x + y$$

can be locally described by bihomogenous polynomials. More precisely, an *addition law* $\mathfrak{p}$ of bidegree (m, n) on $\iota(A) \subset \mathbb{P}^r$ is the data of $r+1$ polynomials

$$p_0, \ldots, p_r \in \mathbb{k}[X_0, \ldots, X_r]/I_1 \otimes \mathbb{k}[Y_0, \ldots, Y_r]/I_2,$$

not all zero, bihomogeneous of degree m in $X_0, \ldots, X_r$ and degree n in $Y_0, \ldots, Y_r$ such that we have

$$\iota \circ \mu\,(x, y) = \Big(p_0\big(\iota(x), \iota(y)\big) : \ldots : p_r\big(\iota(x), \iota(y)\big)\Big)$$

for all points $(x, y) \in (A \times A)(\overline{\mathbb{k}})$ where these polynomials are not all zero. The set of points where an addition law is not defined is called its *exceptional subset.* We also define the *zero addition law* which is independent of the bidegree and given by zero polynomials. It is denoted by 0 and its exceptional subset is $(A \times A)(\overline{\mathbb{k}})$. We use in Section 2 the structure of $\mathbb{k}$-vector space of the set of addition laws having fixed bidegree together with 0, as shown below in Propostion 1.5.

In this paper we are interested in the construction of a single addition law which describes the group morphism μ on $(A \times A)(\mathbb{k})$ where $\mathbb{k}$ is a finite field of characteristic different from two and $A/\mathbb{k}$ the Jacobian of a genus 2 curve embedded in $\mathbb{P}^{15}$.

DEFINITION 1.4. A set S of addition laws is said to be $\mathbb{k}$-*complete* if for any point $(x, y) \in (A \times A)(\mathbb{k})$ there is an addition law in S defined on an open subset containing (x, y). The set S is said to be *complete* if the previous property is true over $\overline{\mathbb{k}}$. If $S = \{\mathfrak{p}\}$ is a singleton, we say the addition law $\mathfrak{p}$ is $\mathbb{k}$-*complete*, or *complete* when $\mathbb{k} = \overline{\mathbb{k}}$.

The following proposition interprets the addition laws of bidegree $(2, 2)$, said *biquadratic*, as global sections of a certain line bundle $\mathcal{M}$.

PROPOSITION 1.5. [**LR85**, Lemma 2.1] *Let $\pi_1, \pi_2 : A \times A \to A$ be the projection maps on the first and second factor. There is a biquadratic addition law, respectively a complete set of biquadratic addition laws, on A with respect to the embedding in $\mathbb{P}^r$ determined by the line bundle $\mathcal{L}$ if, and only if,*

$$H^0(A \times A, \mathcal{M}) \neq 0,$$

respectively the complete linear system $|\mathcal{M}|$ is base-point free, where

$$\mathcal{M} := \mu^*\mathcal{L}^{-1} \otimes \pi_1^*\mathcal{L}^2 \otimes \pi_2^*\mathcal{L}^2.$$

More precisely, following [**AKR11**, Section 2], there is the bijective map

$$\text{(2)} \qquad \left\{ \begin{array}{c} \text{nonzero biquadratic addition laws} \\ \text{on } \operatorname{Jac}(\mathcal{C}) \text{ with respect to } \iota \end{array} \right\} \Big/ \mathbb{k}^* \begin{array}{ccc} \leftrightarrow & |\mathcal{M}| \\ \overline{\mathfrak{p}} \quad \mapsto & \operatorname{div}(w), \end{array}$$

where w is a nonzero global section of $\mathcal{M}$ such that $\mathfrak{p}$ is defined outside $\operatorname{div}(w)$. The next lemma gives a nice description of this line bundle $\mathcal{M}$. We recall that a line bundle $\mathcal{L} = \mathcal{L}(D)$, defined by a divisor D, is said *symmetric* if $\mathcal{L} \cong [-1]^*\mathcal{L}$, or equivalently $D \sim [-1]^*D$.

LEMMA 1.6. [**LR85**, Propositions 2.2 and 2.3] *Let $\mathcal{L}$ be an ample line bundle on A and $\delta : A \times A \to A$ be the difference map $(x, y) \mapsto x - y$.*

1) if $\mathcal{L}$ is not symmetric then $H^0(A \times A, \mathcal{M}) = 0$.

2) if $\mathcal{L}$ is symmetric, then $\mathcal{M} \cong \delta^*\mathcal{L}$. Moreover $\mathcal{M}$ is base-point free and $h^0(A \times A, \mathcal{M}) = h^0(A, \mathcal{L})$.

This Lemma tells that, if $\mathcal{L}(D)$ is symmetric, then it is necessary and sufficient to find a divisor in $|\mathcal{L}(D)|$ without $\Bbbk$-rational points to get a $\Bbbk$-complete biquadratic addition law. We end this section with the statement of the existence of such an addition law.

PROPOSITION 1.7. [**AKR11**, Statement and Proof of Theorem 4.8] *Let $\Bbbk = \mathbb{F}_q, q \geq 7$, be a finite field and $\mathcal{C}/\mathbb{F}_q$ be a genus 2 curve. There exists an $\mathbb{F}_q$-complete biquadratic addition law on the embedding of* Jac$(\mathcal{C})$ *in $\mathbb{P}^{15}$ by 4Θ.*

Moreover, it is proved in the reference of the previous proposition that the exceptional subset of the addition law is explicitely determined.

2. Construction

2.1. A basis of biquadratic addition laws on $Jac(\mathcal{C}) \hookrightarrow P^{15}$. Riemann's addition formulæ are widely known and common in the literature. We use the general formulæ given by Baily [**Bai62**] and apply it to obtain the following formulæ for theta function of level 4.

PROPOSITION 2.1. [**Bai62**, Section 2.2, Formulæ (9)] *Let $a_k, b_l \in \frac{1}{2}\mathbb{Z}^2/\mathbb{Z}^2$, $k, l = 1, \ldots, 4$. Assume we have*

$$-a_1 + a_2 + a_3 + a_4 = 2a, \qquad -b_1 + b_2 + b_3 + b_4 = 2b,$$

with a and b in $\frac{1}{2}\mathbb{Z}^2/\mathbb{Z}^2$ then for all z_1, z_2 in $\mathbb{C}^2$ we have

$$4\,\vartheta\begin{bmatrix} a_1 \\ b_1 \end{bmatrix}(z_1 + z_2)\,\vartheta\begin{bmatrix} a_2 \\ b_2 \end{bmatrix}(z_1 - z_2)\,\vartheta\begin{bmatrix} a_3 \\ b_3 \end{bmatrix}(0)\,\vartheta\begin{bmatrix} a_4 \\ b_4 \end{bmatrix}(0) =$$

$$\sum_{\alpha,\beta \in \frac{1}{2}\mathbb{Z}^2/\mathbb{Z}^2} \vartheta\begin{bmatrix} a_1+a+\alpha \\ b_1+b+\beta \end{bmatrix}(z_2)\,\vartheta\begin{bmatrix} a_2-a+\alpha \\ b_2-b+\beta \end{bmatrix}(z_2)\,\vartheta\begin{bmatrix} a_3-a+\alpha \\ b_3-b+\beta \end{bmatrix}(z_1)\,\vartheta\begin{bmatrix} a_4-a+\alpha \\ b_4-b+\beta \end{bmatrix}(z_1).$$

For all a_1, a_2, b_1, b_2 in $\frac{1}{2}\mathbb{Z}^2/\mathbb{Z}^2$, there exists a_3, a_4, b_3, b_4 in $\frac{1}{2}\mathbb{Z}^2/\mathbb{Z}^2$ verifying the condition of the proposition and such that the product of theta constants $\vartheta\begin{bmatrix} a_3 \\ b_3 \end{bmatrix}(0)\,\vartheta\begin{bmatrix} a_4 \\ b_4 \end{bmatrix}(0)$ is non zero. We now go back to Notation 1.1.

Remark 2.2. The embedding $\mathbf{Jac}(\mathcal{C}) \hookrightarrow \mathbb{P}^{15}$ is given by the line bundle $\mathcal{L} = \mathcal{L}(4\Theta)$. From now on we consider the functions ϑ_i as global sections of this line bundle. We also use the same notation for points on $\mathbf{Jac}(\mathcal{C})$ and their image by ι to make the expressions more readable. Recall that we also denote the theta constants $\vartheta_i(0)$ as ϑ_i, but the distinction is clear in the context where they are used.

Remark 2.3. The formulæ above express for $i, j = 1, \ldots, 16$ the product $\vartheta_i(z_1 + z_2)\vartheta_j(z_1 - z_2)$ as a biquadratic bihomogeneous polynomial in the level 4 theta functions $(\vartheta_1(z_1), \ldots, \vartheta_{16}(z_1))$ and $(\vartheta_1(z_2), \ldots, \vartheta_{16}(z_2))$. Note also that they are defined over the field of definition of the theta constants, denoted $\mathbb{K}$. As we want the morphism δ to be $\Bbbk$-rational we assume that the theta constants are $\Bbbk$-rational. This is not a theoretical problem because, if they were not $\Bbbk$-rational, any $\mathbb{K}$-complete addition law will be $\Bbbk$-complete.

Next, fixing the index j, if z_1, z_2 are such that $\vartheta_j(z_1 - z_2) \neq 0$, there exists biquadratic bihomogeneous polynomials $p_{i,j}$ such that

$$\begin{aligned}(3)\qquad \vartheta_i(z_1+z_2)\,\vartheta_j(z_1-z_2) &= \\ &p_{i,j}\big((\vartheta_1(z_1),\ldots,\vartheta_{16}(z_1)),(\vartheta_1(z_2),\ldots,\vartheta_{16}(z_2))\big).\end{aligned}$$

This allows us to construct an addition law $\mathfrak{p}_j = (p_{1,j}, \ldots, p_{16,j})$ defined outside the exceptional subset $\delta^*(\vartheta_j)_0$, where $(\vartheta_j)_0$ is the divisor of the theta function ϑ_j. Note that this divisor can be described by the points whose j-th coordinate in $\mathbb{P}^{15}$ is zero. Indeed, let

$$x_k = \big(\vartheta_1(z_k) : \cdots : \vartheta_{16}(z_k)\big) \in \mathbf{Jac}(\mathcal{C}), \quad k = 1, 2,$$

be two points such that $x_1 - x_2 \notin (\vartheta_j)_0$, *i.e.* satisfying $\vartheta_j(z_1 - z_2) \neq 0$. We have

$$\begin{aligned}\iota\circ\mu\,(x_1,x_2) &= \big(\vartheta_1(z_1+z_2) : \cdots : \vartheta_{16}(z_1+z_2)\big)\\ &= \big(\vartheta_j(z_1-z_2)\vartheta_1(z_1+z_2) : \cdots : \vartheta_j(z_1-z_2)\vartheta_{16}(z_1+z_2)\big)\\ &= \mathfrak{p}_j(x_1,x_2).\end{aligned}$$

Notation 2.4. For $j = 1, \ldots, 16$, we denote $\mathfrak{p}_j$ the addition law on $\mathbf{Jac}(\mathcal{C})$ whose exceptional subset is $\delta^*(\vartheta_j)_0$ presented above.

Clearly the set of addition laws $\{\mathfrak{p}_1, \ldots, \mathfrak{p}_{16}\}$ is complete because the functions ϑ_i define an embedding of $\mathbf{Jac}(\mathcal{C})$ so have no common zero. But in fact only three of them are sufficient to get a complete set.

THEOREM 2.5. *There exists a complete set of biquadratic addition laws of cardinality three on* $\mathbf{Jac}(\mathcal{C}) \hookrightarrow \mathbb{P}^{15}$.

PROOF. We pick two addition laws $\mathfrak{p}_i, \mathfrak{p}_j$ in $\{\mathfrak{p}_{11}, \ldots, \mathfrak{p}_{16}\}$, then the divisor of the theta functions they are associated with are translates of Θ by some $(A_i) - (\infty)$ and $(A_j) - (\infty)$ with $i \neq j \in \{1, \ldots, 5, \infty\}$. Their intersection is reduced to the two 2-torsion points $\{0, (A_i) - (A_j)\}$ [**Mum75**, p. 275]. But each image in $\mathbb{P}^{15}$ by ι of a 2-torsion point of $\mathbf{Jac}(\mathcal{C})$ has six zero coordinates, so there exists a theta function ϑ_k whose divisor does not contain these two points. Finally the set $\{\mathfrak{p}_i, \mathfrak{p}_j, \mathfrak{p}_k\}$ is complete. □

Moreover we have the following proposition:

PROPOSITION 2.6. *Let* $\mathcal{C}$ *be a genus* 2 *curve. The set* $\{\mathfrak{p}_1, \ldots, \mathfrak{p}_{16}\}$ *is a basis of the space of biquadratic addition laws on* $\mathbf{Jac}(\mathcal{C}) \hookrightarrow \mathbb{P}^{15}$.

PROOF. We have $\dim_{\mathbb{k}}\big(\mathcal{L}(4\Theta)\big) = 16$, so by Lemma 1.6 case 2) we only need to show that the family is free. Let assume there exists a linear relation

$$(4)\qquad \sum \lambda_j \mathfrak{p}_j = 0.$$

Let denote by O_J the neutral element of $\mathbf{Jac}(\mathcal{C})$. Then for all points $x = \big(\vartheta_1(z) : \cdots : \vartheta_{16}(z)\big) \in \mathbf{Jac}(\mathcal{C})$, the relation

$$\sum \lambda_j \mathfrak{p}_j(x, O_J) = 0$$

in $\mathbb{P}^{15}$ gives $\sum \lambda_j p_{i,j}(x, O_J) = 0$ for all $i = 1, \ldots, 16$. Moreover there exists a k_0 such that $\vartheta_{k_0}(z) \neq 0$, so

$$0 = \sum \lambda_j p_{k_0,j}(x, O_J) = \sum \lambda_j \vartheta_{k_0}(z+0)\vartheta_j(z-0) = \vartheta_{k_0}(z) \sum \lambda_j \vartheta_j(z).$$

The dependance in k_0 being eliminated we finally get

$$\sum \lambda_j \vartheta_j = 0$$

which gives a contradiction because the family $\{\vartheta_j,\ j = 1, \ldots, 16\}$ is a basis for the theta functions of level 4. Hence the assumption of the existence of the relation (4) is not true, and $\{\mathfrak{p}_1, \ldots, \mathfrak{p}_{16}\}$ is a free family. □

2.2. Idea of the construction. From now on and without mention of the contrary we assume that $\Bbbk = \mathbb{F}_q$, where q is greater than 7 and is odd. In the previous subsection we have built a basis for the space of addition laws we are interested in. Now we want to construct the addition law announced in Proposition 1.7. We denote it by $\mathfrak{p}$. According to Lemma 1.6 case 2) its exceptional subset is of the form $\delta^* D$ with $D \in \mathbf{Div}_{\Bbbk}\left(\mathbf{Jac}(\mathcal{C})\right)$ and $D \sim 4\Theta$. Let us construct this divisor. The assumption $q \geq 7$ implies the existence of a degree 4 closed point of the form $\left\{P_0, P_0^\sigma, \overline{P_0}, \overline{P_0}^\sigma\right\}$, with σ the Frobenius over $\mathbb{F}_q$. Let define $\alpha_0 := (P_0) + (P_0^\sigma) - 2(\infty)$ and $\alpha_{l+1} := \alpha_l^\sigma$ for $l = 0, 1, 2$. Then the divisor $D = \sum \Theta_{\alpha_i}$ has the desired properties to induce an $\mathbb{F}_q$-complete biquadratic addition law, where Θ_{α_l}, $l = 0, \ldots, 3$, is the translation by α_l of the theta divisor Θ on $\mathbf{Jac}(\mathcal{C})$. In the sequel, D is taken of this form.

Our aim is to find a projective solution $(\lambda_1 : \cdots : \lambda_{16})$ for the relation

$$\mathfrak{p} = \sum \lambda_j \mathfrak{p}_j \tag{5}$$

using an interpolation method. To get this, let $x \in D$. As we want $\mathfrak{p}$ not to be defined on $\delta^* D$, in particular at (x, O_J), we search for solutions of the linear system

$$0 = \sum \lambda_j \mathfrak{p}_j(x, O_J).$$

Varying x in D, we want to get a linear system of rank 15. Indeed, the solution is projective because the exceptional subset of an addition law defines it up to scalar multiplications. Note that although x is a non rational point, we solve the system over the base field $\Bbbk$.

The following proposition allows to avoid the computation of the last six coefficients, corresponding to the odd theta functions, and then to reduce significantly the running time. Recall, following Notation 2.4, that the addition laws $\mathfrak{p}_j$ have $\delta^*(\vartheta_j)_0$ as exceptional subset.

THEOREM 2.7. *Assume $\Bbbk = \mathbb{F}_q$, with $q \geq 7$ and $(2, q) = 1$. Let $\mathfrak{p}$ be the addition law introduced above and $\mathfrak{p} = \sum \lambda_j \mathfrak{p}_j$ the desired linear relation. We have $\lambda_{11} = \cdots = \lambda_{16} = 0$.*

PROOF. Remark that by construction for all $x \in D$ we have $-x \in D$. We have

$$\forall x \in D, \qquad \mathfrak{p}(x, O_J) = \mathfrak{p}(O_J, x) = 0. \tag{6}$$

Using the parity of the theta functions ϑ_j, the second equality becomes

$$\mathfrak{p}(O_J, x) = \sum \lambda_j \mathfrak{p}_j(O_J, x) = \sum_{j=1}^{10} \lambda_j \mathfrak{p}_j(x, O_J) - \sum_{j=11}^{16} \lambda_j \mathfrak{p}_j(x, O_J).$$

We use it in the formulæ (6) and are led to consider the next two equations

$$\forall x \in D, \qquad \sum_{j=1}^{10} \lambda_j \mathfrak{p}_j(x, O_J) = 0, \text{ and } \sum_{j=11}^{16} \lambda_j \mathfrak{p}_j(x, O_J) = 0. \tag{7}$$

Let us define the two biquadratic addition laws appearing here

$$\mathfrak{q}_1 := \sum_{j=1}^{10} \lambda_j \mathfrak{p}_j, \qquad \mathfrak{q}_2 := \sum_{j=11}^{16} \lambda_j \mathfrak{p}_j.$$

We want to prove that $\mathfrak{q}_2$ is zero. Let D_1, D_2 be two effective divisors on $\mathbf{Jac}(\mathcal{C})$ such that $\delta^* D_1, \delta^* D_2$ are the respective exceptional subsets of $\mathfrak{q}_1$ and $\mathfrak{q}_2$. By Formula (2), they verify for $k = 1, 2$ either $D_k \sim 4\Theta$ or $\mathfrak{q}_k = 0$. The formulæ (7) imply $D \leq D_k$, hence, for both k, either $D_k = D$ or $\mathfrak{q}_k = 0$. But we point out that $\mathfrak{q}_2(O_J, O_J) = 0$ because the theta constants involved in (3) are zero. As (O_J, O_J) is a $\Bbbk$-rational point of $\mathbf{Jac}(\mathcal{C}) \times \mathbf{Jac}(\mathcal{C})$, it does not belong to D. Hence $D < D_2$, and the second addition law $\mathfrak{q}_2$ is zero. A fortiori $\lambda_{11} = \cdots = \lambda_{16} = 0$. □

Remark 2.8. We do not get more information on the coefficients λ_j using that $\mathfrak{p}(-x, O_J) = \mathfrak{p}(O_J, -x) = 0$ for $x \in D$.

Points on the divisor $D = \sum \Theta_{\alpha_i}$ are easily generated in Mumford coordinates: they are of the form $x = (P) - (\infty) + \alpha_i$ where $P \in \mathcal{C}(\overline{\Bbbk})$. First, note that by construction, we know the Mumford coordinates (u_i, v_i) of α_i. Then, take a random x_P in $\overline{\Bbbk}$, find a y_P such that $y_P^2 = f(x_P)$. Hence $(x - x_P, y_P)$ are the Mumford coordinates of $(P) - (\infty)$ and it remains to use Cantor algorithm to find the Mumford coordinates of x. This cannot be done easily using only theta coordinates. Now that we have the Mumford coordinates of x, we use the morphisms to compute its theta coordinates.

2.3. Numerical results. AVISOGENIES is a MAGMA package for working with genus 2 curves, and more generally with abelian varieties, using theta functions[1]. Using some already implemented functions, we wrote code to compute the coefficients λ_i given an hyperelliptic curves. This code is now part of the AVIsogenies package.

Example 2.9. Consider the curve

$$\mathcal{C}: \quad y^2 = f(x) = x^5 + 5782x^4 + 2517x^3 + 2312x^2 + 9402x$$

defined over $\Bbbk = \mathbb{F}_{10007}$. We chose this example to have rational theta constants, which are

$$\begin{array}{llll} \vartheta_1 = 1, & \vartheta_2 = 5242, & \vartheta_3 = 7727, & \vartheta_4 = 678, \\ \vartheta_5 = 3926, & \vartheta_6 = 7092, & \vartheta_7 = 5628, & \vartheta_8 = 7556, \\ \vartheta_9 = 3666, & \vartheta_{10} = 904. & & \end{array}$$

Let

$$K = \mathbb{F}_{10007}[X]/X^2 + 1 \simeq \mathbb{F}_{10007^2}$$

and $x_0 = 8310 + 2164\sqrt{-1}$. The point $P_0 = \left(x_0, \sqrt{f(x_0)}\right)$ is a point of the curve $\mathcal{C}(\mathbb{F}_{10007^4})$ which doesn't belong to $\mathcal{C}(\mathbb{F}_{10007^2})$. We define the divisor D

[1] It can be found at `http://avisogenies.gforge.inria.fr/`.

on $\mathbf{Jac}(\mathcal{C})$ as in the preceeding Subsection and search for the addition law $\mathfrak{p}$. The corresponding nonzero λ_i are given by

$$\begin{array}{llll} \lambda_1 = 1, & \lambda_2 = 6924, & \lambda_3 = 1940, & \lambda_4 = 9380, \\ \lambda_5 = 5155, & \lambda_6 = 1278, & \lambda_7 = 7239, & \lambda_8 = 1761, \\ \lambda_9 = 6859, & \lambda_{10} = 5891. & & \end{array}$$

This compution took less than a minute. It is possible to check that the addition law is $\mathbb{k}$-complete by an exhaustive computation. Note that it is enough to check $\mathfrak{p}(D, O_J) = D$ for all divisor D of $\mathbf{Jac}(\mathcal{C})(\mathbb{k})$. This verification took almost a week.

Concerning the efficiency of these addition laws, it is clearly not to their advantage when we look at $\mathfrak{p}_1$ expressed below in Example A.1. One verifies in this Appendix that the total cost to compute the desired addition law $\mathfrak{p}$ is

$$736\mathbf{m} + 32\mathbf{s} + 160\mathbf{m}_\vartheta,$$

where the number of operations in the base field $\mathbb{k}$ are denoted by $\mathbf{m}$ for a multiplication, $\mathbf{s}$ for a squaring and $\mathbf{m}_\vartheta$ for a multiplication by a coefficient, that only depends on the theta constants, which can be precomputed.

In comparison, the classical representation of points in $\mathbf{Jac}(\mathcal{C})$ as elements of the divisor class group of $\mathcal{C}$ and the use of Mumford's representation and Cantor's algorithm provides extremly cheaper costs, *e.g.* $47\mathbf{m} + 4\mathbf{s}$ for a general addition in even characteristic [**Lan05**]. There also exist pseudo-addition laws on the Kummer surface of the variety that can be computed much faster [**Duq04, Gau07**].

Remark 2.10. The relative simplicity of the addition laws $\mathfrak{p}_i$ may be related to the fact that the set $\{\mathfrak{p}_i\}$ is G_2-complete with G the subgroup of 2-torsion of $\mathbf{Jac}(\mathcal{C})$ in the sense defined by Kohel [**Koh11**, Section 7] but none of them is $\mathbb{k}$-complete. An open question is whether more efficient addition laws, considering an action of a torsion subgroup on them, can be constructed ?

Appendix A. Operation count

Let $x = (X_1 : \cdots : X_{16})$ and $y = (Y_1 : \cdots : Y_{16})$ be two points on $\mathbf{Jac}(\mathcal{C})$. We point out that the addition laws $\mathfrak{p}_i$ with $1 \le i \le 10$, have a particular form. Ideed, for $i \ne j$, each polynomial $p_{i,j}$ is the product of a constant and a polynomial with coefficients in $\{\pm 1\}$ defined by eight monomials. Likewise, the monomials defining $p_{i,i}$ are $\pm X_i^2 Y_i^2$. We give the addition law $\mathfrak{p}_1$ in the Example A.1 below to illustrate our reasoning. The other addition laws have a similar combinatorial aspect. We assume that the coefficients λ_j are precomputed. To calculate $\lambda_j \mathfrak{p}_j$ we first need $160\mathbf{m}_\vartheta$.Then we detail the computation of the monomials. Note that we do not take into account additions or sign changes costs. We first compute all the products $X_i X_j$ and $Y_i Y_j$, this costs $240\mathbf{m} + 32\mathbf{s}$ and the products $X_i X_j Y_i Y_j$ in $256\mathbf{m}$. These monomials are exactly the one included in the ten first polynomials of the addition laws $\mathfrak{p}_j$ (see Example A.1), so we do not need any other multiplication to calculated and the $p_{i,j}$ for $1 \le i, j \le 10$. For the remaining $p_{i,j}$ with $11 \le i \le 16$ and $1 \le j \le 10$, we point out that if a monomial $X_{i_0} X_{j_0} Y_{k_0} Y_{l_0}$ appears, so does $X_{k_0} X_{l_0} Y_{i_0} Y_{j_0}$ with the same sign. We then use the relation

$$X_{i_0}X_{j_0}Y_{k_0}Y_{l_0} + X_{k_0}X_{l_0}Y_{i_0}Y_{j_0} = (X_{i_0}X_{j_0} + X_{k_0}X_{l_0})(Y_{i_0}Y_{j_0} + Y_{k_0}Y_{l_0}) - X_{i_0}X_{j_0}Y_{i_0}Y_{j_0} - X_{k_0}X_{l_0}Y_{k_0}Y_{l_0}$$

to calculate each $p_{i,j}$ with $\mathbf{4m}$. Hence, the addition laws $\lambda_1\mathfrak{p}_1, \dots, \lambda_{10}\mathfrak{p}_{10}$ can be computed in $\mathbf{736m} + \mathbf{32s} + \mathbf{160m}_\vartheta$.

Example A.1. As an illustrative example, we present the addition law $\mathfrak{p}_1$.

$p_{1,1} = \frac{1}{\vartheta_1^2}(X_1^2Y_1^2 + X_2^2Y_2^2 + X_3^2Y_3^2 + X_4^2Y_4^2 + X_5^2Y_5^2 + X_6^2Y_6^2 + X_7^2Y_7^2 + X_8^2Y_8^2 + X_9^2Y_9^2 + X_{10}^2Y_{10}^2 + X_{11}^2Y_{11}^2 + X_{12}^2Y_{12}^2 + X_{13}^2Y_{13}^2 + X_{14}^2Y_{14}^2 + X_{15}^2Y_{15}^2 + X_{16}^2Y_{16}^2)$,

$p_{2,1} = \frac{2}{\vartheta_1\vartheta_2}(X_1X_2Y_1Y_2 + X_3X_4Y_3Y_4 + X_5X_{15}Y_5Y_{15} + X_6X_{13}Y_6Y_{13} + X_7X_{12}Y_7Y_{12} + X_8X_{10}Y_8Y_{10} + X_9X_{11}Y_9Y_{11} + X_{14}X_{16}Y_{14}Y_{16})$,

$p_{3,1} = \frac{2}{\vartheta_1\vartheta_3}(X_1X_3Y_1Y_3 + X_2X_4Y_2Y_4 + X_5X_{13}Y_5Y_{13} + X_6X_{15}Y_6Y_{15} + X_7X_9Y_7Y_9 + X_8X_{14}Y_8Y_{14} + X_{11}X_{12}Y_{11}Y_{12} + X_{10}X_{16}Y_{10}Y_{16})$,

$p_{4,1} = \frac{2}{\vartheta_1\vartheta_4}(X_1X_4Y_1Y_4 + X_2X_3Y_2Y_3 + X_5X_6Y_5Y_6 + X_7X_{11}Y_7Y_{11} + X_8X_{16}Y_8Y_{16} + X_9X_{12}Y_9Y_{12} + X_{10}X_{14}Y_{10}Y_{14} + X_{13}X_{15}Y_{13}Y_{15})$,

$p_{5,1} = \frac{2}{\vartheta_1\vartheta_5}(X_1X_5Y_1Y_5 - X_2X_{15}Y_2Y_{15} - X_3X_{13}Y_3Y_{13} + X_4X_6Y_4Y_6 + X_7X_8Y_7Y_8 - X_9X_{14}Y_9Y_{14} - X_{10}X_{12}Y_{10}Y_{12} + X_{11}X_{16}Y_{11}Y_{16})$,

$p_{6,1} = \frac{2}{\vartheta_1\vartheta_6}(X_1X_6Y_1Y_6 - X_2X_{13}Y_2Y_{13} - X_3X_{15}Y_3Y_{15} + X_4X_5Y_4Y_5 + X_7X_{16}Y_7Y_{16} + X_8X_{11}Y_8Y_{11} - X_9X_{10}Y_9Y_{10} - X_{12}X_{14}Y_{12}Y_{14})$,

$p_{7,1} = \frac{2}{\vartheta_1\vartheta_7}(X_1X_7Y_1Y_7 - X_2X_{12}Y_2Y_{12} + X_3X_9Y_3Y_9 - X_4X_{11}Y_4Y_{11} + X_5X_8Y_5Y_8 - X_6X_{16}Y_6Y_{16} - X_{10}X_{15}Y_{10}Y_{15} + X_{13}X_{14}Y_{13}Y_{14})$,

$p_{8,1} = \frac{2}{\vartheta_1\vartheta_8}(X_1X_8Y_1Y_8 + X_2X_{10}Y_2Y_{10} - X_3X_{14}Y_3Y_{14} - X_4X_{16}Y_4Y_{16} + X_5X_7Y_5Y_7 - X_6X_{11}Y_6Y_{11} - X_9X_{13}Y_9Y_{13} + X_{12}X_{15}Y_{12}Y_{15})$,

$p_{9,1} = \frac{2}{\vartheta_1\vartheta_9}(X_1X_9Y_1Y_9 - X_2X_{11}Y_2Y_{11} + X_3X_7Y_3Y_7 - X_4X_{12}Y_4Y_{12} + X_5X_{14}Y_5Y_{14} - X_6X_{10}Y_6Y_{10} + X_8X_{13}Y_8Y_{13} - X_{15}X_{16}Y_{15}Y_{16})$,

$p_{10,1} = \frac{2}{\vartheta_1\vartheta_{10}}(X_1X_{10}Y_1Y_{10} + X_2X_8Y_2Y_8 - X_3X_{16}Y_3Y_{16} - X_4X_{14}Y_4Y_{14} + X_5X_{12}Y_5Y_{12} - X_6X_9Y_6Y_9 + X_7X_{15}Y_7Y_{15} - X_{11}X_{13}Y_{11}Y_{13})$,

$p_{11,1} = \frac{2}{\vartheta_8\vartheta_6}(X_1X_{11}Y_6Y_8 + X_2X_9Y_{10}Y_{13} + X_3X_{12}Y_{14}Y_{15} + X_4X_7Y_5Y_{16} + X_5X_{16}Y_4Y_7 + X_6X_8Y_1Y_{11} + X_{10}X_{13}Y_2Y_9 + X_{14}X_{15}Y_3Y_{12})$,

$p_{12,1} = \frac{2}{\vartheta_7\vartheta_2}(X_1X_{12}Y_2Y_7 + X_2X_7Y_1Y_{12} + X_3X_{11}Y_4Y_9 + X_4X_9Y_3Y_{11} + X_5X_{10}Y_8Y_{15} + X_6X_{14}Y_{13}Y_{16} + X_8X_{15}Y_5Y_{10} + X_{13}X_{16}Y_6Y_{14})$,

$p_{13,1} = \frac{2}{\vartheta_6\vartheta_2}(X_1X_{13}Y_2Y_6 + X_2X_6Y_1Y_{13} + X_3X_5Y_4Y_{15} + X_4X_{15}Y_3Y_5 - X_7X_{14}Y_{12}Y_{16} - X_8X_9Y_{10}Y_{11} - X_{10}X_{11}Y_8Y_9 - X_{12}X_{16}Y_7Y_{14})$,

$p_{14,1} = \frac{2}{\vartheta_5\vartheta_9}(X_1X_{14}Y_5Y_9 - X_2X_{16}Y_{11}Y_{15} + X_3X_8Y_7Y_{13} - X_4X_{10}Y_6Y_{12} + X_5X_9Y_1Y_{14} - X_6X_{12}Y_4Y_{10} + X_7X_{13}Y_8Y_3 - X_{11}X_{15}Y_2Y_{16})$,

$p_{15,1} = \frac{2}{\vartheta_5\vartheta_2}(X_1X_{15}Y_2Y_5 + X_2X_5Y_1Y_{15} + X_3X_6Y_4Y_{13} + X_4X_{13}Y_3Y_6 + X_7X_{10}Y_8Y_{12} + X_8X_{12}Y_7Y_{10} + X_9X_{16}Y_{11}Y_{14} + X_{11}X_{14}Y_9Y_{16})$,

$p_{16,1} = \frac{2}{\vartheta_3\vartheta_{10}}(X_1X_{16}Y_3Y_{10} - X_2X_{14}Y_4Y_8 + X_3X_{10}Y_1Y_{16} - X_4X_8Y_2Y_{14} - X_5X_{11}Y_{12}Y_{13} + X_6X_7Y_9Y_{15} + X_9X_{15}Y_6Y_7 - X_{12}X_{13}Y_5Y_{11})$.

References

[AKR11] C. Arene, D. Kohel, and C. Ritzenthaler. Complete addition laws on abelian varieties. eprint arXiv:1102.2349, February 2011. `http://arxiv.org/abs/1102.2349`.

[Bai62] W. L. Baily, Jr. On the Theory of θ-Functions, the Moduli of Abelian Varieties, and the Moduli of Curves. *Ann. of Math. (2)*, 75:342–381, March 1962. MR0162799 (29:103)

[BKL09] D. J. Bernstein, D. Kohel, and T. Lange. Twisted Hessian curves. Preprint, 2009.

[BL95] W. Bosma and H.W. Lenstra, Jr. Complete systems of two addition laws for elliptic curves. *J. Number Theory*, 53(2):229–240, 1995. MR1348761 (96f:11079)

[BL04] C. Birkenhake and H. Lange. *Complex abelian varieties*, volume 302 of *Grundlehren der Mathematischen Wissenschaften [Fundamental Principles of Mathematical Sciences]*. Springer-Verlag, Berlin, second edition, 2004. MR2062673 (2005c:14001)

[BL07] D. J. Bernstein and T. Lange. Faster Addition and Doubling on Elliptic Curves. In K. Kurosawa, editor, *Advances in Cryptology - ASIACRYPT 2007*, volume 4833 of *Lecture Notes in Comput. Sci.*, pages 29–50. Springer Berlin / Heidelberg, 2007. MR2565722 (2011d:11125)

[Cos11] R. Cosset. *Applications des fonctions thêta à la cryptographie sur courbes hyperelliptiques*. PhD thesis, Université Henri-Poincarré, Nancy 1, France, 2011.

[CR11] R. Cosset and D. Robert. Computing (ℓ, ℓ)-isogenies in polynomial time on jacobians of genus 2 curves. Cryptology ePrint Archive, Report 2011/143, March 2011. `http://eprint.iacr.org/2011/143`.

[Duq04] S. Duquesne. Montgomery scalar multiplication for genus 2 curves. In D. Buell, editor, *ANTS-VI*, volume 3076 of *Lecture Notes in Comput. Sci.*, pages 153–168. Springer–Verlag, 2004. MR2137351 (2006c:11074)

[Edw07] H. M. Edwards. A normal form for elliptic curves. *Bull. Amer. Math. Soc.*, 44:393–422, April 2007. MR2318157 (2008b:14052)

[FJ10] R. Farashahi and M. Joye. Efficient Arithmetic on Hessian Curves. In P. Nguyen and D. Pointcheval, editors, *Public Key Cryptography - PKC 2010*, volume 6056 of *Lecture Notes in Comput. Sci.*, pages 243–260. Springer Berlin / Heidelberg, 2010. MR2660746

[Gau07] P. Gaudry. Fast genus 2 arithmetic based on theta functions. *J. Math. Cryptol.*, 1(3):243–265, 2007. MR2372155 (2009f:11156)

[Koh11] D. Kohel. Addition law structure of elliptic curves. *J. Number Theory*, 131(5):894–919, 2011. MR2772478 (2012c:14072)

[Lan05] T. Lange. Formulae for arithmetic on genus 2 hyperelliptic curves. *Appl. Algebra Engrg. Comm. Comput.*, 15(5):295–328, 2005. MR2122308 (2005j:14082)

[LM05] T. Lange and P. K. Mishra. SCA resistant parallel explicit formula for addition and doubling of divisors in the Jacobian of hyperelliptic curves of genus 2. In *Progress in cryptology—INDOCRYPT 2005*, volume 3797 of *Lecture Notes in Comput. Sci.*, pages 403–416. Springer, Berlin, 2005. MR2237799 (2007c:94145)

[LR85] H. Lange and W. Ruppert. Complete systems of addition laws on abelian varieties. *Invent. Math.*, 79:603–610, 1985. MR782238 (86f:14029)

[Mum66] D. Mumford. On the equations defining abelian varieties. I. *Invent. Math.*, 1:287–354, 1966. MR0204427 (34:4269)

[Mum67a] D. Mumford. On the equations defining abelian varieties. II. *Invent. Math.*, 3:75–135, 1967. MR0219541 (36:2621)

[Mum67b] D. Mumford. On the equations defining abelian varieties. III. *Invent. Math.*, 3:215–244, 1967. MR0219542 (36:2622)

[Mum70] D. Mumford. *Abelian varieties*. Tata Institute of Fundamental Research Studies in Mathematics, No. 5. Published for the Tata Institute of Fundamental Research, Bombay, 1970. MR0282985 (44:219)

[Mum75] D. Mumford. *Curves and their Jacobians*. The University of Michigan Press, Ann Arbor, Mich., 1975. MR0419430 (54:7451)

[Mum83] D. Mumford. *Tata lectures on theta. I*, volume 28 of *Progr. Math.* Birkhäuser Boston Inc., Boston, MA, 1983. With the assistance of C. Musili, M. Nori, E. Previato and M. Stillman. MR2352717 (2008h:14042)

[Mum84] D. Mumford. *Tata lectures on theta. II*, volume 43 of *Progr. Math.* Birkhäuser Boston Inc., Boston, MA, 1984. Jacobian theta functions and differential equations, With the collaboration of C. Musili, M. Nori, E. Previato, M. Stillman and H. Umemura. MR742776 (86b:14017)

[Tho70] J. Thomae. Beitrag zur Bestimmung von $\vartheta\,(0, 0, \ldots 0)$ durch die Klassbmodulb algebraischer Functionnen. *J. Reine Angew. Math.*, 70:201–222, 1870.

[vW98] P. van Wamelen. Equations for the Jacobian of a hyperelliptic curve. *Trans. Amer. Math. Soc.*, 350(8):3083–3106, 1998. MR1432144 (98k:14038)

Institut de Mathématiques de Luminy, UMR 6206, Campus de Luminy, Case 907, 13288 Marseille, France.

E-mail address: arene.christophe@gmail.com

Laboratoire Lorrain de Recherche en Informatique et ses Applications, UMR 7503, Campus scientifique, BP 239, 54506 Vandoeuvre-lès-Nancy, France.

E-mail address: romain.cosset@loria.fr

Contemporary Mathematics
Volume **574**, 2012
http://dx.doi.org/10.1090/conm/574/11427

Number of points in an Artin-Schreier covering

Régis Blache

ABSTRACT. We improve the bounds for the number of rational points of an Artin-Schreier covering of a curve defined over a finite field, following ideas of Rojas Leon and Wan. This number of points is classically written as a sum of exponential sums, and we are able to find some cancellation in the outer sum using the theory of ℓ-adic sheaves and their Fourier transform. Precisely, we study the local and global monodromies of an ℓ-adic sheaf; under certain generic hypotheses, these are big enough to improve the known bounds by a factor $\sqrt{q}$, q being the cardinality of the field of definition of the Artin-Schreier covering.

Introduction

Let C denote a (projective, smooth, geometrically irreducible) curve of genus g, defined over the field $k := \mathbb{F}_q$, and with function field $K(C)$. For any $r \geq 1$, let $k_r := \mathbb{F}_{q^r}$ be the degree r extension of k inside a fixed algebraic closure $\overline{k}$, and denote by $N_r(C)$ the number of k_r-rational points of the curve C. Let $f \in K(C)$ be a non constant function over C, with pole divisor

$$(f)_\infty := \sum_{i=1}^{n} d_i P_i,$$

where we assume each d_i prime to the characteristic p, and the P_i are in $C(\overline{k})$. We denote by P_f :=Supp$(f)_\infty$ the set of poles of f.

We denote by C_f the curve whose function field is the extension $K(C)(y)$ of $K(C)$, where y satisfies the equation $y^q - y = f(x)$. We have constructed an Artin-Schreier covering of the curve C [**11**, Section 3.7].

This is a wildly ramified cover at the poles of the function f, and from the Riemann-Hurwitz formula, the curve C_f has genus g_f satisfying (*cf* [**7**])

$$2g_f - 2 = q(2g-2) + (q-1)\sum_{i=1}^{n}(d_i+1).$$

Moreover the number of points of the curve C_f over the field k_r can be expressed from exponential sums over $C \backslash P_f$. Let ψ be a non trivial additive character of k:

2010 *Mathematics Subject Classification.* Primary: 11M38, 11T23, 14G15.
Key words and phrases. Number of points of curves over a finite field, Artin-Schreier covering.

we define the sums

$$S_r(C, \psi, tf) := \sum_{P \in C(k_r) \setminus P_f(k_r)} \psi\left(t \mathrm{Tr}_r(f(P))\right)$$

where Tr_r denotes the trace from k_r to k. In terms of these sums, we have the following expression, valid for any $r \geq 1$

$$N_r(C_f) = N_r(C) + \sum_{t \in k^\times} S_r(C, \psi, tf). \tag{0.1}$$

As a consequence, the zeta function of the curve C_f can be written in the following way

$$Z(C_f; T) = Z(C; T) \prod_{t \in k^\times} L(C, \psi, tf; T), \tag{0.2}$$

where the L-functions are the ones associated to the curve C and the functions tf, defined as usual by

$$L(C, \psi, tf; T) := \exp\left(\sum_{r \geq 1} S_r(C, \psi, tf) \frac{T^r}{r}\right).$$

From the work of Weil, it is known that the functions $L(C, tf; T)$ are polynomials with degree $D := 2g - 2 + \sum_{i=1}^n (d_i + 1)$ (*cf* [**1, 7**]). Moreover all their reciprocal roots are algebraic integers having complex absolute value $\sqrt{q}$. We get the following estimate for the difference between the numbers of k_r-rational points of the curves C and C_f

$$|N_r(C_f) - N_r(C)| \leq (q-1)D\sqrt{q^r} = O(q^{\frac{r}{2}+1}).$$

The aim of this work is to get an improvement for this bound in certain (generic) cases, and to find a bound with magnitude $O(q^{\frac{r+1}{2}})$. Note that one cannot hope such a bound to be always true: it is well-known that some maximal curves (whose number of rational points attains the Weil upper bound), as the Hermitian one, can be written as Artin-Schreier coverings of the projective line (see for instance [**4**] for more details on this).

We follow closely, and adapt to the above situation, the ideas already used in [**9**] and [**10**]. We try to find cancellation in the sum in the formula (0.1): this is the same phenomenon as for the sum $S_1(C, \psi, f)$, which is the sum of approximately q complex numbers of moduli 1 and has order of magnitude $O(q^{\frac{1}{2}})$ instead of $O(q)$. Let us first describe the case $r = 1$.

We construct a lisse ℓ-adic sheaf $\mathcal{H}$ over $\mathbb{G}_m$, pure of weight 1, such that the trace of the action of Frobenius on its stalk at a geometric point over $t \in k^\times$ is the sum $S_1(C, \psi, tf)$. From Grothendieck trace formula, we can express the difference between the numbers of k-rational points of the curves C and C_f, $N_1(C_f) - N_1(C)$, from the traces of the action of Frobenius on the cohomology spaces of $\mathcal{H}$ with compact support; actually this number is the degree 1 coefficient of the power series expansion around 0 of the rational function

$$L(\mathcal{H}; T) = \frac{\det(1 - \mathrm{Frob}_k T | \mathrm{H}^1_c(\mathbb{G}_{m,\bar{k}}, \mathcal{H}))}{\det(1 - \mathrm{Frob}_k T | \mathrm{H}^2_c(\mathbb{G}_{m,\bar{k}}, \mathcal{H}))}.$$

The sheaf $\mathcal{H}$ can be seen as (the restriction to $\mathbb{G}_m$ of) the ℓ-adic Fourier transform of a sheaf over $\mathbb{A}^1$; this interpretation, joint with Laumon's stationary phase principle

[**8**] gives us a fairly complete description of the local monodromies of the sheaf $\mathcal{H}$. Moreover, results by Katz [**5**] give, under certain hypotheses, the whole geometric monodromy, i.e. the action of the group $\pi_1(\mathbb{G}_{m,\bar{k}})$ on the generic fiber of $\mathcal{H}$. With these results at hand, we are able to show that (in many cases) the denominator above is trivial, and from the fundamental theorem of [**2**], we ensure that the cohomology spaces appearing in the numerator are mixed with weights less than 2.

When $r > 1$, we interpret the sum $S_r(C, \psi, tf)$ as the action of Frobenius on the stalk at t of the (virtual) sheaf

$$[\mathcal{H}]^r := \sum_{i=0}^{r-1} (-1)^i \mathbb{S}_{\lambda_i} \mathcal{H};$$

(where $\mathbb{S}_{\lambda_i}$ is the Schur functor associated to the partition $\lambda_i(r - i, 1, \ldots, 1)$ of r [**3**]) called the *r-th Adams operation of $\mathcal{H}$*; it is a lisse virtual sheaf over $\mathbb{G}_m$, pure of weight r. It has been used in [**6**] and in [**12**, Lemma 4.2] with slightly different coefficients, and in [**9**, Proposition 3.4] just as above. Here the number $N_r(C_f) - N_r(C)$ can be seen as the degree 1 coefficient of the power series expansion around 0 of the rational function

$$L([\mathcal{H}]^r; T) = \frac{\det(1 - \mathrm{Frob}_k T | \mathrm{H}^1_c(\mathbb{G}_{m,\bar{k}}, [\mathcal{H}]^r))}{\det(1 - \mathrm{Frob}_k T | \mathrm{H}^2_c(\mathbb{G}_{m,\bar{k}}, [\mathcal{H}]^r))}.$$

The description of $\mathcal{H}$ can be used to give information on $[\mathcal{H}]^r$, and to show that under certain (generic) conditions, the denominator above is trivial, and the action of Frobenius on the cohomology space appearing at the numerator is mixed of weights less than $r + 1$. We can also give an upper bound for the absolute value of the Euler characteristic of this sheaf, from which we deduce the

THEOREM 0.1. *Notations being as above, assume the differential df has only simple zeroes on C. Assume moreover that either*

i/ the number r is odd;

ii/ the hypersurface in C^r defined by the equation $f(P_1) + \ldots + f(P_r) = 0$ is non-singular;

iii/ we have $p > 2D + 1$, the function f is supermorse, its critical values $\{s_1, \ldots, s_D\}$ are such that the differences $s_i - s_j$, $1 \leq i, j \leq D$, $i \neq j$ are pairwise distinct, and we do not have D even, $s = s_1 + \ldots + s_D = 0$ and $r = D$.

Then we have the following estimate

$$|N_r(C_f) - N_r(C)| \leq C_{D,r} q^{\frac{r+1}{2}},$$

where we have set

$$C_{D,r} := \sum_{i=0}^{r-1} \binom{D - 1 + r - i}{r} \binom{r - 1}{i}.$$

The paper is organized as follows: in section 1, we recall some facts about certain coverings of curves whose Galois group is a p-group and give a first general bound for the number $|N_r(C_f) - N_r(C)|$ studied above. Then we try to improve it under the hypotheses described in Theorem 0.1; in section 2, we define precisely the sheaves presented above, and give some information about their geometric local monodromies, using Laumon's stationnary phase principle for Fourier transform of

ℓ-adic sheaves. In section 3, we give a cohomological expression for the number $N_r(C_f) - N_r(C)$ from the L-function of one of the above sheaves; then we use the information about local monodromies to show the theorem in cases *i/ ,ii/*. Finally we deal with case *iii/* in section 4, determining the geometric monodromy group of the sheaf $\mathcal{H}$.

REMARK 0.2. Note that in the case $r = 1$, this estimate can be deduced in a very elementary way: for a fixed $P \in C(k)$, the equation $y^q - y = f(P)$ has q solutions if $f(P) = 0$, and no solution otherwise. Since the places in $P_f(k)$ are totally ramified in the extension $K(C_f)/K(C)$, we get the following exact formula

$$N_1(C_f) - N_1(C) = q\#\text{Supp}(f)_0(k) - \#(C - P_f)(k).$$

REMARK 0.3. Note that one assumption here is that the curve $y^q - y = f(P)$ is defined over the field k; as remarked in [**10**], there should be a similar improvement when it is defined only over the larger field k_r. But we were not able to prove it in this way.

REMARK 0.4. We thank Antonio Rojas Leon for pointing out that another way to obtain similar estimates would be to show that the sheaf $\mathcal{F}$ defined in section 2 satisfies the requirements of [**9**, Proposition 6.4]. We have chosen to stay closer to the ideas of [**10**] in order to make the exposition clearer and more natural.

1. Number of points in Artin-Schreier coverings: Weil bound

In this section, we consider a (geometrically irreducible, smooth) curve C defined over the field $k = \mathbb{F}_q$, with function field K_C, and a non constant function $f \in K_C$ with pole divisor $(f)_\infty := \sum_{i=1}^n d_i P_i$; as in the introduction, we assume that the d_i are prime to p. Let $A(x) := \sum_{i=0}^s a_i x^{p^i} \in k\{x\}$ be a separable (i.e. $a_0 \neq 0$) additive polynomial. We define the covering $C_{f,A}$ as a smooth curve whose function field is the extension $K_C(y)$ defined by the equation $A(y) = f$. Finally, let ψ be the additive character of k defined by $\psi(x) = \exp\left(\frac{2i\pi}{p}\text{Tr}_{k/\mathbb{F}_p}(x)\right)$.

The aim of this section is to give an expression for the difference between the numbers of rational points in the covering using character sums, and to deduce a bound for it from Weil's theorem. These results must be very classical, and we include them here just because we did not find an appropriate reference.

DEFINITION 1.1. Let A be a separable additive polynomial, of degree p^s; we define its *adjoint polynomial* as $A^*(x) := \sum_{i=0}^s (a_i x)^{p^{s-i}}$.

LEMMA 1.2. *Let $x_0 \in k_r$; the number of solutions of the equation $A(y) = x_0$ in k_r is*

$$N_{A,r}(x_0) = \sum_{a \in k_r, A^*(a)=0} \psi(ax_0).$$

PROOF. The map $y \mapsto A(y)$ is a linear endomorphism of the $\mathbb{F}_p$-vector space k_r; as a consequence we have

$$N_{A,r}(x_0) = \begin{cases} \#\text{Ker}A(k_r) & \text{if} \quad x_0 \in \text{Im } A(k_r) \\ 0 & \text{else} \end{cases}$$

Using the perfect pairing $(x, y) \mapsto \psi(xy)$, we can rewrite

$$N_{A,r}(x_0) = \sum_{a \in (\text{Im}A(k_r))^\perp} \psi(ax_0).$$

Now we have $a \in (\text{Im } A(k_r))^\perp$ if and only if for any $x \in k_r$, $\psi(aA(x)) = 1$. An easy calculation shows that $\psi(aA(x)) = \psi(xA^*(a))$, and since the pairing is non degenerate, this is equivalent to $A^*(a) = 0$. This is the desired result. □

As a consequence, we can give an exact formula for the number of rational points of $C_{f,A}$ from the number of rational points of C and certain character sums.

PROPOSITION 1.3. We have the following formula for the difference of the numbers of points in the covering above

$$N_r(C_{f,A}) - N_r(C) = \sum_{a \in k_r^\times,\ A^*(a)=0} S_r(C, \psi, af)$$

PROOF. On one hand, the k_r-rational poles of f –the elements of $P_f(k_r)$– are totally ramified in the covering, and each one gives rise to a unique k_r-rational point of $C_{f,A}$. If x is a point in $C(k_r)$ which is not a pole of f, then it gives rise to $N_{A,r}(f(x))$ points in $C_{f,A}(k_r)$. Summing up, and using the result of the above Lemma, we get

$$\begin{aligned} N_r(C_{f,A}) &= \#P_f(k_r) + \sum_{a \in k_r,\ A^*(a)=0} S_r(C, \psi, af) \\ &= N_r(C) + \sum_{a \in k_r^\times,\ A^*(a)=0} S_r(C, \psi, af). \end{aligned}$$

since the sum $S_r(C, \psi, 0f)$ is exactly the number of k_r-rational points of C which are not poles of f. □

It remains to use Weil bound, as in [**7**], to give an upper bound for the difference above

COROLLARY 1.4. Notations are as in the introduction: we have the following bound

$$N_r(C_{f,A}) - N_r(C) \leq \#\text{Ker}A^*(k_r)D\sqrt{q^r}.$$

REMARK 1.5. Note that Hurwitz formula (see for instance [**11**]) gives the following for the genus of $C_{f,A}$

$$2\,(g(C_{f,A}) - g(C)) = (\deg A - 1)D.$$

As a consequence, when $\text{Ker}A \not\subseteq k_r$, the above corollary gives an improvement on Weil's bound (of order at least p since $\text{Ker}A^*(k_r)$ is a sub vector space of k_r).

However the cases considered in the remaining of this paper cannot be treated in this way, since we will always have $\text{Ker}A \subseteq k_r$ from now.

2. Study of an etale sheaf attached to a family of exponential sums

Here we shall define the sheaf $\mathcal{H}$ (briefly) described in the introduction as the restriction to $\mathbb{G}_{m,k}$ of the Fourier transform of a sheaf $\mathcal{F}$ over $\mathbb{A}^1_k$. Then we compute the local monodromies of the sheaf $\mathcal{H}$ at its points of non-lissity. We begin by describing the sheaf $\mathcal{F}$.

We consider the function f as a morphism of degree $d := \sum_{i=1}^n d_i$ from $C \backslash P_f$ to $\mathbb{A}^1_k$. Let $\mathcal{F}$ be the constructible sheaf on $\mathbb{A}^1_k$ defined as the kernel of the surjective trace map $f_*\overline{\mathbb{Q}}_\ell \to \overline{\mathbb{Q}}_\ell$. It has rank $d-1$.

It is well known that this sheaf is lisse away from the critical values of the function f. For this reason we let $(df)_0 := \sum_{i=0}^{d_0'} m_i Q_i$ be the divisor of zeroes of the differential of f; we denote its support by $\mathcal{Q}$; the points in $\mathcal{Q}$ are the critical points of f. In the following we assume that each integer $m_i + 1$ is prime to p.

Let $S = \{s_1, \dots, s_t\} = \{f(Q_1), \dots, f(Q_{d_0'})\}$ be the set of critical values of f. We order the points Q_i in such a way that the critical points above s_i are the $Q_{j_i}, \dots, Q_{j_{i+1}-1}$. In other words, we can write the zero divisor of $f - s_i$ in the following way

$$(f - s_i)_0 = \sum_{k=j_i}^{j_{i+1}-1} (m_k + 1) Q_k + D_i,$$

where D_i is a divisor with all points of multiplicity one, whose support is disjoint with $\{Q_{j_i}, \dots, Q_{j_{i+1}-1}\}$.

We list some properties of the sheaf $\mathcal{F}$; we will use them when we study the Fourier transform of $\mathcal{F}$.

In the following the letter χ will always denote some (multiplicative) character of $\overline{k}^\times$. Moreover, we denote by $\mathcal{L}_\psi$ (*resp.* $\mathcal{L}_\chi$) the lisse rank one $\overline{\mathbb{Q}}_\ell$-sheaf on $\mathbb{A}^1$ (*resp.* on $\mathbb{G}_m = \mathbb{A}^1 \backslash \{0\}$) obtained from the Artin-Schreier (*resp.* Kummer) covering by extension of the structural group by ψ (*resp.* by χ) [**1**].

LEMMA 2.1. *The sheaf $\mathcal{F}$ has the following properties*

i/ it is lisse over the open $U := \mathbb{A}_k^1 \backslash S$;

ii/ the trace of the action of Frobenius at $t \in \mathbb{A}_k^1(k_r)$ is

$$Tr(Frob_t | \mathcal{F}_t) = \#(f^{-1}(t))(k_r) - 1;$$

iii/ the action of the inertia group I_{s_i}, modulo its invariants, on the generic fibre $\mathcal{F}_{\overline{\eta}}$ is given by

$$\mathcal{F}(s_i)/\mathcal{F}_{s_i} \simeq \bigoplus_{k=j_i}^{j_{i+1}-1} \bigoplus_{\chi_k^{m_k+1} = \mathbf{1},\ \chi_k \neq \mathbf{1}} \mathcal{L}_{\chi_k};$$

iv/ the action of the inertia group I_∞ at infinity on the generic fibre $\mathcal{F}_{\overline{\eta}}$ is given by

$$\mathcal{F}(\infty) \simeq \overline{\mathbb{Q}}_\ell^{n-1} \bigoplus_{k=1}^{n} \bigoplus_{\chi_k^{d_k} = \mathbf{1},\ \chi_k \neq \mathbf{1}} \mathcal{L}_{\chi_k};$$

v/ it is a Fourier sheaf in the sense of [**5**, 7.3.5].

PROOF. Most of the above properties follow directly from the definition of the sheaf $\mathcal{F}$ as a direct factor of the sheaf $f_* \overline{\mathbb{Q}}_\ell$; we justify them briefly. Assertion *i/* comes from the fact that the map f induces an etale covering of degree d from $C \backslash \mathcal{Q}$ to U. Now remark that $\mathrm{Tr}(\mathrm{Frob}_t | \mathcal{F}_t) = \mathrm{Tr}(\mathrm{Frob}_t | (f_* \overline{\mathbb{Q}}_\ell)_t) - 1$; this proves assertion *ii/*.

From [**5**, Lemma 7.3.2], the sheaf $f_* \overline{\mathbb{Q}}_\ell$ is a middle extension on $\mathbb{A}_k^1$ in the sense of [**5**, 7.3]; thus its direct factor $\mathcal{F}$ shares this property. As a consequence, we have $\mathcal{F}_t = \mathcal{F}_{\overline{\eta}}^{I_t}$ for any $t \in \mathbb{A}_k^1$.

In order to show the last three statement, which are of a geometric nature, we make a suitable extension of scalar, so that all points become rational over the new base field. At a critical value s_i, if u_k is an uniformizing parameter for C at Q_k,

$j_i \leq k \leq j_{i+1}-1$, then locally at Q_k we have $f - s_i = a_k u_k^{m_k+1} g_k$ for some g_k in $\mathcal{O}_{C,Q_i}^\times$, $a_k \in k^\times$, and since we assumed m_k+1 prime to p, the map $f - s_i$ from the strict henselization of C at Q_i is isomorphic to the map $u \mapsto u_k^{m_k+1}$. Thus under the action of the inertia group at s_i, the generic stalk of $f_*\overline{\mathbb{Q}}_\ell$ is isomorphic to

$$\left(\bigoplus_{k=j_i}^{j_{i+1}-1} \bigoplus_{\chi_k^{m_k+1}=\mathbf{1}} \mathcal{L}_{\chi_k}\right) \oplus \overline{\mathbb{Q}}_\ell^{\deg D_i}.$$

As a consequence, we get $\mathcal{F}_{s_i} \simeq \overline{\mathbb{Q}}_\ell^{\deg D_i - j_i + j_{i+1}}$ for the space of invariants of I_{s_i}, and assertion *iii/* follows. Reasoning the same way, we also get assertion *iv/* since we assumed the d_i prime to p.

To prove the last assertion, we follow [**5**, Lemma 7.10.2.1]; the only difference is that we do not assume $\deg f < p$ here. But from above, the representation $\mathcal{F}(\infty)$ is tame, and the vanishing $\mathrm{H}^0(\mathbb{A}^1, \mathcal{F} \otimes \mathcal{L}_{\psi_t}) = \mathrm{H}_c^2(\mathbb{A}^1, \mathcal{F} \otimes \mathcal{L}_{\psi_t}) = 0$ persits when $t \neq 0$. For $t = 0$ the argument is the same as in *loc. cit.* □

We describe now the naive Fourier transform of $\mathcal{F}$ in the sense of [**5**], i.e. the sheaf defined in terms of the two projections π_1, π_2 from $\mathbb{A}^2$ with coordinates (x, y) to $\mathbb{A}^1$, and the multiplication map $m : \mathbb{A}^1 \times \mathbb{A}^1 \to \mathbb{A}^1$ by

$$\mathcal{G} := \mathcal{R}^1 \pi_{1!}(\pi_2^* \mathcal{F} \otimes m^* \mathcal{L}_\psi).$$

From the work of Laumon [**8**] (see also [**5**]), we can give a rather precise description of the sheaf $\mathcal{G}$.

LEMMA 2.2. *The sheaf $\mathcal{G}$ has the following properties*

i/ it is lisse over $\mathbb{G}_{m,k}$, pure of weight 1 *and has rank D;*

ii/ the trace of the action of Frobenius at $t \in \mathbb{G}_{m,k}(k_r)$ is

$$\mathit{Tr}(\mathit{Frob}_t | \mathcal{G}_t) = -S_r(C, \psi, tf);$$

iii/ the fiber at 0 *of $\mathcal{G}$ is*

$$\mathcal{G}_0 = \mathcal{G}_{\overline{\eta}}^{I_0} = H_c^1(\mathbb{A}_{\overline{k}}^1, \mathcal{F}) \simeq H_c^1(C_{\overline{k}} - P_{f,\overline{k}}, \overline{\mathbb{Q}}_\ell),$$

a $\overline{\mathbb{Q}}_\ell$-vector space of dimension $2g + n - 1$, and the trace of the action of Frobenius is $\mathit{Tr}(\mathit{Frob}_0^r | \mathcal{G}_0) = q^r - \#(C - P_f)(k_r)$;

iv/ the local monodromy at infinity of $\mathcal{G}$ is given by

$$\mathcal{G}(\infty) \simeq \bigoplus_{i=1}^{t} \left(\bigoplus_{k=j_i}^{j_{i+1}-1} \bigoplus_{\substack{\chi_k^{m_k+1}=\mathbf{1} \\ \chi_k \neq \mathbf{1}}} \mathcal{L}_{\chi_k}\right) \otimes \mathcal{L}_{\psi_{s_i}}.$$

PROOF. The first three assertions follow from the properties of Fourier transform (see for instance [**5**, Theorem 7.3.8, Lemma 7.3.9]), except the isomorphism in the third one, and the assertion on the dimension.

In order to show that, we just remark that in the long cohomology exact sequence coming from the definition of $\mathcal{F}$, we have $\mathrm{H}_c^1(\mathbb{A}_{\overline{k}}^1, \mathcal{F}) \simeq \mathrm{H}_c^1(\mathbb{A}_{\overline{k}}^1, f_*\overline{\mathbb{Q}}_\ell)$; thus this space is isomorphic to $\mathrm{H}_c^1(C_{\overline{k}} - P_{f,\overline{k}}, \overline{\mathbb{Q}}_\ell)$. Moreover, the excision sequence for the partition $C = (C - P_f) \coprod P_f$ gives

$$0 \to \mathrm{H}_c^0(C_{\overline{k}}, \overline{\mathbb{Q}}_\ell) = \overline{\mathbb{Q}}_\ell \to \mathrm{H}_c^0(P_{f,\overline{k}}, \overline{\mathbb{Q}}_\ell) = \overline{\mathbb{Q}}_\ell^n \to \mathrm{H}_c^1(C_{\overline{k}} - P_{f,\overline{k}}, \overline{\mathbb{Q}}_\ell) \to \mathrm{H}_c^1(C_{\overline{k}}, \overline{\mathbb{Q}}_\ell) \to 0,$$

and the last space has dimension $2g$, giving the dimension of $\mathcal{G}_0$.

The last assertion is the consequence of Laumon's Theorem of stationary phase: the monodromy action at infinity on $\mathcal{G}$ can be described from local Fourier transforms as

$$LFT_\psi(\infty,\infty)(\mathcal{F}(\infty))\bigoplus\left(\oplus_{i=1}^t LFT_\psi(s_i,\infty)(\mathcal{F}(s_i)/\mathcal{F}_{s_i})\right).$$

From [**5**, Theorem 7.5.4], the slope >1 part of $\mathcal{G}(\infty)$ comes from the slope >1 part of $\mathcal{F}(\infty)$ via $LFT_\psi(\infty,\infty)$, and it vanishes. Thus the monodromy action at infinity on $\mathcal{F}$ comes from the description of the monodromy action on $\mathcal{F}$ at its points of non-lissity, and *loc. cit.* gives the desired result.

□

REMARK 2.3. One can be more precise for the monodromy action at 0 on $\mathcal{G}$: from [**5**, Theorem 7.4.3], we have an exact sequence

$$0\to \mathrm{H}^1_c(\mathbb{A}^1_{\overline{k}},\mathcal{F})\to\mathcal{G}(0)\to LFT_\psi(\infty,0)(\mathcal{F}(\infty))\to 0,$$

whose last term we can precise from [**5**, Theorem 7.5.4]

$$LFT_\psi(\infty,0)(\mathcal{F}(\infty))\simeq\overline{\mathbb{Q}}_\ell^{n-1}\bigoplus_{k=1}^{n}\bigoplus_{\chi_k^{d_k}=\mathbf{1},\ \chi_k\neq\mathbf{1}}\mathcal{L}_{\chi_k}.$$

We are ready to define the sheaf $\mathcal{H}$, that we shall use (together with its Adams operations) in the next sections to show our main result.

DEFINITION 2.4. Let ι denote the inclusion $\iota:\mathbb{G}_{m,k}\to\mathbb{A}^1_k$; we define the sheaf $\mathcal{H}$ on $\mathbb{G}_{m,k}$ as $\mathcal{H}:=\iota^*\mathcal{G}$.

For any $r\geq 1$, we define the *r-th Adams operation of* $\mathcal{H}$ as the (virtual) sheaf

$$[\mathcal{H}]^r:=\sum_{i=0}^{r-1}(-1)^i\cdot\mathbb{S}_{\lambda_i}\mathcal{H};$$

where $\mathbb{S}_{\lambda_i}$ is the Schur functor associated to the partition $\lambda_i(r-i,1,\dots,1)$ of r.

REMARK 2.5. One can find different decompositions of the virtual sheaf $[\mathcal{H}]^r$ as a sum of lisse sheaves in the literature; the first one seems to be

$$[\mathcal{H}]^r:=\sum_{i=0}^{r}(-1)^{i-1}(i-1)\cdot\mathrm{Sym}^{r-i}\mathcal{H}\otimes\wedge^i\mathcal{H}$$

from [**12**]; in [**6**] one can find

$$[\mathcal{H}]^r:=\sum_{i=1}^{r}(-1)^{i-1}i\cdot\mathrm{Sym}^{r-i}\mathcal{H}\otimes\wedge^i\mathcal{H}.$$

The one we choose here comes from [**9**, Proposition 3.4]; it reflects the decomposition of the sheaves $\mathrm{Sym}^{r-i}\mathcal{H}\otimes\wedge^i\mathcal{H}$ into irreducible representations of the symmetric group S_r acting on $\mathcal{H}^{\otimes r}$, and presents much less cancellation than the two other expressions. For this reason, the coefficients are smaller and give better bounds.

Recall that we use the sheaf $[\mathcal{H}]^r$ in order to evaluate the difference between the numbers of points in the Artin-Schreier covering C_f/C. This is motivated by the following lemma, which comes immediately from the equality $\mathrm{Tr}(\mathrm{Frob}_t|[\mathcal{H}]^r)=\mathrm{Tr}(\mathrm{Frob}_t^r|\mathcal{H})$ (*cf* [**9**, Proposition 3.4]) and lemma 2.2 *ii/*.

LEMMA 2.6. *Let $t \in k^\times$; then we have*

$$Tr(Frob_t|[\mathcal{H}]^r_t) = -S_r(C, \psi, tf).$$

3. Cohomology spaces, and the proof of the theorem

In this section we give some information on the L-functions of the sheaves $[\mathcal{H}]^r$; although we will not be as precise as [**10**], where the arithmetic action is also considered, this will be sufficient to prove Theorem 0.1.

First, from Grothendieck's trace formula (note that since the sheaf $[\mathcal{H}]^r$ is lisse over $\mathbb{G}_m$ the space H^0_c vanishes), we have for any $r \geq 1$

$$\begin{aligned} L([\mathcal{H}]^r, T) &= \frac{\det(1 - \mathrm{Frob}_k T | \mathrm{H}^1_c(\mathbb{G}_{m,\bar{k}}, [\mathcal{H}]^r))}{\det(1 - \mathrm{Frob}_k T | \mathrm{H}^2_c(\mathbb{G}_{m,\bar{k}}, [\mathcal{H}]^r))} \\ &= \frac{\prod_{i=0}^{r-1} \det(1 - \mathrm{Frob}_k T | \mathrm{H}^1_c(\mathbb{G}_{m,\bar{k}}, \mathbb{S}_{\lambda_i}\mathcal{H}))^{(-1)^i}}{\prod_{i=0}^{r-1} \det(1 - \mathrm{Frob}_k T | \mathrm{H}^2_c(\mathbb{G}_{m,\bar{k}}, \mathbb{S}_{\lambda_i}\mathcal{H}))^{(-1)^i}} \end{aligned}$$

From Deligne's main theorem, since the sheaf $[\mathcal{H}]^r$ is pure of weight r, the action of Frobenius on $\mathrm{H}^1_c(\mathbb{G}_{m,\bar{k}}, \mathbb{S}_{\lambda_i}\mathcal{H})$ is mixed of weights $\leq r+1$; in order to prove Theorem 0.1, we just have to show that the denominators above are trivial, and to give an upper bound for the dimensions of the first cohomology spaces.

LEMMA 3.1. *Assume that the differential df has only simple zeroes on C, and one of the following conditions hold*

i/ r is odd,

ii/ the hypersurface in C^r defined by the equation $f(P_1) + \ldots + f(P_r) = 0$ is non-singular;

then for any i we have $H^2_c(\mathbb{G}_{m,\bar{k}}, \mathbb{S}_{\lambda_i}\mathcal{H}) = 0$.

PROOF. From Poincaré duality, we are reduced to show that for any i, we have $\mathrm{H}^0(\mathbb{G}_{m,\bar{k}}, \mathcal{J}) = 0$, where $\mathcal{J} = (\mathbb{S}_{\lambda_i}\mathcal{H})^\vee(-1)$ is a Tate twisted dual, i.e. that the action of $\pi_1(\mathbb{G}_{m,\bar{k}})$ on this space has no invariant.

Denote by $\mathcal{F}_-$, $\mathcal{G}_-$ and $\mathcal{H}_-$ the sheaves defined as in section 2, but from the function $-f$. The sheaf $\mathcal{F}$ is self-dual; since it is a Fourier sheaf, from [**5**, Theorem 7.3.8], the dual of $\mathcal{G}$ is the sheaf $\mathcal{G}^\vee = NFT_\psi([-1]^*\mathcal{F})(1) = NFT_\psi(\mathcal{F}_-)(1) = \mathcal{G}_-(1)$. As a consequence, its restriction $\mathcal{H}$ has dual $\mathcal{H}^\vee = \mathcal{H}_-(1)$, and for any $0 \leq i \leq r$, the sheaf $\mathbb{S}_{\lambda_i}\mathcal{H}$ has dual $\mathbb{S}_{\lambda_i}\mathcal{H}_-(r)$. We conclude that $\mathcal{J} = \mathbb{S}_{\lambda_i}\mathcal{H}_-(r-1)$.

Since all representations $\mathbb{S}_{\lambda_i}\mathcal{H}_-$ are subrepresentations of the tensor representation $\mathcal{H}_-^{\otimes r}$, it suffices to show that the action of $\pi_1(\mathbb{G}_{m,\bar{k}})$ on this last one has no invariant. Actually it is sufficient to show that the action of I_∞ on the vector space $(\mathcal{G}_-)^{\otimes r}_{\bar{\eta}}$ has this property.

The hypothesis on the zeroes of df, joint with Lemma 2.2 *iv/* ensures that all multiplicative characters appearing in the representation $\mathcal{G}_-(\infty)$ are equal to the quadratic one. We denote it by ρ from now on. Thus the action of I_∞ on $(\mathcal{G}_-)^{\otimes r}_{\bar{\eta}}$ splits as a direct sum of representations of the form $\mathcal{L}_{\rho^r} \otimes \mathcal{L}_{\psi_{s_{i_1}+\ldots+s_{i_r}}}$ for some $s_{i_j} \in S$.

If r is odd, we obtain a non trivial action (as a reflection) of an element of order 2 in I_∞, and there is no non-zero invariant. Assume r is even; then the action of I_∞ on $(\mathcal{G}_-)^{\otimes r}_{\bar{\eta}}$ splits as a direct sum of representations of the form $\mathcal{L}_{\psi_{s_{i_1}+\ldots+s_{i_r}}}$

for some $s_{i_j} \in S$. Then the action of the wild inertia group is non trivial, except when we have $s_{i_1} + \ldots + s_{i_r} = 0$. Since the $s_i = f(Q_{k_i})$ are critical values, this last equality is verified if and only if the point $(Q_{k_1}, \ldots, Q_{k_r})$ is a singular point on the hypersurface $f(P_1) + \ldots + f(P_r) = 0$.

We have shown that, under either hypothesis, the action of I_∞ on the vector space $(\mathcal{G}_-)_{\overline{\eta}}^{\otimes r}$ has no non-zero invariants. This concludes the proof of the Lemma. □

LEMMA 3.2. *We have the following upper bound on the dimension of the first cohomology space of the sheaf* $\mathbb{S}_{\lambda_i}\mathcal{H}$

$$\dim H^1_c(\mathbb{G}_{m,\overline{k}}, \mathbb{S}_{\lambda_i}\mathcal{H}) \leq \binom{D-1+r-i}{r}\binom{r-1}{i}.$$

PROOF. We follow the proof of [**10**, Corollary 4]. From the splitting of the action of the inertia group at infinity on the generic fiber of the sheaf presented in the proof of Lemma 3.1, all slopes at infinity of the sheaf $\mathbb{S}_{\lambda_i}\mathcal{H}$ are 0 or 1; thus we can bound from above the Swan conductor at infinity (which is the dimension of H^1_c by the Grothendieck-Neron-Ogg-Shafarevic formula) by the generic rank of this sheaf. This is the claim. □

We are ready to show cases *i/* and *ii/* of Theorem 0.1 from the results above

PROOF. *of Theorem* 0.1, *cases i/ and ii/* From Lemma 2.6, the difference $N_r(C_f) - N_r(C)$ is the degree one coefficient of the power series expansion of $L([\mathcal{H}]^r; T)$ around 0. Now we have the cohomological expression

$$L([\mathcal{H}]^r; T) = \prod_{i=0}^{r} \det(1 - \mathrm{Frob}_k T | \mathrm{H}^1_c(\mathbb{G}_{m,\overline{k}}, \mathbb{S}_{\lambda_i}\mathcal{H}))^{(-1)^i}.$$

From Deligne's Theorem [**2**, Theorem 1], the $\mathrm{H}^1_c(\mathbb{G}_{m,\overline{k}}, \mathbb{S}_{\lambda_i}\mathcal{H})$ are mixed of weight $\leq r+1$, and considering the degree one coefficients, we get

$$|N_r(C_f) - N_r(C)| \leq \left(\sum_{i=0}^{r} \dim \mathrm{H}^1_c(\mathbb{G}_{m,\overline{k}}, \mathbb{S}_{\lambda_i}\mathcal{H})\right) q^{\frac{r+1}{2}}.$$

The proof is now complete from Lemma 3.2. □

4. Global monodromy, and a refinement of the theorem

In this section, we deal with the whole geometric monodromy group of the sheaf $\mathcal{H}$; in the preceding section, we needed only consider the action of the inertia group at infinity to get the vanishing of the second cohomology space. Here we shall use computations of Katz [**5**] which apply here under a more restrictive hypothesis on the characteristic and the critical values of the function f. In turn, we will be able to get the bound in case *iii/* of Theorem 0.1, which is valid without restriction on r. We assume $p > 2D+1$ in this section.

Recall that the geometric monodromy group of the sheaf $\mathcal{H}$ is the Zariski closure of the action of $\pi_1(\mathbb{G}_{m,\overline{k}})$ on the generic fiber $\mathcal{H}_{\overline{\eta}}$: it is a closed subgroup of the general linear group $\mathbf{GL}(\mathcal{H}_{\overline{\eta}})$. We denote it by G. Note that from the construction of $\mathcal{H}$, it is the same as the geometric monodromy group of the sheaf $\mathcal{G}$. We shall determine this last group, building on results of Katz.

As in the preceding section, we assume that the differential df has only simple zeroes on C, moreover we assume that f separates these zeroes. In the terminology of Katz [**5**, Section 7.10], f is a supermorse function. From Lemma 2.1 *iii/*, the inertia group at any critical value of f (i.e. at any point where $\mathcal{F}$ is not lisse) acts on the generic fiber of $\mathcal{F}$ by (tame) reflections. Moreover, from [**5**, Lemma 7.10.2.3], the sheaf $\mathcal{F}$ is geometrically irreducible. From [**5**, Theorem 7.9.4], the sheaf $\mathcal{G}$ remains geometrically irreducible, and the connected component of identity G_0 is a semisimple algebraic group [**2**, Corollaire 1.3.9]. If we assume moreover that the numbers $s_i - s_j$, $1 \leq i, j \leq D$, $i \neq j$, are pairwise distinct, it follows from [**5**, Theorem 7.9.6] that G_0 must contain $\mathbf{SL}(\mathcal{H}_{\overline{\eta}})$, thus we have $G_0 = \mathbf{SL}(\mathcal{H}_{\overline{\eta}})$.

In order to determine G, it remains (by the same result of Deligne) to determine its image by the determinant. From Lemma 2.2 and Remark 2.3, the determinant $\det \mathcal{G}$ is a lisse rank one sheaf over $\mathbb{G}_m$, with action of I_0 isomorphic to $\mathcal{L}_{\rho^N}$, where N is the number of poles of f with even multiplicity in $C(\overline{k})$. Since f is a supermorse function, we have $t = \#S = D$, and the local monodromy at infinity can be expressed in the following, simpler way

$$\mathcal{G}(\infty) \simeq \bigoplus_{i=1}^{D} \mathcal{L}_\rho \otimes \mathcal{L}_{\psi_{s_i}};$$

As a consequence, as a representation of the inertia group at infinity, the sheaf $\det \mathcal{G}$ is isomorphic to $\mathcal{L}_{\rho^D} \otimes \mathcal{L}_{\psi_s}$, where $s = s_1 + \ldots + s_D$. Since the sheaves $\det \mathcal{G}$ and $\mathcal{L}_{\rho^D} \otimes \mathcal{L}_{\psi_s}$ carry the same representations of the inertia groups at 0 and ∞ (note that $N \equiv D \mod 2$), they give rise to the same representation of $\pi_1(\mathbb{G}_{m,\overline{k}})$, and we conclude that they are geometrically isomorphic.

We summarize the hypotheses we made, and the results we deduced

LEMMA 4.1. *Assume $p > 2D+1$, f is a supermorse function, and the differences $s_i - s_j$, $1 \leq i, j \leq D$, $i \neq j$, are pairwise distinct. For V a vector space, denote by $\mathbf{GL}_m(V)$ the group of automorphisms of V whose determinant is an m-th root of unity. Set $s = s_1 + \ldots + s_D$. Then the geometric monodromy group of $\mathcal{H}$ is*

i/ $\mathbf{SL}(\mathcal{H}_{\overline{\eta}})$ *if $s = 0$ and D is even ;*
ii/ $\mathbf{GL}_2(\mathcal{H}_{\overline{\eta}})$ *if $s = 0$ and D is odd ;*
iii/ $\mathbf{GL}_p(\mathcal{H}_{\overline{\eta}})$ *if $s \neq 0$ and D is even ;*
iv/ $\mathbf{GL}_{2p}(\mathcal{H}_{\overline{\eta}})$ *if $s \neq 0$ and D is odd.*

It is now possible to conclude that the denominator in the cohomological expression of $L([\mathcal{H}]^r, T)$ is (almost) trivial in many more cases, and to deduce the last case of the main theorem.

THEOREM 4.2. *Assume that f satisfies the requirements of Lemma 4.1. Then we have the estimate*

$$|N_r(C_f) - N_r(C)| \leq C_{D,r} q^{\frac{r+1}{2}},$$

unless D is even, $s = 0$ and $r = D$, in which case there exists some $\beta \in \{\pm 1\}$ such that

$$|N_D(C_f) - N_D(C) - \beta q^{\frac{D}{2}+1}| \leq C_{D,D} q^{\frac{D+1}{2}}.$$

PROOF. We adapt the proof of [**10**, Corollary 5] to our presentation of the sheaf $[\mathcal{H}]^r$; the idea is to show that the second cohomology group vanishes, or is one dimensional in the second case, by computing the invariants of the action of the global monodromy group.

First note that from [**9**, Proposition 3.4], the sheaf $\mathbb{S}_{\lambda_i}\mathcal{H}$ is a subsheaf of $\mathrm{Sym}^{r-i}\mathcal{H} \otimes \wedge^i\mathcal{H}$. In the course of the proof of [**10**, Corollary 5], it is shown that the action of $\mathbf{SL}(\mathcal{H}_{\overline{\eta}})$ has no non-zero invariant on this last sheaf unless we have $i = r-1$ (or $i = r$ but from the definition of $\mathcal{H}$ we do not have to consider this case here). The action of $\mathbf{SL}(\mathcal{H}_{\overline{\eta}})$ on $\mathbb{S}_{\lambda_{r-1}}\mathcal{H} = \wedge^r\mathcal{H}$ has no non-zero invariant except for $r = D$. Then the geometric monodromy group of $\mathcal{H}$ contains the map ζId for any mD-th root of unity ζ; choose some primitive one. The last map acts on $\wedge^D\mathcal{H}$ as multiplication by ζ^D, and as long as $m \neq 1$, we get no non-zero invariant.

In the remaining case, $\wedge^D\mathcal{H}_{\overline{\eta}}$ is the one dimensional invariant space under the action of the geometric monodromy group on $[\mathcal{H}^D]_{\overline{\eta}}$. Since this group is $\mathbf{SL}(\mathcal{H}_{\overline{\eta}})$, the sheaf $\det\mathcal{H}$ is geometrically trivial; since $\mathcal{H}$ is pure of weight 1 and has rank D, $\det\mathcal{H}$ must be of the form $(q^{\frac{D}{2}}\beta)^{\deg}$, where β has absolute value 1. As representations of the Galois group $\mathrm{Gal}(\overline{k}/k)$, we have the isomorphism $\mathrm{H}^2_c(\mathbb{G}_{m,\overline{k}}, \wedge^D\mathcal{H}) \simeq \det\mathcal{H}(-1)$. This allows us to rewrite the denominator in the following way

$$\det(1 - \mathrm{Frob}_k T|\mathrm{H}^2_c(\mathbb{G}_{m,\overline{k}}, \wedge^D\mathcal{H})) = \det(1 - \mathrm{Frob}_k T|\det\mathcal{H}(-1)) = 1 - q^{\frac{D}{2}+1}\beta T.$$

Finally, the functions f and $-f$ give rise to the same L-function $L([\mathcal{H}]^D, T)$, thus the set of its roots and poles is stable under complex conjugation; since $q^{\frac{D}{2}+1}\beta$ is the only element of this set of weight $D+2$, we must have $\beta \in \{\pm 1\}$. This completes the proof of the last assertion. □

References

[1] P. Deligne, *Sommes trigonométriques*, in SGA $4\frac{1}{2}$, Lecture Notes in Math. **569** (1977), Springer. MR0463174 (57:3132)

[2] P. Deligne, *La conjecture de Weil : II*, Publ. Math. I.H.E.S. **132** (1980), 137-252. MR601520 (83c:14017)

[3] W. Fulton, J. Harris, *Representation theory: a first course*, Grad. Texts in Math. **129** (1991), Springer-Verlag. MR1153249 (93a:20069)

[4] A. Garcia, S. Tafazolian, *On additive polynomials and certain maximal curves*, J. Pure Appl. Alg. **212** (2008), 2513-2521. MR2440263 (2009e:11121)

[5] N. Katz, *Exponential sums and differential equations*, Annals of Math. Studies **124** (1990), PUP. MR1081536 (93a:14009)

[6] N. Katz, *Frobenius-Schur indicator and the ubiquity of Brock-Granville quadratic excess*, Finite Fields and Appl. **7** (2001), 45-69. MR1803935 (2002d:11069)

[7] G. Lachaud, *Artin-Schreier curves, exponential sums, and the Carlitz-Uchiyama bound for geometric codes*, J. Number Th. **39** (1991), 18-40. MR1123166 (93c:11108)

[8] G. Laumon, *Transformation de Fourier, constantes d'équations fonctionnelles et conjectures de Weil*, Publ. Math. I.H.É.S. **65** (1987), 131-210. MR908218 (88g:14019)

[9] A. Rojas Leon, *Rationality of trace and norm L-functions*, preprint avalaible at `http://arxiv.org/abs/1007.5324`.

[10] A. Rojas Leon, D. Wan, *Improvement of the Weil bound for Artin-Schreier curves*, Math. Ann. **351** (2011), 417–442. MR2836667

[11] H. Stichtenoth, *Algebraic function fields and codes*, Springer-Verlag, Berlin (1993). MR1251961 (94k:14016)

[12] D. Wan, *Dwork's conjecture on unit root zeta function*, Ann. Math. **150** (1999), 867-927. MR1740990 (2001a:11108)

Équipe LAMIA, IUFM de la Guadeloupe
Current address: Équipe LAMIA, IUFM de la Guadeloupe
E-mail address: `rblache@iufm.univ-ag.fr`

Contemporary Mathematics
Volume **574**, 2012
http://dx.doi.org/10.1090/conm/574/11423

Some more functions that are not APN infinitely often. The case of Gold and Kasami exponents

Eric Férard, Roger Oyono, and François Rodier

ABSTRACT. We prove a necessary condition for some polynomials of Gold and Kasami degree to be APN over $\mathbb{F}_{q^n}$ for large n.

1. Introduction

The vector Boolean functions are used in cryptography to construct block ciphers and an important criterion on these functions is high resistance to differential cryptanalysis.

Let $q = 2^n$ for some positive integer n. A function $f : \mathbb{F}_q \longrightarrow \mathbb{F}_q$ is said to be *almost perfect nonlinear* (APN) on $\mathbb{F}_q$ if the number of solutions in $\mathbb{F}_q$ of the equation

$$f(x + a) + f(x) = b$$

is at most 2, for all $a, b \in \mathbb{F}_q$, $a \neq 0$. This kind of function has a good resistance to differential cryptanalysis as was proved by Nyberg in [**20**].

So far, the study of APN functions has focused on power functions. Recently it was generalised to other functions, particularly quadratic polynomials (Edel, Kyureghyan and Pott [**11**], or Budaghyan, Carlet, Felke and Leander [**4**]) or polynomials on small fields (Dillon [**9**]). On the other hand, several authors (Berger, Canteaut, Charpin and Laigle-Chapuy [**2**], Byrne and McGuire [**5**], Jedlicka [**17**], Rodier [**21**], or Férard and Rodier [**12, 13**]) showed that APN functions did not exist in certain cases.

There are many classes of function for which it can be shown that each function is APN for at most a finite number of extensions [**23, 21**]. So we fix a finite field $\mathbb{F}_q$ and a function $f : \mathbb{F}_q \to \mathbb{F}_q$ given by a polynomial in $\mathbb{F}_q[x]$ and we set the question of whether this function can be APN for an infinite number of extensions of $\mathbb{F}_q$.

In this approach, Hernando and McGuire [**14**] showed a result on the classification of APN monomials which has been conjectured for 40 years: the only exponents such that the monomial x^d are APN over infinitely many extension of $\mathbb{F}_2$ are of the form $2^i + 1$ or $4^i - 2^i + 1$. One calls these exponents *exceptional*

2010 *Mathematics Subject Classification.* Primary 11T06, 12E05, 14Q10, 11T71.
Key words and phrases. APN functions, algebraic surfaces, finite fields, absolute irreducible polynomials.

exponents. Then it is natural to formulate for polynomial functions the following conjecture.

CONJECTURE 1.1 (Aubry, McGuire and Rodier). *A polynomial on $\mathbb{F}_q$ can be APN for an infinity of extensions of $\mathbb{F}_q$ only if it is CCZ equivalent (as was defined by Carlet, Charpin and Zinoviev in* [**7**]*) to a monomial x^t where t is an exceptional exponent.*

A means to prove this conjecture is to remark that the APN property is equivalent to the fact that the rational points of the algebraic surface X in a 3-dimensional space defined by

$$\phi(x,y,z)=\frac{f(x)+f(y)+f(z)+f(x+y+z)}{(x+y)(x+z)(y+z)}$$

(which is a polynomial in $\mathbb{F}_q[x,y,z]$) are all in a surface V made of the three planes $x+y=0, x+z=0, y+z=0$.

Some cases of this conjecture have been studied already, in particular the case of polynomials of odd degree, not Gold or Kasami [**1**]. It is also true for polynomials of degree <13 (see [**1**] and [**21**]). Some partials results have been obtain in case of polynomials of Gold degree or of even degree [**1, 22**]. We recall them in Section 3 (see Theorems 3.1 to 3.5). Nevertheless, in characteristic 3, there exists polynomials, not equivalent to monomials, which are PN for an infinity of extensions of $\mathbb{F}_q$ (see [**18**, théorème 3.3.7] or [**8, 10**]). In this paper, we will study polynomials of Kasami degree. The proofs happen to be somehow the same as in Gold degree, with a few changes anyway.

In Section 5, we study the special case of binomials of Gold and Kasami degree. For instance, we prove that any binomial of Gold degree could not be APN on infinitely many extensions of $\mathbb{F}_q$.

2. Preliminaries

We define

$$\phi(x,y,z)=\frac{f(x)+f(y)+f(z)+f(x+y+z)}{(x+y)(x+z)(y+z)}$$

which is a polynomial in $\mathbb{F}_q[x,y,z]$. This polynomial defines a surface X in the three dimensional affine space $\mathbb{A}^3$.

If X is absolutely irreducible (or has an absolutely irreducible component defined over $\mathbb{F}_q$) then f is not APN on $\mathbb{F}_{q^n}$ for all n sufficiently large. As shown in [**21**], this follows from the Lang-Weil bound for surfaces, which guarantees many $\mathbb{F}_{q^n}$-rational points on the surface for all n sufficiently large.

We call $\phi_j(x,y,z)$ the ϕ function associated to the monomial x^j. The function $\phi_j(x,y,z)$ is homogeneous of degree $j-3$.

We recall a result due to Janwa, Wilson, [**15**, Theorem 5] about Kasami exponents.

THEOREM 2.1. *If $f(x)=x^{2^{2k}-2^k+1}$ then*

$$\phi(x,y,z)=\prod_{\alpha\in\mathbb{F}_{2^k}-\mathbb{F}_2}p_\alpha(x,y,z) \tag{2.1}$$

where for each α, $p_\alpha(x,y,z)$ is an absolutely irreducible polynomial of degree 2^k+1 on $\mathbb{F}_{2^k}$ such that $p_\alpha(x,0,1)=(x-\alpha)^{2^k+1}$.

3. Some functions that are not APN infinitely often

The best known examples of APN functions are the Gold functions x^{2^k+1} and the Kasami-Welch functions $x^{4^k-2^k+1}$. These functions are defined over $\mathbb{F}_2$, and are APN on any field $\mathbb{F}_{2^m}$ where $\gcd(k,m)=1$. For other odd degree polynomial functions, we can state a general result.

THEOREM 3.1 (Aubry, McGuire and Rodier, [**1**]). *If the degree of the polynomial function f is odd and not a Gold or a Kasami-Welch number then f is not APN over $\mathbb{F}_{q^n}$ for all n sufficiently large.*

In the even degree case, we can state the result when half of the degree is odd, with an extra minor condition.

THEOREM 3.2 (Aubry, McGuire and Rodier, [**1**]). *If the degree of the polynomial function f is $2e$ with e odd, and if f contains a term of odd degree, then f is not APN over $\mathbb{F}_{q^n}$ for all n sufficiently large.*

In [**22**] we have some results for the case of polynomials of degree $4e$ where e is odd.

THEOREM 3.3. *If the degree of the polynomial function f is even such that $\deg(f)=4e$ with $e\equiv 3 \pmod 4$, and if the polynomials of the form*

$$(x+y)(y+z)(z+x)+P$$

with

$$P(x,y,z)=c_1(x^2+y^2+z^2)+c_4(xy+xz+zy)+b_1(x+y+z)+d \tag{3.1}$$

for $c_1,c_4,b_1,d\in\mathbb{F}_{q^3}$, do not divide ϕ then f is not APN over $\mathbb{F}_{q^n}$ for n large.

We have more precise results for polynomials of degree 12.

THEOREM 3.4. *If the degree of the polynomial f defined over $\mathbb{F}_q$ is 12, then either f is not APN over $\mathbb{F}_{q^n}$ for large n or f is CCZ equivalent to the Gold function x^3. In this case f is of the form*

$$L(x^3)+L_1 \text{ or } (L(x))^3+L_1$$

where L is a linearized polynomial

$$x^4+x^2(c^{1+q}+c^{1+q^2}+c^{q+q^2})+xc^{1+q+q^2},$$

c is an element of $\mathbb{F}_{q^3}$ such that $c+c^q+c^{q^2}=0$ and L_1 is a q-affine polynomial of degree at most 8 (that is a polynomial whose monomials are of degree 0 or a power of 2).

We have some results on the polynomials of Gold degree $d=2^k+1$.

THEOREM 3.5 (Aubry, McGuire and Rodier, [**1**]). *Suppose $f(x)=x^d+g(x)$ where $\deg(g)\le 2^{k-1}+1$. Let $g(x)=\sum_{j=0}^{2^{k-1}+1}a_jx^j$. Suppose moreover that there exists a nonzero coefficient a_j of g such that $\phi_j(x,y,z)$ is absolutely irreducible (where $\phi_i(x,y,z)$ denote the polynomial $\phi(x,y,z)$ associated to x^i). Then f is not APN over $\mathbb{F}_{q^n}$ for all n sufficiently large.*

4. Polynomials of Kasami degree

Suppose the degree of f is a Kasami number $d = 2^{2k} - 2^k + 1$. Set d to be this value for this section. Write f as $f(x) = x^d + g(x)$ with $\deg(g) \leq d - 1$.

Then the degree of ϕ is $d - 3 = 2^{2k} - 2^k - 2$. We will prove the absolute irreducibility for a certain type of f.

4.1. The case $\deg(g) \leq 2^{2k-1} - 2^{k-1} + 1$.

THEOREM 4.1. *Suppose $f(x) = x^d + g(x)$ where $\deg(g) \leq 2^{2k-1} - 2^{k-1} + 1$. Let $g(x) = \sum_{j=0}^{2^{2k-1}-2^{k-1}+1} a_j x^j$. Suppose moreover that there exists a nonzero coefficient a_j of g such that $\phi_j(x, y, z)$ is absolutely irreducible. Then $\phi(x, y, z)$ is absolutely irreducible.*

Proof: Suppose $\phi(x, y, z) = P(x, y, z)Q(x, y, z)$ with $\deg P \geq \deg Q$. Write each polynomial as a sum of homogeneous parts:

$$\sum_{j=3}^{d} a_j \phi_j(x, y, z) = (P_s + P_{s-1} + \cdots + P_0)(Q_t + Q_{t-1} + \cdots + Q_0) \tag{4.1}$$

where P_j, Q_j are homogeneous of degree j. Then from the Theorem 2.1 we get

$$P_s Q_t = \prod_{\alpha \in \mathbb{F}_{2^k} - \mathbb{F}_2} p_\alpha(x, y, z).$$

In particular this implies that P_s and Q_t are relatively prime as the product is made of distinct irreducible factors.

The homogeneous terms of degree less than $d-3$ and greater than $2^{2k-1} - 2^{k-1}$ are 0, by the assumed bound on the degree of g. Equating terms of degree $s+t-1$ in the equation (4.1) gives $P_s Q_{t-1} + P_{s-1} Q_t = 0$. Hence P_s divides $P_{s-1} Q_t$ which implies P_s divides P_{s-1} because $\gcd(P_s, Q_t) = 1$, and we conclude $P_{s-1} = 0$ as $\deg P_{s-1} < \deg P_s$. Then we also get $Q_{t-1} = 0$. Similarly, $P_{s-2} = 0 = Q_{t-2}$, $P_{s-3} = 0 = Q_{t-3}$, and so on until we get the equation

$$P_s Q_0 + P_{s-t} Q_t = 0$$

since we suppose that $s \geq t$. This equation implies P_s divides $P_{s-t} Q_t$, which implies P_s divides P_{s-t}, which implies $P_{s-t} = 0$. Since $P_s \neq 0$ we must have $Q_0 = 0$.

We now have shown that $Q = Q_t$ is homogeneous. In particular, this means that $\phi_j(x, y, z)$ is divisible by $p_\alpha(x, y, z)$ for some $\alpha \in \mathbb{F}_{2^k} - \mathbb{F}_2$ and for all j such that $a_j \neq 0$. We are done if there exists such a j with $\phi_j(x, y, z)$ irreducible. Since $\phi_j(x, y, z)$ is defined over $\mathbb{F}_2$ it implies that $p_\alpha(x, y, z)$ also, which is a contradiction with the fact that α is not in $\mathbb{F}_2$.

□

REMARK 4.1. As in Theorem 3.5, the above theorem and the next corollary are true with the weaker hypothesis that there exists a nonzero coefficient a_j such that ϕ_j is prime to ϕ_d. We give in Section 5 some criterion about $j < d$ satisfying $\gcd(\phi_j, \phi_d) = 1$.

COROLLARY 4.1. *Suppose $f(x) = x^d + g(x)$ where g is a polynomial in $\mathbb{F}_q[x]$ such that $\deg(g) \leq 2^{2k-1} - 2^{k-1} + 1$. Let $g(x) = \sum_{j=0}^{2^{2k-1}-2^{k-1}+1} a_j x^j$. Suppose* *moreover that there exists a nonzero coefficient a_j of g such that $\phi_j(x,y,z)$ is absolutely irreducible. Then the polynomial f is not APN on infinitely many extensions of $\mathbb{F}_q$.*

REMARK 4.2. It is quite possible that, for a polynomial f as in Corollary 4.1, there is no extension of $\mathbb{F}_q$ where f is APN. This is an interesting but non trivial problem.

4.2. Next step: The case $\deg(g) = 2^{2k-1} - 2^{k-1} + 2$.

If we jump one degree more we need other arguments to prove irreducibility.

THEOREM 4.2. *Let $q = 2^n$. Suppose $f(x) = x^d + g(x)$ where $g(x) \in \mathbb{F}_q[x]$ and $\deg(g) = 2^{2k-1} - 2^{k-1} + 2$. Let $k \geq 3$ be odd and relatively prime to n. If $g(x)$ does not have the form $ax^{2^{2k-1}-2^{k-1}+2} + a^2x^3$ then ϕ is absolutely irreducible, while if $g(x)$ does have the form $ax^{2^{2k-1}-2^{k-1}+2} + a^2x^3$ then either ϕ is irreducible or ϕ splits into two absolutely irreducible factors which are both defined over $\mathbb{F}_q$.*

Proof: Suppose $\phi(x,y,z) = P(x,y,z)Q(x,y,z)$ with $\deg P \geq \deg Q$ and let

$$g(x) = \sum_{j=0}^{2^{2k-1}-2^{k-1}+2} a_j x^j.$$

Write each polynomial as a sum of homogeneous parts:

$$\sum_{j=3}^{d} a_j\phi_j(x,y,z) = (P_s + P_{s-1} + \cdots + P_0)(Q_t + Q_{t-1} + \cdots + Q_0).$$

Then

$$P_sQ_t = \prod_{\alpha\in\mathbb{F}_{2^k}-\mathbb{F}_2} p_\alpha(x,y,z).$$

In particular this means P_s and Q_t are relatively prime as in the previous theorem.

Since $s \geq t$, we have $s \geq 2^{2k-1} - 2^{k-1} - 1$. Comparing each degree gives $P_{s-1} = 0 = Q_{t-1}$, $P_{s-2} = 0 = Q_{t-2}$, and so on until we get the equation of degree $s+1$

$$P_sQ_1 + P_{s-t+1}Q_t = 0$$

which implies $P_{s-t+1} = 0 = Q_1$.

If $s \neq t$ then $s \geq 2^{2k-1} - 2^{k-1}$. Note then that $a_{s+3}\phi_{s+3} = 0$. The equation of degree s is

$$P_sQ_0 + P_{s-t}Q_t = a_{s+3}\phi_{s+3} = 0.$$

This means that $P_{s-t} = 0$, so $Q_0 = 0$. We now have shown that $Q = Q_t$ is homogeneous. In particular, this means that $\phi(x,y,z)$ is divisible by $p_\alpha(x,y,z)$ for some $\alpha \in \mathbb{F}_{2^k} - \mathbb{F}_2$, which is impossible, as we will show. Indeed, since the leading coefficient of g is not 0, the polynomial $\phi_{2^{2k-1}-2^{k-1}+2}$ occurs in ϕ; as

$$\phi_{2^{2k-1}-2^{k-1}+2} = \phi^2_{2^{2k-2}-2^{k-2}+1}(x+y)(y+z)(z+x), \tag{4.2}$$

this polynomial is prime to ϕ, because if $p_\alpha(x,y,z)$ occurs in the polynomials $\phi_{2^{2k-1}-2^{k-1}+2}$, then it will occur in $\phi_{2^{2k-2}-2^{k-2}+1}$. If that is the case, the polynomial $p_\alpha(x,0,1) = (x-\alpha)^{2^k+1}$ would divide $\phi_{2^{2k-2}-2^{k-2}+1}(x,0,1)$. One has

$$(x+y)(y+z)(z+x)\phi_{2^{2k-2}-2^{k-2}+1}(x,y,z)$$
$$= x^{2^{2k-2}-2^{k-2}+1} + y^{2^{2k-2}-2^{k-2}+1} + z^{2^{2k-2}-2^{k-2}+1} + (x+y+z)^{2^{2k-2}-2^{k-2}+1}$$

hence

$$x(x+1)\phi_{2^{2k-2}-2^{k-2}+1}(x,0,1) = x^{2^{2k-2}-2^{k-2}+1} + 1 + (x+1)^{2^{2k-2}-2^{k-2}+1}.$$

Let $u = x - \alpha$. We have, for some polynomial R:

$$\begin{aligned}
& x(x+1)\phi_{2^{2k-2}-2^{k-2}+1}(x,0,1) \\
= \; & (u+\alpha)(u+\alpha+1)\phi_{2^{2k-2}-2^{k-2}+1}(u+\alpha,0,1) \\
= \; & (u+\alpha)^{2^{2k-2}-2^{k-2}+1} + 1 + (u+\alpha+1)^{2^{2k-2}-2^{k-2}+1} \\
= \; & \alpha^{2^{2k-2}-2^{k-2}+1} + u\alpha^{2^{2k-2}-2^{k-2}} + u^{2^{k-2}}\alpha^{2^{2k-2}-2^{k-1}+1} + 1 + \\
& +(\alpha+1)^{2^{2k-2}-2^{k-2}+1} + u(\alpha+1)^{2^{2k-2}-2^{k-2}} \\
& +u^{2^{k-2}}(\alpha+1)^{2^{2k-2}-2^{k-1}+1} + u^{2^{k-2}+1}R(u).
\end{aligned}$$

As $\alpha^{2^k-1} = 1$ we have $\alpha^{2^{2k-2}-2^{k-2}} = \alpha^{2^{k-2}(2^k-1)} = 1$. So

$$\begin{aligned}
& x(x+1)\phi_{2^{2k-2}-2^{k-2}+1}(x,0,1) \\
= \; & \alpha + u + u^{2^{k-2}}\alpha^{1-2^{k-2}} + 1 + (\alpha+1) + u + u^{2^{k-2}}(\alpha+1)^{1-2^{k-2}} + u^{2^{k-2}+1}R(u) \\
= \; & u^{2^{k-2}}(\alpha^{1-2^{k-2}} + (\alpha+1)^{1-2^{k-2}}) + u^{2^{k-2}+1}R(u)
\end{aligned}$$

which is a contradiction.

Suppose next that $s = t = 2^{2k-1} - 2^{k-1} - 1$ in which case the degree s equation is

$$P_sQ_0 + P_0Q_s = a_{s+3}\phi_{s+3}.$$

If $Q_0 = 0$, then

$$\phi(x,y,z) = \sum_{j=3}^{d} a_j\phi_j(x,y,z) = (P_s + P_0)Q_s$$

which implies that

$$\phi(x,y,z) = a_d\phi_d(x,y,z) + a_{2^{2k-1}-2^{k-1}+2}\phi_{2^{2k-1}-2^{k-1}+2}(x,y,z) = P_sQ_t + P_0Q_t$$

and $P_0 \neq 0$, since $g \neq 0$. So one has $\phi_{2^{2k-1}-2^{k-1}+2}$ divides $\phi_d(x,y,z)$ which is impossible by (4.2).

We may assume then that $P_0 = Q_0$. Then we have

$$\phi(x,y,z) = (P_s + P_0)(Q_s + Q_0) = P_sQ_s + P_0(P_s + Q_s) + P_0^2. \tag{4.3}$$

Note that this implies $a_j = 0$ for all j except $j = 3$ and $j = s + 3$. This means

$$f(x) = x^d + a_{s+3}x^{s+3} + a_3x^3.$$

So if $f(x)$ does not have this form, this shows that ϕ is absolutely irreducible.

If on the contrary ϕ splits as $(P_s + P_0)(Q_s + Q_0)$, the factors $P_s + P_0$ and $Q_s + Q_0$ are irreducible, as can be shown by using the same argument.

Assume from now on that $f(x) = x^d + a_{s+3}x^{s+3} + a_3x^3$ and that (4.3) holds. Then $a_3 = P_0^2$, so clearly $P_0 = \sqrt{a_3}$ is defined over $\mathbb{F}_q$. We claim that P_s and Q_s are actually defined over $\mathbb{F}_2$.

We know from (2.1) that P_sQ_s is defined over $\mathbb{F}_2$.

Also $P_0(P_s + Q_s) = a_{s+3}\phi_{s+3}$, so $P_s + Q_s = (a_{s+3}/\sqrt{a_3})\phi_{s+3}$. On the one hand, $P_s + Q_s$ is defined over $\mathbb{F}_{2^k}$ by Theorem 2.1. On the other hand, since ϕ_{s+3} is defined over $\mathbb{F}_2$ we may say that $P_s + Q_s$ is defined over $\mathbb{F}_q$. Because $(k,n) = 1$ we may conclude that $P_s + Q_s$ is defined over $\mathbb{F}_2$. Note that the leading coefficient of

$P_s + Q_s$ is 1, so $a_{s+3}^2 = a_3$. Whence if this condition is not true, then ϕ is absolutely irreducible.

Let σ denote the Galois automorphism $x \mapsto x^2$. Then $P_s Q_s = \sigma(P_s Q_s) = \sigma(P_s)\sigma(Q_s)$, and $P_s + Q_s = \sigma(P_s + Q_s) = \sigma(P_s) + \sigma(Q_s)$. This means σ either fixes both P_s and Q_s, in which case we are done, or else σ interchanges them. In the latter case, σ^2 fixes both P_s and Q_s, so they are defined over $\mathbb{F}_4$. Because they are certainly defined over $\mathbb{F}_{2^k}$ by Theorem 2.1, and k is odd, they are defined over $\mathbb{F}_{2^k} \cap \mathbb{F}_4 = \mathbb{F}_2$.

Finally, we have now shown that X either is irreducible, or splits into two absolutely irreducible factors defined over $\mathbb{F}_q$. □

REMARK 4.3. For $k = 3$, the polynomial ϕ corresponding to $f(x) = x^{57} + ax^{30} + a^2x^3$ where $a \in \mathbb{F}_q$ is irreducible. Indeed if it were not, we would have P_{27} and Q_{27} defined over $\mathbb{F}_2$, so by Theorem 2.1, $P_{27} = p_\beta(x,y,z)p_{\beta^2}(x,y,z)p_{\beta^4}(x,y,z)$ and $Q_{27} = p_{\beta^3}(x,y,z)p_{\beta^5}(x,y,z)p_{\beta^6}(x,y,z)$ for some $\beta \in \mathbb{F}_8 - \mathbb{F}_2$. So, up to inversion, we would check that $P_{27}(x,0,1) = (1+x+x^3)^9$ and $Q_{27}(x,0,1) = (1+x^2+x^3)^9$, hence $P_{27}(x,0,1) + Q_{27}(x,0,1) = (1+x+x^3)^9 + (1+x^2+x^3)^9$, and one can check that this is not equal to $\phi_{30}(x,0,1)$ as it should be.

5. Binomials that are not APN infinitely often

Another class of functions which are known not to be APN on infinitely many extensions of $\mathbb{F}_q$ comes from certain binomials:

THEOREM 5.1 (Voloch[**23**]). *Let $f(x) = x^m + cx^r$, where $c \in \mathbb{F}_{2^n}^*$, $3 \le r < m$ are coprime integers, not both even, neither a power of two and such that $(m-1, r-1)$ is a power of two. Then f is not APN on infinitely many extension of $\mathbb{F}_{2^n}$.*

We note that the assumption m, r coprime could be omitted as mentioned in [**21**]. In the following we will look at binomials of Kasami degree or Gold degree, i.e. binomials of the form $f(x) = x^d + ax^{d'}$ with $d = 2^{2k} - 2^k + 1$ or $d = 2^k + 1$, $d' < d$. We will restrict to such d' which are not a power of 2 since the class of APN functions is invariant by addition of a q-affine polynomial.

5.1. Binomials of Gold degree. Let $d = 2^k + 1$ be a Gold exponent, $a \in \mathbb{F}_q^*$ and $d' < d$ an integer not a power of two. We deduce immediately from Voloch's theorem the following result:

THEOREM 5.2. *Let $k \ge 1$, $d = 2^k + 1$, $a \in \mathbb{F}_q^*$ and $d' < d$ an integer not a power of two. If $f(x) = x^d + ax^{d'} \in \mathbb{F}_q[x]$, then the polynomial $\phi(x,y,z)$ is absolutely irreducible, and f is in particular not APN on infinitely many extensions of $\mathbb{F}_q$.*

5.2. Binomials of Kasami degree. Let $k \ge 3$ an integer and $d = 2^{2k} - 2^k + 1$ a Kasami exponent. In what follows, we will prove:

THEOREM 5.3. *Let $k \ge 3$, $d = 2^{2k} - 2^k + 1$ and $a \in \mathbb{F}_q^*$. Let $d' < d$ an integer not a power of two and not of the form $2^v(2^k\ell + 1)$ where ℓ is an integer such that $\gcd(\ell, 2^k - 1) \ne 1$. If $f(x) = x^d + ax^{d'} \in \mathbb{F}_q[x]$, then the polynomial $\phi(x,y,z)$ is absolutely irreducible, and f is in particular not APN on infinitely many extensions of $\mathbb{F}_q$.*

To prove Theorem 5.3, we will first derive the following lemma:

LEMMA 5.4. *Let $3 \leq t < d$ be an odd integer. If there is some $\alpha \in \mathbb{F}_{2^k} - \mathbb{F}_2$ such that p_α divides ϕ_t, then $t = 2^k\ell+1$ where ℓ is an integer not coprime to $2^k - 1$.*

Proof: Suppose that there is some $\alpha \in \mathbb{F}_{2^k} - \mathbb{F}_2$ such that p_α divides ϕ_t. Then, $t - 3 \geq 2^k + 1$ and there is a polynomial $r(x, y, z)$ such that

$$(x+y)(x+z)(y+z)p_\alpha(x,y,z)r(x,y,z) = x^t + y^t + z^t + (x+y+z)^t.$$

Evaluating the above equality at $y = 0$ and $z = 1$ yields

$$x(x+1)p_\alpha(x,0,1)r(x,0,1) = x^t + 1 + (x+1)^t.$$

Let $u = x + \alpha$. Then

$$(u+\alpha)(u+\alpha+1)u^{2^k+1}r(u+\alpha,0,1) = (u+\alpha)^t + 1 + (u+\alpha+1)^t,$$

and thus

$$(u+\alpha)^t + (u+\alpha+1)^t + 1 = 0 \mod u^{2^k+1}.$$

On the other hand, we know that

$$\begin{aligned} 1 + (u+\alpha)^t + (u+\alpha+1)^t &= 1 + \sum_{i=0}^{t} \binom{t}{i} \left(\alpha^{t-i} + (\alpha+1)^{t-i}\right) u^i \\ &= 1 + \alpha^t + (\alpha+1)^t + (\alpha^{t-1} + (\alpha+1)^{t-1})u \\ &\quad + \sum_{i=2}^{2^k} \binom{t}{i} \left(\alpha^{t-i} + (\alpha+1)^{t-i}\right) u^i \mod u^{2^k+1}. \end{aligned}$$

Now, if $\alpha^t + (\alpha+1)^t + 1 = \alpha^{t-1} + (\alpha+1)^{t-1} = 0$, we then have $(\alpha+1)^t = \alpha^{t-1}(\alpha+1)$ and thus

$$1 = \alpha^t + (\alpha+1)^t = \alpha^t + \alpha^{t-1}(\alpha+1) = \alpha^{t-1}.$$

From $\alpha^{t-1} = 1$ and $\alpha^{2^k-1} = 1$ we conclude that the order of $\alpha \neq 1$ should divide $\gcd(t-1, 2^k-1)$, and thus $\gcd(t-1, 2^k-1) > 1$.

Recall that $\alpha^t + (\alpha+1)^t + 1 = \alpha^{t-1} + (\alpha+1)^{t-1} = 0$. For any integer j:

$$\begin{aligned} \alpha^{t-1-2^j} + (\alpha+1)^{t-1-2^j} &= \frac{\alpha^{t-1}}{\alpha^{2^j}} + \frac{(\alpha+1)^{t-1}}{(\alpha+1)^{2^j}} \\ &= \frac{\alpha^{t-1}}{\alpha^{2^j}} + \frac{\alpha^{t-1}}{(\alpha+1)^{2^j}} = \frac{\alpha^{t-1}}{\alpha^{2^j}(1+\alpha^{2^j})} \neq 0. \end{aligned}$$

From this, we conclude that $\binom{t}{1+2^j} \equiv 0 \mod 2$ for any $j \in [1, k-1]$, which is equivalent, by Lucas theorem's [**19**][p. 230], to say that the $(j+1)$-th bit of t is 0 for j in the range $[1, k-1]$ i.e. 2^k divides $t-1$ since t is odd. In particular there is some integer ℓ such that $t - 1 = 2^k\ell$ and thus $\gcd(t-1, 2^k-1) = \gcd(\ell, 2^k-1) > 1$. □

We will use the following Lemma 5.1 from [**21**] to prove that ϕ is absolutely irreducible.

LEMMA 5.5. *Let $\phi(x,y,z) \in \mathbb{F}_q[x,y,z]$ be the sum of two homogeneous polynomials, i.e. $\phi = \phi_r + \phi_d$ where ϕ_i is a homogeneous polynomial of degree i, $r < d$. Suppose that $\gcd(\phi_d, \phi_r) = 1$ and either ϕ_d or ϕ_r factors into distinct factors over $\overline{\mathbb{F}}_q$. Then ϕ is absolutely irreducible.*

Proof of Theorem 5.3: According to the definition of ϕ and from $f(x) = x^d + ax^{d'}$ we get

$$\phi(x,y,z) = \phi_d(x,y,z) + a\phi_{d'}(x,y,z),$$

where ϕ_d (resp. $\phi_{d'}$) is a homogeneous polynomial of degree $d-3$ (resp. $d'-3$).

Let denote by v the 2-adic valuation of d' and t odd such that $d' = 2^v \cdot t$ where t is ≥ 3.

From Theorem 2.1 we know that ϕ_d factors over $\overline{\mathbb{F}}_q$ as $\prod_{\alpha\in\mathbb{F}_{2^k}-\mathbb{F}_2} p_\alpha(x,y,z)$. In order to prove that ϕ is absolutely irreducible, and as a consequence of Lemma 5.5, it is then sufficient to show that $\gcd(\phi_d, \phi_{d'}) = 1$. Suppose that $\gcd(\phi_d, \phi_{d'}) \neq 1$, i.e. there is an $\alpha \in \mathbb{F}_{2^k} - \mathbb{F}_2$ such that p_α divides $\phi_{d'}$. Since $d' = 2^v t$, we have

$$\phi_{d'}(x,y,z) = \phi_t^{2^v}(x,y,z) \cdot ((x+y)(x+z)(y+z))^{2^v-1},$$

and in particular p_α divides ϕ_t which is impossible by Lemma 5.4.

References

1. Y. Aubry, G. McGuire, F. Rodier, A Few More Functions That Are Not APN Infinitely Often, Finite Fields: Theory and applications, Ninth International conference Finite Fields and Applications, McGuire et al. editors, Contemporary Math. n°518, AMS, Providence (RI), USA, 2010. Available on arXiv: n° 0909.2304. MR2648536 (2011j:14061)
2. T. Berger, A. Canteaut, P. Charpin and Y. Laigle-Chapuy, On almost perfect nonlinear functions over F_2^n, IEEE Trans. Inform. Theory, 52(9), pp. 4160–4170, 2006. MR2298539 (2008a:94107)
3. C. Bracken, E. Byrne, N. Markin, G. McGuire, New families of quadratic almost perfect nonlinear trinomials and multinomials, Finite Fields and their Applications, 14 (2008), pp. 703–714. MR2435056 (2009e:11224)
4. L. Budaghyan, C. Carlet, P. Felke and G. Leander, An infinite class of quadratic APN functions which are not equivalent to power mappings, Cryptology ePrint Archive, 2005/359.
5. E. Byrne and G. McGuire On the Non-Existence of Quadratic APN and Crooked Functions on Finite Fields, WCC 2005.
6. A. Canteaut, Differential cryptanalysis of Feistel ciphers and differentially δ-uniform mappings, In Selected Areas on Cryptography, SAC'97, pp. 172–184, Ottawa, Canada, 1997.
7. C. Carlet, P. Charpin and V. Zinoviev, Codes, bent functions and permutations suitable for DES-like cryptosystems, Designs, Codes and Cryptography, 15(2), pp. 125–156, 1998. MR1658423 (99k:94030)
8. R. S. Coulter and R. W. Matthews, Planar functions and planes of Lenz-Barlotti class II, Des. Codes Cryptogr., 10(2), pp. 167–184, 1997. MR1432296 (97j:51010)
9. J. F. Dillon, APN Polynomials: An update, Fq9 International conference on finite fields and their Applications, July 2009.
10. C. Ding and J. Yuan, A family of skew Hadamard difference sets, J. Combin. Theory Ser. A, 113(7), pp. 1526–1535, 2006. MR2259075 (2008c:05020)
11. Y. Edel, G. Kyureghyan and A. Pott, A new APN function which is not equivalent to a power mapping. IEEE Trans. Inform. Theory 52, (2006), n°2, pp. 744–747. MR2236189 (2007e:94055)
12. E. Férard and F. Rodier, Non linéarité des fonctions booléennes données par des polynômes de degré binaire 3 définies sur $\mathbb{F}_{2^m}$ avec m pair [Nonlinearity of Boolean functions given by polynomials of binary degree 3 defined on $\mathbb{F}_{2^m}$ with m even]. Arithmetic, geometry, cryptography and coding theory 2009, pp. 41–53, Contemp. Math., 521, Amer. Math. Soc., Providence, RI, 2010. MR2744032 (2011m:94090)
13. E. Férard and F. Rodier, Non linéarité des fonctions booléennes données par des traces de polynômes de degré binaire 3 [Nonlinearity of Boolean functions given by traces of polynomials of binary degree 3]. Algebraic geometry and its applications, pp. 388–409, Ser. Number Theory Appl., 5, World Sci. Publ., Hackensack, NJ, 2008. MR2484066 (2009k:94199)
14. F. Hernando and G. McGuire, Proof of a conjecture on the sequence of exceptional numbers, classifying cyclic codes and APN functions, Available on *arXiv:0903.2016v3* [cs.IT], 2009. MR2824545

15. H. Janwa and R. M. Wilson, Hyperplane sections of Fermat varieties in P^3 in char. 2 and some applications to cyclic codes, *Applied Algebra, Algebraic Algorithms and Error-Correcting Codes, Proceedings AAECC-10 (G Cohen, T. Mora and O. Moreno Eds.)*, pp. 180-194, Lecture Notes in Computer Science, Vol. 673, Springer-Verlag, NewYork/Berlin 1993. MR1251978 (95b:11119)
16. H. Janwa, G. McGuire and R. M. Wilson, Double-error-correcting cyclic codes and absolutely irreducible polynomials over GF(2), *Applied J. of Algebra*, 178, pp. 665–676 (1995). MR1359909 (97a:94007)
17. D. Jedlicka, APN monomials over $GF(2^n)$ for infinitely many n, Finite Fields Appl. 13 (2007), n°4, pp. 1006–1028. MR2360537 (2008i:11144)
18. E. Leducq, *Autour des codes de Reed-Muller généralisés*, PhD thesis, Université Paris 7, 2011. Available on http://www.math.jussieu.fr/~elodieleducq/these.pdf
19. E. Lucas, Théorie des fonctions numériques simplement périodiques, *American Journal of Mathematics*, vol. 1, pp. 197-240 and pp. 289–321, 1878. MR1505176
20. K. Nyberg, Differentially uniform mappings for cryptography, *Advances in cryptology—Eurocrypt '93* (Lofthus, 1993), pp. 55–64, Lecture Notes in Comput. Sci., Vol. 765, Springer, Berlin, 1994. MR1290329 (95e:94039)
21. F. Rodier, Bornes sur le degré des polynômes presque parfaitement non-linéaires, in Arithmetic, Geometry, Cryptography and Coding Theory, G. Lachaud, C. Ritzenthaler and M.Tsfasman editors, Contemporary Math. n°487, AMS, Providence (RI), USA, pp. 169-181, 2009. Available on *arXiv:math/0605232v3* [math.AG]. MR2555993 (2010h:94195)
22. F. Rodier, Functions of degree $4e$ that are not APN Infinitely Often, Cryptography and Communications, 1–14, 2011.
23. F. Voloch, Symmetric cryptography and Algebraic curves, *Algebraic geometry and its applications, Proceedings of the first SAGA conference*, Ser. Number theory and its applications, World Sci. Publ., Hackensack, NJ, pp. 135–141, 2008. MR2484052 (2009m:14036)

ÉQUIPE GAATI, UNIVERSITÉ DE LA POLYNÉSIE FRANÇAISE
E-mail address: eric.ferard@upf.pf

ÉQUIPE GAATI, UNIVERSITÉ DE LA POLYNÉSIE FRANÇAISE
E-mail address: roger.oyono@upf.pf

INSTITUT DE MATHÉMATIQUES DE LUMINY, CNRS, UNIVERSITÉ DE LA MÉDITERRANÉE, MARSEILLE
E-mail address: rodier@iml.univ-mrs.fr

Contemporary Mathematics
Volume **574**, 2012
http://dx.doi.org/10.1090/conm/574/11420

Rational curves with many rational points over a finite field

Satoru Fukasawa, Masaaki Homma, and Seon Jeong Kim

ABSTRACT. We study a particular plane curve over a finite field whose normalization is of genus 0. The number of rational points of this curve achieves the Aubry-Perret bound for rational curves. The configuration of its rational points and a generalization of the curve are also presented.

1. Introduction

Let C be a curve of degree d in projective plane $\mathbb{P}^2$ over a finite field $\mathbb{F}_q$. We are interested in the number $N_q(C)$ of the set $C(\mathbb{F}_q)$ of $\mathbb{F}_q$-points. Especially we want to give a good upper bound for $N_q(C)$ in terms of d and q for curves C in a certain class. In 2010, the second and the third authors proved a fact of this direction [**9**].

THEOREM. *If a plane curve C of degree $d \geq 2$ over $\mathbb{F}_q$ has no $\mathbb{F}_q$-linear components, then*

$$N_q(C) \leq (d-1)q + 1 \tag{1}$$

unless $d = q = 4$ and C is projectively equivalent to

$$K : (X+Y+Z)^4 + (XY+YZ+ZX)^2 + XYZ(X+Y+Z) = 0$$

over $\mathbb{F}_4$. In this exceptional case, $N_4(K) = 14$.

Since the upper bound (1) was originally conjectured by Sziklai [**15**], we refer it as the Sziklai bound. Note that the bound (1) makes sense only if $2 \leq d \leq q+2$ because it is worse than the obvious bound:

$$N_q(C) \leq {}^{\#}\mathbb{P}^2(\mathbb{F}_q) = q^2 + q + 1$$

if $d > q + 2$.

The study of curves that attain the Sziklai bound is still under way [**8, 10**], but we have not yet met any example of a curve with singularities which attains it. We guess that there will be a better bound for curves with singularities, which is our motivation for focusing on them.

2010 *Mathematics Subject Classification.* Primary 14H50, Secondary 14G15, 14G05, 14N10.

The first author was supported in part by Grant-in-Aid for for Young Scientists (22740001), JSPS.

The second author was supported in part by Grant-in-Aid for Scientific Research (21540051), JSPS.

The third author was supported in part by Basic Science Research Program through the National Research Foundation of Korea (NRF) funded by the Ministry of Education, Science and Technology (2010-0028027).

Let C' be an irreducible curve of degree d in $\mathbb{P}^2$ over $\mathbb{F}_q$ whose normalization is $\mathbb{P}^1$. Since the morphism $\mathbb{P}^1 \to C'$ given by the normalization is defined over $\mathbb{F}_q$, the number of nonsingular $\mathbb{F}_q$-points of C' is at most $q+1$. Therefore

$$N_q(C') \le q + 1 + \frac{1}{2}(d-1)(d-2), \tag{2}$$

because the number of singularities of C' is at most $\frac{1}{2}(d-1)(d-2)$. This bound is a special case of a result of Aubry and Perret [**1**, Prop. 2.3]. So we refer this bound as the Aubry-Perret bound for rational curves. The bound (2) is, of course, better than (1) in the meaningful range of d.

Let B be the rational plane curve over $\mathbb{F}_q$ defined by the image of

$$\mathbb{P}^1 \ni (s,t) \to (s^{q+1}, s^q t + s t^q, t^{q+1}) \in \mathbb{P}^2. \tag{3}$$

B is of degree $q+1$ and $N_q(B)$ actually attains the Aubry-Perret bound for $d = q+1$. This curve over an algebraically closed field containing $\mathbb{F}_q$ appeared in Ballico and Hefez's classification list [**2**, Th. 1] of non-reflexive plane curves of degree $q+1$ with second-order q in a different parametrization

$$(s,t) \to (t^{q+1}, s(s-t)^q, s^q(s-t)) \in \mathbb{P}^2$$

from (3). Actually those parametrizations are equivalent over $\mathbb{F}_{q^2}$, but not over $\mathbb{F}_q$. However we refer the curve B parametrized by (3) as the Ballico-Hefez curve. Recently, in [**3**] the first author has studied the Galois points of B by using a parametrization

$$(s,t) \to (s^{q+1}, (s+t)^{q+1}, t^{q+1}) \in \mathbb{P}^2$$

equivalent to (3) over $\mathbb{F}_q$, and has found that the constellation of Galois points of B is described in a similar way that the second author did in [**7**] for Hermitian curves.

As Hermitian curves have many interesting properties in finite geometry including coding theory, Ballico-Hefez curves also might have lovely properties because of this similarity[1].

In Section 2, we verify that the number of $\mathbb{F}_q$-points of B is actually $q+1+\frac{1}{2}q(q-1)$, in other words B is a singular maximal curve, and show that the zeta function of B is given by $\frac{(1+T)^{\frac{q^2-q}{2}}}{(1-T)(1-qT)}$. In Section 3, we consider the case q is odd. We give combinatorial characterizations of the set of $\mathbb{F}_q$-points of B, and compute the τ_i's, where τ_i is the number of $\mathbb{F}_q$-lines that are i-secants to $B(\mathbb{F}_q)$. In Section 4, we compute the τ_i's for q even. By using those results, we compute parameters of codes coming from $B(\mathbb{F}_q)$ in Section 5. Some of them have the largest minimum distance under fixed length and dimension. In the last section, we propose a generalization of the Ballico-Hefez curve, which is a rational curve in $\mathbb{P}^n$ parametrized by elementary symmetric polynomials in $t, t^q, \dots, t^{q^{n-1}}$. We give a formula of the number of $\mathbb{F}_q$-points of this curve.

2. Arithmetic of the curve B

We study the arithmetic properties of the Ballico-Hefez curve B. We prepare some notations. Φ denotes the morphism (3), that is, $\Phi(s,t) = (s^{q+1}, s^q t + s t^q, t^{q+1})$, and φ denotes $\Phi|\mathbb{A}^1$, where $\mathbb{A}^1 = \{s \neq 0\}$, that is, $\varphi(t) = (1, t+t^q, t^{q+1})$. For a point $P = (\alpha, \beta) \in \mathbb{P}^1$, P^q denotes the point $(\alpha^q, \beta^q) \in \mathbb{P}^1$. $\operatorname{Sing} B$ denotes the set of singularities of B, and $\operatorname{Flex} B$ the set of inflection points of B. The precise

[1] For other similarity, see Remark 2.5.

definition of an "inflection point" in our case will be given in 2.1 below. We also take interest in the order-sequence and the q-Frobenius order-sequence for Φ. The notion of q-Frobenius order-sequence was introduced by Stöhr and Voloch [**14**] as a tool to bound the number of $\mathbb{F}_q$-points of curves. Here we give the definition of them in a little more general setting than our original one.

2.1. *Order-sequence and q-Frobenius order-sequence.* Let $f : \tilde{C} \to \mathbb{P}^n$ be a morphism from a nonsingular curve $\tilde{C}$ over $\overline{\mathbb{F}}_q$, where $\overline{\mathbb{F}}_q$ denotes the algebraic closure of $\mathbb{F}_q$. In a neighborhood of an assigned point $P \in \tilde{C}$, f can be represented by $(n+1)$-tuple regular functions $(f_0, \dots, f_n)$ one of which is the constant function 1. Let t be a local parameter at P. Then regular functions around P can be embedded into the formal power series ring $\overline{\mathbb{F}}_q[[t]]$ via the identification with the completion of the local ring at $P \in \tilde{C}$. The i-th Hasse derivation $D^{(i)}$ on $\overline{\mathbb{F}}_q[[t]]$ is given by $D^{(i)}t^k = \binom{k}{i}t^{k-i}$. Then $\{(\varepsilon_0, \dots, \varepsilon_n) \mid \varepsilon_0 < \dots < \varepsilon_n,\ \det\left((D^{(\varepsilon_i)}f_j)(P)\right)_{i,j} \neq 0\}$ is nonempty. The minimum $(n+1)$-tuple in the above set by the lexicographic order is called the Hermitian P-invariant for f. The Hermitian P-invariant is constant if P is in a certain nonempty open subset of $\tilde{C}$. The Hermitian P-invariant of a point in this open subset is called the order-sequence for f. For details, consult [**6**, Ch.7].

Let $C = f(\tilde{C}) \subset \mathbb{P}^n$, and $\varepsilon_0 < \dots < \varepsilon_n$ the order-sequence for f. For a nonsingular point $P' = f(P) \in C$, the linear subspace spanned by $\nu + 1$ vectors $\{((D^{(\varepsilon_i)}f_0)(P), \dots, (D^{(\varepsilon_i)}f_n)(P)) | i = 0, 1, \dots, \nu\}$ in $\mathbb{P}^n$ is the tangent ν-plane at P', which is denoted by $T^{(\nu)}_{P'}C$.

If the image of P' by q-Frobenius map lies on $T^{(\nu)}_{P'}C$ with some $\nu < n$ for almost all $P' \in C$, the curve C is said to be q-Frobenius nonclassical, and the minimum number ν having the above property is called the q-Frobenius index of C (see, [**4**, Prop. 2]). Let ν be the q-Frobenius index of C. The sequence $\{\varepsilon_0, \varepsilon_1, \dots, \varepsilon_n\} \setminus \{\varepsilon_\nu\}$ is called the q-Frobenius order-sequence of C.

For a plane curve $C = f(\tilde{C}) \subset \mathbb{P}^2$, a nonsingular point $P' \in C$ is an inflection point if $i(C.T^{(1)}_{P'}(C); P') > \varepsilon_2$.

Now we go back to our original setting.

THEOREM 2.2. (i) *For any $P \in \mathbb{P}^1$, the map induced by Φ on tangent spaces*

$$d\Phi_P : T_{P,\mathbb{P}^1} \to T_{\Phi(P),\mathbb{P}^2}$$

is injective.

(ii) $\operatorname{Sing} B$ *consists of* $\frac{q^2-q}{2}$ *ordinary double points.*

(iii) *For* $P \in \mathbb{P}^1$, $\Phi(P) \in \operatorname{Sing} B$ *if and only if* $P \in \mathbb{P}^1(\mathbb{F}_{q^2}) \setminus \mathbb{P}^1(\mathbb{F}_q)$. *In this case,* $\Phi^{-1}(\Phi(P)) = \{P, P^q\}$.

(iv) *The order-sequence for Φ is $\{0, 1, q\}$, and B is q-Frobenius nonclassical.*

(v) $\Phi(P)$ *is an inflection point of B if and only if* $P \in \mathbb{P}^1(\mathbb{F}_q)$.

(vi) $B(\mathbb{F}_q) = \operatorname{Flex} B \cup \operatorname{Sing} B$.

PROOF. For the assertion (i), by the symmetry of Φ in s and t, it is enough to show that $d\varphi_P : T_{P,\mathbb{P}^1} \to T_{\varphi(P),\mathbb{P}^2}$ is injective. The injectivity holds true because

$$\frac{d\varphi}{dt} = (0, 1, t^q). \tag{4}$$

Since no tangent vector collapses by Φ, $\Phi(P)$ is a singular point if and only if $^{\#}\Phi^{-1}(\Phi(P)) > 1$. We want to determine such points. For $P_\infty = (0, 1)$, obviously

$\Phi^{-1}(\Phi(P_\infty)) = \{P_\infty\}$. For two distinct points $P_\alpha = (1, \alpha)$ and $P_\beta = (1, \beta) \in \mathbb{A}^1$, $\Phi(P_\alpha) = \Phi(P_\beta)$ if and only if $\varphi(\alpha) = \varphi(\beta)$, that is,

$$\begin{cases} \alpha + \alpha^q &= \beta + \beta^q \\ \alpha^{q+1} &= \beta^{q+1}. \end{cases}$$

Hence $\Phi(P_\alpha) = \Phi(P_\beta)$ implies that $(X - \alpha)(X - \alpha^q) = (X - \beta)(X - \beta^q)$. Since $\alpha \neq \beta$, we have $\alpha = \beta^q$ and $\alpha^q = \beta$. Hence $P_\alpha, P_\beta \in \mathbb{P}^1(\mathbb{F}_{q^2}) \setminus \mathbb{P}^1(\mathbb{F}_q)$ with $P_\alpha^q = P_\beta$. Conversely, this condition obviously leads to $\Phi(P_\alpha) = \Phi(P_\beta)$. Therefore $\Phi^{-1}(\operatorname{Sing} B) = \mathbb{P}^1(\mathbb{F}_{q^2}) \setminus \mathbb{P}^1(\mathbb{F}_q)$ and $\mathbb{P}^1(\mathbb{F}_{q^2}) \setminus \mathbb{P}^1(\mathbb{F}_q) \xrightarrow{\Phi} \operatorname{Sing} B$ is a 2-to-1 map. Moreover, since the embedded tangent line to B at the branch corresponding to $P_\alpha \in \mathbb{P}^1$ is

$$\begin{vmatrix} X & Y & Z \\ 1 & \alpha + \alpha^q & \alpha^{q+1} \\ 0 & 1 & \alpha^q \end{vmatrix} = 0 \quad \text{(see (4))}, \tag{5}$$

$\Phi(P_\alpha) = \Phi(P_{\alpha^q}) \in \operatorname{Sing} B$ is an ordinary double point, that is, those two branches have different tangent lines. So (ii) and (iii) have been established, and furthermore we have also established

$$B(\mathbb{F}_q) = \Phi(\mathbb{P}^1(\mathbb{F}_q)) \cup \operatorname{Sing} B \tag{6}$$

because $\alpha + \alpha^q$ and α^{q+1} are the trace and the norm from $\mathbb{F}_{q^2}$ to $\mathbb{F}_q$ respectively.

Next we calculate the order-sequence for Φ. Since $d\Phi \neq 0$, the first two orders are 0 and 1. Let $D_t^{(\nu)}$ be the ν-th Hasse derivation on the function field $\overline{\mathbb{F}}_q(B) = \overline{\mathbb{F}}_q(t)$ with respect to t. Since

$$\begin{aligned} \det \begin{pmatrix} \varphi \\ D_t^{(1)}\varphi \\ D_t^{(\nu)}\varphi \end{pmatrix} &= \det \begin{pmatrix} 1 & t + t^q & t^{q+1} \\ 0 & 1 & t^q \\ 0 & \binom{q}{\nu} t^{q-\nu} & \binom{q+1}{\nu} t^{q+1-\nu} \end{pmatrix} \\ &= \begin{cases} 0 & (q > \nu > 1) \\ t - t^q & (\nu = q) \end{cases}, \end{aligned}$$

the order-sequence for Φ is $\{0, 1, q\}$.

Since the tangent line at $\Phi(P_\alpha)$ is given by (5), $P_\alpha^q = (1, (\alpha^q + \alpha)^q, (\alpha^{q+1})^q)$ lies on it. Hence B is q-Frobenius nonclassical. The set of inflection points of B is given by the support of the Wronskian divisor (or the ramification divisor) [**14**, page 6]:

$$(1+2)(q+1)P_\infty + \operatorname{div}\left(\det\begin{pmatrix} \varphi \\ D_t^{(1)}\varphi \\ D_t^{(q)}\varphi \end{pmatrix}\right) + (0 + 1 + q)\operatorname{div} dt = \sum_{P \in \mathbb{P}^1(\mathbb{F}_q)} P.$$

Hence $\operatorname{Flex} B = \Phi(\mathbb{P}^1(\mathbb{F}_q))$, which, together with (6), implies (vi). □

As was mentioned in Introduction, $N_q(B)$ attains the Aubry-Perret bound for rational curves.

COROLLARY 2.3. *The Ballico-Hefez curve B is of degree $q + 1$, and $N_q(B) = q + 1 + \frac{q(q-1)}{2}$.*

COROLLARY 2.4. *The zeta function of B is $Z_B(T) = \frac{(1+T)^{\frac{q^2-q}{2}}}{(1-T)(1-qT)}$.*

PROOF. Since $\Phi : \mathbb{P}^1 \to B$ is the normalization of B, we can calculate the zeta function of B by [**1**, Th. 2.1] together with the following informations: $\operatorname{Sing} B$ consists of $\frac{q^2-q}{2}$ points that are $\mathbb{F}_q$-rational, and $\Phi^{-1}(\Phi(P)) = \{P, P^q\}$ with $P \in \mathbb{P}^1(\mathbb{F}_{q^2}) \setminus \mathbb{P}^1(\mathbb{F}_q)$. □

REMARK 2.5. Corollaries 2.3 and 2.4 suggest another similarity between the Ballico-Hefez curve and the Hermitian curve. In general, if C' is an irreducible curve over $\mathbb{F}_q$ with the normalization $\mathbb{P}^1 \overset{\pi}{\to} C'$, then the zeta function of C' is of the form $\frac{L_{C'}(T)}{(1-T)(1-qT)}$ where $L_{C'}(T)$ is a polynomial of degree $\#\pi^{-1}(\operatorname{Sing} C') - \#\operatorname{Sing} C'$ (say, $\Delta_{C'}$). Let $\{\beta_1, \dots, \beta_{\Delta_{C'}}\}$ be the set of reciprocal roots of $L_{C'}$. Then $|\beta_i| = 1$ $(i = 1, \dots, \Delta_{C'})$ and $N_{q^r}(C') = q^r + 1 - \sum_{i=1}^{\Delta_{C'}} \beta_i^r$ (see, [**1**]). Especially, $N_q(C') \leq q+1+\Delta_{C'}$, and if equality holds, then we have $\Delta_{C'} \leq \frac{q^2-q}{2}$ by Ihara's argument [**11**]. Actually equality holds in both inequalities for the Ballico-Hefez curve. Comparing with Rück and Stichtenoth's characterization of Hermitian curves [**12**], we might expect Ballico-Hefez curves to be characterized among rational curves by those two properties.

The following lemma will be used later.

LEMMA 2.6. *For each $\mathbb{F}_q$-point Q of $\mathbb{P}^2$ which does not lie on B, there are two points $P_1, P_2 \in B(\mathbb{F}_q) \setminus \operatorname{Sing} B$ such that $T_{P_1} B \cap T_{P_2} B = \{Q\}$. Moreover the pair $\{P_1, P_2\}$ is uniquely determined by Q.*

PROOF. Since $i(B.T_{P_i}B; P_i) = q+1$ by Theorem 2.2, $T_{P_i}B \cap B = \{P_i\}$. Hence the map

$$\begin{array}{ccc} S^2(B(\mathbb{F}_q) \setminus \operatorname{Sing} B) \setminus \Delta & \to & \mathbb{P}^2(\mathbb{F}_q) \setminus B(\mathbb{F}_q) \\ \{P_1, P_2\} & \mapsto & T_{P_1} B \cap T_{P_2} B \end{array}$$

is well-defined, where $S^2(B(\mathbb{F}_q) \setminus \operatorname{Sing} B)$ denotes the symmetric product of $B(\mathbb{F}_q) \setminus \operatorname{Sing} B$ and Δ its diagonal subset. Since the source and the target of this map have the same cardinality $\frac{q(q+1)}{2}$, it is enough to show the following fact; *Let $P_i' = (\alpha_i, \beta_i) \in \mathbb{P}^1$ $(i = 1, 2, 3)$ be three distinct points. Then three embedded tangent lines at branches corresponding to $P_i' \in \mathbb{P}^1$ are not concurrent.* In fact, the embedded tangent line at the branch corresponding to $P_i' \in \mathbb{P}^1$ is

$$\begin{vmatrix} X & Y & Z \\ 1 & \frac{\beta_i}{\alpha_i} + (\frac{\beta_i}{\alpha_i})^q & (\frac{\beta_i}{\alpha_i})^{q+1} \\ 0 & 1 & (\frac{\beta_i}{\alpha_i})^q \end{vmatrix} = 0,$$

if $\alpha_i \neq 0$. More simply, the tangent line is given by the equation

$$\beta_i^{2q} X - \alpha_i^q \beta_i^q Y + \alpha_i^{2q} Z = 0, \tag{7}$$

which is also valid for $(\alpha_i, \beta_i) = (0, 1)$. Since

$$\begin{vmatrix} \beta_1^{2q} & -\alpha_1^q \beta_1^q & \alpha_1^{2q} \\ \beta_2^{2q} & -\alpha_2^q \beta_2^q & \alpha_2^{2q} \\ \beta_3^{2q} & -\alpha_3^q \beta_3^q & \alpha_3^{2q} \end{vmatrix} = -\prod_{i<j} (\alpha_i \beta_j - \alpha_j \beta_i)^q,$$

those three lines are not concurrent. □

3. Geometry of B with $\mathbb{F}_q$-lines, for q odd

The projective plane of $\mathbb{F}_q$-lines in $\mathbb{P}^2$ is denoted by $\check{\mathbb{P}}^2(\mathbb{F}_q)$. An $\mathbb{F}_q$-line l is an i-line if $^\#(B(\mathbb{F}_q) \cap l) = i$. The cardinality of $\{l \in \check{\mathbb{P}}^2(\mathbb{F}_q) | l \text{ is an } i\text{-line}\}$ is denoted by τ_i. Since $\deg B = q+1$, only $q+2$ numbers $\{\tau_i | 0 \leq i \leq q+1\}$ make sense. The purpose of this and next sections is to determine the exact values of the τ_i's.

In this section, we assume q is odd.

LEMMA 3.1. *Let C_0 be the conic defined by $4XZ - Y^2 = 0$. Then $B \cap C_0 = B(\mathbb{F}_q) \cap C_0(\mathbb{F}_q) = B(\mathbb{F}_q) \setminus \operatorname{Sing} B$, and $T_P(B) = T_P(C_0)$ for any $P \in B \cap C_0$.*

PROOF. Let $\Phi(\alpha, \beta) = P \in B(\mathbb{F}_q) \setminus \operatorname{Sing} B$. Then $(\alpha, \beta) \in \mathbb{P}^1(\mathbb{F}_q)$. Hence we may suppose $\alpha, \beta \in \mathbb{F}_q$, and have $P = (\alpha^{q+1}, \alpha^q\beta + \alpha\beta^q, \beta^{q+1}) = (\alpha^2, 2\alpha\beta, \beta^2)$, which lies on C_0. So

$$B(\mathbb{F}_q) \setminus \operatorname{Sing} B \subseteq C_0(\mathbb{F}_q) \cap B(\mathbb{F}_q) \subseteq B \cap C_0.$$

$T_P B$ is given by $\beta^{2q}X - \alpha^q\beta^qY + \alpha^{2q}Z = 0$ (see (7)). On the other hand, $T_P C_0$ is defined by $4\beta^2X - 4\alpha\beta Y + 4\alpha^2 Z = 0$. Since $\alpha, \beta \in \mathbb{F}_q$, $T_P B = T_P C_0$. Hence

$$2(q+1) = (B.C_0) = \sum_{P \in B(\mathbb{F}_q) \setminus \operatorname{Sing} B} i(B.C_0; P) \geq 2(q+1),$$

which means $B \cap C_0 = B(\mathbb{F}_q) \setminus \operatorname{Sing} B$. □

Now we give geometric characterization of the set of rational points $B(\mathbb{F}_q)$ of the Ballico-Hefez curve for odd q. The configuration of $B(\mathbb{F}_q)$ is more or less known. We define two subsets $\mathcal{S}$ and $\mathcal{T}$ in $\mathbb{P}^2(\mathbb{F}_q)$.

We use some terms in finite geometry. For a given plane conic[2] D over $\mathbb{F}_q$, a point $P \in \mathbb{P}^2(\mathbb{F}_q)$ is *internal* to D if there is no tangent $\mathbb{F}_q$-line to D passing through P. An $\mathbb{F}_q$-point P is *external* to D if P does not lie on D and there is a tangent $\mathbb{F}_q$-line to D passing through P. Since q is odd, there are exactly two tangent $\mathbb{F}_q$-lines to D passing through P if P is external to D. Especially, the set of external points to D is

$$\{T_P D \cap T_Q D \mid P, Q \in D(\mathbb{F}_q) \text{ and } P \neq Q\}.$$

A subset of $\mathbb{P}^2(\mathbb{F}_q)$ is an *arc* if no 3 points of the subset are collinear, and an arc is an *n-arc* if the arc consists of n points.

DEFINITION 3.2. (1) Let $l_1, \ldots, l_{q+1}$ be $q+1$ $\mathbb{F}_q$-lines that form an arc in $\check{\mathbb{P}}^2(\mathbb{F}_q)$, that is, no three of the $q+1$ lines are concurrent. $\mathcal{S}(l_1, \ldots, l_{q+1})$ is the set $\mathbb{P}^2(\mathbb{F}_q) \setminus \bigcup_{i<j} l_i \cap l_j$.

If $q+1$ $\mathbb{F}_q$-lines $l'_1, \ldots, l'_{q+1}$ form another $q+1$-arc in $\check{\mathbb{P}}^2(\mathbb{F}_q)$, then $\mathcal{S}(l_1, \ldots, l_{q+1})$ and $\mathcal{S}(l'_1, \ldots, l'_{q+1})$ are projectively equivalent, which can be proved by using a theorem of Segre [**13**]. $\mathcal{S}$ denotes one of the $\mathcal{S}(l_1, \ldots, l_{q+1})$'s, which makes sense up to projective equivalence.

(2) Let D be a conic over $\mathbb{F}_q$. $\mathcal{T}_D$ denotes the set of internal points to D together with $D(\mathbb{F}_q)$, which appeared in [**5**, Example 12.6 (3)].

Since two conics D_1 and D_2 over $\mathbb{F}_q$ are projectively equivalent over $\mathbb{F}_q$, so are $\mathcal{T}_{D_1}$ and $\mathcal{T}_{D_2}$. $\mathcal{T}$ denotes one of the $\mathcal{T}_D$'s.

[2]In our context, the word "conic" connotes that it is absolutely irreducible. Since an absolutely irreducible conic cannot have a singular point by Bézout's theorem, our conic is also nonsingular.

THEOREM 3.3. *The three subsets $B(\mathbb{F}_q)$, $\mathcal{S}$ and $\mathcal{T}$ in $\mathbb{P}^2(\mathbb{F}_q)$ are projectively equivalent over $\mathbb{F}_q$.*

PROOF. First we show that $\mathcal{S}$ can be constructed by the same way of constructing $\mathcal{T}$. By a theorem of Segre [**13**], $l_1, \ldots, l_{q+1}$ lie on a conic D' in $\check{\mathbb{P}}^2(\mathbb{F}_q)$. Then the dual D of D' in $\mathbb{P}^2(\mathbb{F}_q)$ is also a conic, and each l_i tangents to D. Hence the point set $\{l_i \cap l_j | 1 \le i < j \le q+1\}$ is the set of external points to D.

Next we consider the case for $B(\mathbb{F}_q)$ and $\mathcal{T}$. By Lemmas 2.6 and 3.1, $\mathbb{P}^2(\mathbb{F}_q) \setminus B(\mathbb{F}_q)$ is the external points to C_0. Therefore $B(\mathbb{F}_q)$ coincides with $\mathcal{T}_{C_0}$.

Since any two conics over $\mathbb{F}_q$ are projectively equivalent, so are those three sets. □

COROLLARY 3.4. (i) *If $\tau_i \neq 0$, then $i = 1$ or $\frac{q+1}{2}$ or $\frac{q+3}{2}$.*
(ii) $\tau_1 = q+1$; $\tau_{\frac{q+1}{2}} = \frac{q(q-1)}{2}$; $\tau_{\frac{q+3}{2}} = \frac{q(q+1)}{2}$.

PROOF. We count these numbers by using the configuration $\mathcal{T}$ in Definition 3.2. Let l be an $\mathbb{F}_q$-line which tangents to the conic D at $P \in D(\mathbb{F}_q)$. For each point $Q \in l(\mathbb{F}_q) \setminus \{P\}$, there exists another tangent $\mathbb{F}_q$-line passing through Q, that is, Q is an external point to D. Therefore any tangent line is 1-line. For an $\mathbb{F}_q$-line which does not tangent to D at any $\mathbb{F}_q$-points, there are exactly $\frac{q+1}{2}$ external points if the line does not meet $D(\mathbb{F}_q)$, or $\frac{q-1}{2}$ if it meets $D(\mathbb{F}_q)$, because there are exactly two $\mathbb{F}_q$-tangent lines passing through an external point to D. Hence such a line is a $\frac{q+1}{2}$-line or a $\frac{q+3}{2}$-line.

For (ii), $\tau_1 = {}^{\#}D(\mathbb{F}_q) = q+1$; $\tau_{\frac{q+3}{2}}$ is the same as the number of $\mathbb{F}_q$-lines joining two distinct points of $D(\mathbb{F}_q)$, which is $\frac{q(q+1)}{2}$. Hence $\tau_{\frac{q+1}{2}} = q^2+q+1-\tau_1-\tau_{\frac{q+3}{2}} = \frac{q(q-1)}{2}$. □

REMARK 3.5. The proof of Corollary 3.4 tells us a characterization of i-lines ($i = 1, \frac{q+1}{2}, \frac{q+3}{2}$) for $\mathcal{T}$. Let l be an $\mathbb{F}_q$-line of $\mathbb{P}^2$. Then l is a 1-line for $\mathcal{T}$ if and only if ${}^{\#}(l \cap D(\mathbb{F}_q)) = 1$; a $\frac{q+1}{2}$-line for $\mathcal{T}$ if and only if ${}^{\#}(l \cap D(\mathbb{F}_q)) = 0$; a $\frac{q+3}{2}$-line for $\mathcal{T}$ if and only if ${}^{\#}(l \cap D(\mathbb{F}_q)) = 2$.

4. Geometry of B with $\mathbb{F}_q$-lines, for q even

The aim of this section is to determine the τ_i's for $B(\mathbb{F}_q)$ with q even, which corresponds to Corollary 3.4 for odd q.

PROPOSITION 4.1. *Suppose q is a power of 2.*

(i) *The $q+1$ points of $B(\mathbb{F}_q) \setminus \operatorname{Sing} B$ are collinear.*
(ii) *If $\tau_i \neq 0$, then $i = 1$ or $\frac{q}{2}+1$ or $q+1$.*
(iii) $\tau_1 = q+1$; $\tau_{\frac{q}{2}+1} = q^2-1$; $\tau_{q+1} = 1$.

PROOF. Let $\Phi(\alpha, \beta) = P \in B(\mathbb{F}_q) \setminus \operatorname{Sing} B$. Since $(\alpha, \beta) \in \mathbb{P}^1(\mathbb{F}_q)$, $P = (\alpha^2, 0, \beta^2)$, that is, P lies on the line $Y = 0$. Let $l_0 = \{Y = 0\}$. Obviously $l_0(\mathbb{F}_q) = B(\mathbb{F}_q) \setminus \operatorname{Sing} B$ and l_0 is a $(q+1)$-line. For any $\mathbb{F}_q$-line $l \neq l_0$, put $n(l) = {}^{\#}(l \cap \operatorname{Sing} B)$. Since any tangent line to a branch at a singular point P of B is not an $\mathbb{F}_q$-line, $i(B.l; P) = 2$ for any $\mathbb{F}_q$-line l passing through P. Hence $1+2n(l) \le q+1$. Fix a point $P_0 \in l_0(\mathbb{F}_q)$, and count the number of points of $\operatorname{Sing} B$ by using $\mathbb{F}_q$-lines passing through P_0. Note that l_0 and $T_{P_0}B$ never meet $\operatorname{Sing} B$.

So we have

$$\frac{q^2-q}{2} = \sum_{\substack{l \ni P_0 \\ \text{with} \\ l \neq l_0, T_{P_0}B}} n(l) \leq \frac{q}{2}(q-1).$$

Therefore an $\mathbb{F}_q$-line l which neither l_0 nor the tangent line to B at an $\mathbb{F}_q$-point of l_0 is a $(\frac{q}{2}+1)$-line, and the number of such lines is $(q+1)(q-1)$. Obviously the tangent line to B at an $\mathbb{F}_q$-point of l_0 is a 1-line, and the number of such lines is $q+1$. This completes the proof. □

5. Codes from Ballico-Hefez curves

5.1. *Codes from a subset of* $\mathbb{P}^2(\mathbb{F}_q)$. Let S be a subset of $\mathbb{P}^2(\mathbb{F}_q)$ which consists of s elements. For each point $P \in S$, we fix a representative (a_0, a_1, a_2) of its coordinates with $a_0, a_1, a_2 \in \mathbb{F}_q$. Then for any homogeneous polynomial $F(X_0, X_1, X_2)$ over $\mathbb{F}_q$, the value $F(P) \in \mathbb{F}_q$ is determined without ambiguity.

Let $\Gamma(\mathcal{O}(i))$ be the vector space of homogeneous polynomials over $\mathbb{F}_q$ of degree i. Then the image of $\mathbb{F}_q$-linear map

$$\Gamma(\mathcal{O}(i)) \ni F \mapsto (F(P))_{P \in S} \in (\mathbb{F}_q)^s$$

gives a linear codes, which is denoted by $C_L(S, \mathcal{O}(i))$.

PROPOSITION 5.2. *Suppose that q is odd.*

(i) $C_L(B(\mathbb{F}_q), \mathcal{O}(1))$ *is a* $\left[\frac{q^2+q+2}{2}, 3, \frac{q^2-1}{2}\right]_q$*-code, and it achieves the Griesmer bound.*

(ii) *For* $q \geq 5$, $C_L(B(\mathbb{F}_q), \mathcal{O}(2))$ *is a* $\left[\frac{q^2+q+2}{2}, 6, \frac{q^2-q-4}{2}\right]_q$*-code.*

(iii) *For* $q \geq 7$, $C_L(B(\mathbb{F}_q), \mathcal{O}(3))$ *is a* $\left[\frac{q^2+q+2}{2}, 10, \frac{q^2-2q-7}{2}\right]_q$*-code.*

PROOF. (i) The parameters are known by Corollaries 2.3 and 3.4. Since $\frac{q^2-1}{2} = \frac{q-1}{2} \cdot q + \frac{q-1}{2}$,

$$\sum_{i=0}^{2} \left\lceil \left(\frac{q^2-1}{2}\right) / q^i \right\rceil = \frac{q^2-1}{2} + \left(\frac{q-1}{2} + 1\right) + 1 = \frac{q^2+q+2}{2},$$

which means the triple of parameters $\left[\frac{q^2+q+2}{2}, 3, \frac{q^2-1}{2}\right]_q$ achieves the Griesmer bound.

(ii) Let D be a curve over $\mathbb{F}_q$ of degree 2 in $\mathbb{P}^2$. If D is absolutely irreducible, then ${}^{\#}D(\mathbb{F}_q) = q+1$; if it is irreducible over $\mathbb{F}_q$ but not absolutely, then ${}^{\#}D(\mathbb{F}_q) = 1$. Hence ${}^{\#}(B(\mathbb{F}_q) \cap D) \leq q+1$ for those two cases. If D is reducible over $\mathbb{F}_q$, then ${}^{\#}(B(\mathbb{F}_q) \cap D) \leq q+3$ by Corollary 3.4. In particular, no degree-two-curve contains $B(\mathbb{F}_q)$, and $C_L(B(\mathbb{F}_q), \mathcal{O}(2))$ is a $\left[\frac{q^2+q+2}{2}, 6, \geq \frac{q^2-q-4}{2}\right]_q$-code, where $\geq \frac{q^2-q-4}{2}$ at the third parameter means the minimum distance of this code is at least $\frac{q^2-q-4}{2}$. For an external point Q to C_0, which does not lie on $B(\mathbb{F}_q)$ by Theorem 3.3, there are exactly two $\mathbb{F}_q$-lines passing through Q each of which tangents to C_0. Hence there are exactly $\frac{q-1}{2}$ $\mathbb{F}_q$-lines passing through Q each of which meets $C_0(\mathbb{F}_q)$ at two points. Since $q \geq 5$, we can choose two lines from those $\frac{q-1}{2}$ lines, which are

$\frac{q+3}{2}$-lines by Remark 3.5. The union of those two lines gives a codeword of weight $\frac{q^2-q-4}{2}$.

(iii) The minimum distance of $C_L(B(\mathbb{F}_q), \mathcal{O}(3))$ is at least $\frac{q^2-2q-7}{2}$ if and only if

$$\#(B(\mathbb{F}_q) \cap D) \leq \frac{3}{2}(q+3) \tag{8}$$

for any curve D over $\mathbb{F}_q$ of degree 3. If D is absolutely irreducible, then $\#(B(\mathbb{F}_q) \cap D) \leq q+1+2\sqrt{q} \leq \frac{3}{2}(q+3)$. In other cases, one can verify (8) easily by using Corollary 3.4. Equality in (8) holds if one takes three internal lines to C_0 passing through an assigned external point, which is possible because $q \geq 7$. □

When q is a power of 2, $B(\mathbb{F}_q)$ contains $q+1$ collinear points, so we can't expect to obtain good codes from the total set.

PROPOSITION 5.3. *Let $q = 2^e$ with $e > 2$. Then the code $C_L(\operatorname{Sing} B, \mathcal{O}(1))$ is a $\left[\frac{q^2-q}{2}, 3, \frac{q^2-2q}{2}\right]_q$-code, which achieves the Griesmer bound.*

PROOF. From Proposition 4.1 and its proof, an i-line for $\operatorname{Sing} B$ exists if and only if $i = 0$ or $\frac{q}{2}$. Hence we have the first half of the assertion. Since $\frac{q^2-2q}{2} = \frac{q-2}{2} \cdot q$, we have the additional assertion. □

6. Generalization of the curve B

We generalize the Ballico-Hefez curve in $\mathbb{P}^2$ to rational curves in higher dimensional projective spaces. We will discuss only on the number of rational points on them.

NOTATION 6.1. Let $X_0, \dots, X_{n-1}$ be variables. We consider $n+1$ elementary symmetric polynomials:

$$\sigma_k(X_0, \dots, X_{n-1}) := \sum_{i_1 < \cdots < i_k} X_{i_1} \cdots X_{i_k} \quad (k = 0, \dots, n),$$

where we understand $\sigma_0 = 1$. We also consider "homogeneous" elementary symmetric polynomials of degree n:

$$\tilde{\sigma}_k(X_0, \dots, X_{n-1}; Y_0, \dots, Y_{n-1}) := \sum_{\substack{i_1 < \cdots < i_k; j_1 < \cdots < j_{n-k} \\ \text{with} \\ \{i_1, \dots, i_k, j_1, \dots, j_{n-k}\} = \{0, \dots, n-1\}}} X_{i_1} \cdots X_{i_k} Y_{j_1} \cdots Y_{n-k}.$$

DEFINITION 6.2. We define $n+1$ homogeneous polynomials of degree $q^{n-1} + q^{n-2} + \cdots + 1$ in s and t by

$$\tilde{\varphi}_k(s,t) := \tilde{\sigma}_k(s, s^q, \dots, s^{q^{n-1}}; t, t^q, \dots, t^{q^{n-1}}),$$

and inhomogeneous ones by

$$\varphi_k(s) := \tilde{\varphi}_k(s, 1) = \sigma_k(s, s^q, \dots, s^{q^{n-1}}) = \sum_{i_1 < \cdots < i_k} s^{q^{i_1}} \cdots s^{q^{i_k}}$$

for $k = 0, 1, \dots, n$.

B_n denotes a curve in $\mathbb{P}^n$ over $\mathbb{F}_q$ defined by the image of

$$\Phi_n : \mathbb{P}^1 \ni (s,t) \mapsto (\tilde{\varphi}_0(s,t), \tilde{\varphi}_1(s,t), \dots, \tilde{\varphi}_n(s,t)) \in \mathbb{P}^n.$$

Since $B_2 = B$, the curve B_n is a generalization of the Ballico-Hefez curve.

REMARK 6.3. B_n is nondegenerate in $\mathbb{P}^n$, that is, there is no hyperplane of $\mathbb{P}^n$ containing B_n.

In fact, suppose $\sum_{i=0}^n \alpha_i \varphi_i(s) = 0$ for $\alpha_0, \dots, \alpha_n \in \overline{\mathbb{F}}_q$. Let $D_s^{(\nu)}$ be the ν-th Hasse derivation with respect to s. Note that for k and $i_0 < \cdots < i_l$, $D_s^{(1+q+q^2+\cdots+q^k)} s^{q^{i_0}+q^{i_1}+\cdots+q^{i_l}} \neq 0$ if and only if $l \geq k$ and $i_\mu = \mu$ ($\mu = 0, \dots, l$), and in this case, $D_s^{(1+q+q^2+\cdots+q^k)} s^{q^{i_0}+q^{i_1}+\cdots+q^{i_l}} = s^{q^{i_{k+1}}+\cdots+q^{i_l}}$. Therefore

$$0 = D_s^{(1+q+q^2+\cdots+q^k)} \left(\sum_{i=0}^n \alpha_i \varphi_i(s) \right) |_{s=0} = \alpha_k,$$

which means that no hyperplane contains B_n.

THEOREM 6.4. (i) *For $P \in \mathbb{P}^1$, the map arising from Φ_n on tangent spaces*

$$d\Phi_{n,P} : T_{P,\mathbb{P}^1} \to T_{\Phi_n(P),\mathbb{P}^n}$$

is injective.

(ii) *For $P \in \mathbb{P}^1$, $^{\#}\Phi_n^{-1}(\Phi_n(P)) > 1$ if and only if $P \in \mathbb{P}^1(\mathbb{F}_{q^n}) \setminus \mathbb{P}^1(\mathbb{F}_q)$. In this case, if $\mathbb{F}_q(P) = \mathbb{F}_{q^k}$, then*

$$\Phi_n^{-1}(\Phi_n(P)) = \{P, P^q, \dots P^{q^{k-1}}\},$$

and

$$\Phi_n(P) = \Phi_n(P^q) = \cdots = \Phi_n(P^{q^{k-1}}) \in \mathbb{P}^n(\mathbb{F}_q).$$

(iii) $B_n(\mathbb{F}_q) = \Phi_n(\mathbb{P}^1(\mathbb{F}_{q^n}))$.

PROOF. (i) By the symmetry of Φ_n in s and t, it is sufficient to check the statement on the affine piece $t \neq 0$. The local parametrization of Φ_n on this affine piece is given by $\Phi_n(s,1) = (\varphi_0(s), \dots, \varphi_n(s))$. Since $(\frac{d\varphi_0(s)}{ds}, \frac{d\varphi_1(s)}{ds}, \dots, \frac{d\varphi_n(s)}{ds}) = (0, 1, \dots, s^{q+\cdots+q^{n-1}})$, $d\Phi_{n,P}$ is injective.

(ii) We may suppose $t = 1$. For two points $P_\alpha = (\alpha, 1), P_\beta = (\beta, 1) \in \mathbb{P}^1$, $\Phi_n(P_\alpha) = \Phi_n(P_\beta)$ if and only if $\varphi_k(\alpha) = \varphi_k(\beta)$ for $k = 0, 1, \dots, n$. Since $\prod_{i=0}^{n-1}(X - s^{q^i}) = \sum_{k=0}^n (-1)^k \varphi_k(s) X^{n-k}$, those conditions are equivalent to the condition

$$\{\alpha, \alpha^q, \dots, \alpha^{q^{n-1}}\} = \{\beta, \beta^q, \dots, \beta^{q^{n-1}}\}$$

with counting multiplicity. Here we need the following lemma, which completes the proof of (ii).

LEMMA 6.5. *Let two elements α and β of $\overline{\mathbb{F}}_q$ be distinct from each other. If $\{\alpha, \alpha^q, \dots, \alpha^{q^{n-1}}\} = \{\beta, \beta^q, \dots, \beta^{q^{n-1}}\}$ with counting multiplicity, then $\mathbb{F}_q(\alpha) = \mathbb{F}_q(\beta)$ is a subfield of $\mathbb{F}_{q^n}$.*

PROOF. Let $\mathbb{F}_q(\alpha) = \mathbb{F}_{q^k}$. Obviously $\mathbb{F}_q(\beta) = \mathbb{F}_{q^k}$ also. Since $\beta = \alpha^{q^i}$ for some i with $1 \leq i \leq n-1$, $\alpha^{q^n} = (\alpha^{q^i})^{q^{n-i}} = \beta^{q^{n-i}} \in \{\alpha, \alpha^q, \dots, \alpha^{q^{n-1}}\}$. Hence $\alpha^{q^n} = \alpha^{q^j}$ for some j with $0 \leq j \leq n-1$. Hence $k | n-j$, particularly $k \leq n$. Let $n = uk + r$ with $0 \leq r < k$. Suppose that $r > 0$. Then each of the r elements $\{\alpha, \alpha^q, \dots, \alpha^{q^{r-1}}\}$ appears $u+1$ times in the total set $\{\alpha, \alpha^q, \dots, \alpha^{q^{n-1}}\}$, and each of the $k-r$ elements $\{\alpha^{q^r}, \dots, \alpha^{q^{k-1}}\}$ u-times. On the other hand, since $\beta = \alpha^{q^i}$, each of the r elements $\{\alpha^{q^i}, \alpha^{q^{i+1}}, \dots, \alpha^{q^{i+r-1}}\}$ appears $u+1$ times in the total set, and each of the $k-r$ elements $\{\alpha^{q^{i+r}}, \dots, \alpha^{q^{i+k-1}}\}$ u-times. Therefore both $\alpha^{q^{r-1}}$

and $\alpha^{q^{i+r-1}}$ appear $u+1$ times in the total set, but their q-th power u times. Hence those elements must coincide. Hence $\alpha = \alpha^{q^i} = \beta$, which is a contradiction. □

Continuation of the proof of Theorem 6.4. (iii) It is obvious that $\Phi_n(P_\alpha) \in B(\mathbb{F}_q)$ for any $\alpha \in \mathbb{F}_{q^n}$. Conversely, if $\Phi_n(P_\alpha) \in B(\mathbb{F}_q)$, then $\varphi_n(\alpha) = \alpha^{1+q+\cdots+q^{n-1}} \in \mathbb{F}_q$. Hence $1 = (\alpha^{1+q+\cdots+q^{n-1}})^{q-1} = \alpha^{q^n-1}$ if $\alpha \neq 0$, which means $\alpha \in \mathbb{F}_{q^n}$. □

From Theorem 6.4, we know the number $N_q(B_n)$ of $\mathbb{F}_q$-points of B_n.

THEOREM 6.6.

$$N_q(B_n) = \frac{1}{n}q^n + \sum_{\substack{d|n \\ with \\ d\neq n}} \frac{1}{d} \prod_{\substack{l:prime \\ with \\ l|\frac{n}{d}}} \left(1 - \frac{1}{l}\right) q^d + 1.$$

PROOF. By (ii) and (iii) of Theorem 6.4,

$$N_q(B_n) = \sum_{k|n} \frac{1}{k} \#\{P \in \mathbb{P}^1 \mid \mathbb{F}_q(P) = \mathbb{F}_{q^k}\}.$$

Since $\mathbb{P}^1(\overline{\mathbb{F}}_q) = \overline{\mathbb{F}}_q \cup \{(1,0)\}$,

$$\#\{P \in \mathbb{P}^1 \mid \mathbb{F}_q(P) = \mathbb{F}_{q^k}\} = \begin{cases} \#\{\alpha \in \mathbb{F}_{q^k} \mid \mathbb{F}_q(\alpha) = \mathbb{F}_{q^k}\} & \text{if } k > 1 \\ \#\mathbb{F}_q + 1 & \text{if } k = 1. \end{cases}$$

Now we compute $\#\{\alpha \in \mathbb{F}_{q^k} \mid \mathbb{F}_q(\alpha) = \mathbb{F}_{q^k}\}$ exactly. Let $k = l_{k,1}^{e_1} \ldots l_{k,r_k}^{e_{r_k}}$ be the decomposition of k as a product of powers of distinct prime numbers. Here we understand the right hand of the decomposition of 1 to be empty. Since

$$\{\alpha \in \mathbb{F}_{q^k} \mid \mathbb{F}_q(\alpha) = \mathbb{F}_{q^k}\} = \mathbb{F}_{q^k} \setminus \bigcup_{i=1}^{r_k} \mathbb{F}_{q^{k/l_{k,i}}},$$

the cardinality of this set is

$$q^k - \sum_{i=1}^{r_k} q^{k/l_{k,i}} + \sum_{i<j} q^{k/l_{k,i}l_{k,j}} - \cdots = \sum_{s=0}^{r_k} (-1)^s \sum_{i_1<\cdots<i_s} q^{k/l_{k,i_1}\cdots l_{k,i_s}}.$$

When $k = 1$, we understand the above sum to be just q. Hence

$$N_q(B_n) = \sum_{k|n} \frac{1}{k} \sum_{s=0}^{r_k} (-1)^s \sum_{i_1<\cdots<i_s} q^{k/l_{k,i_1}\cdots l_{k,i_s}} + 1.$$

For each d with $d|n$, we gather together the terms in q^d. If $d = n$, the only term is $\frac{1}{n}q^n$. If $d < n$, put $\frac{n}{d} = m_1^{f_1} \cdots m_u^{f_u}$, where $m_1, \ldots, m_u$ are distinct prime numbers with $f_1, \ldots, f_u \geq 1$. It is obvious that for a fixed d,

$$\{k \mid k/l_{k,j_1} \cdots l_{k,j_s} = d \text{ for some } j_1 < \cdots < j_s\} = \{m_{i_1} \cdots m_{i_s} d \mid i_1 < \cdots < i_s\}.$$

Hence the coefficient of q^d is

$$\sum_{s=0}^{u} \sum_{i_1<\cdots<i_s} (-1)^s \frac{1}{m_{i_1} \cdots m_{i_s} d} = \frac{1}{d} \prod_{i=1}^{u} \left(1 - \frac{1}{m_i}\right) = \frac{1}{d} \prod_{\substack{l|\frac{n}{d} \\ l:\text{ prime}}} \left(1 - \frac{1}{l}\right).$$

This completes the proof.

REMARK 6.7. The order-sequence, the q-Frobenius index and the q-Frobenius order-sequence for $\Phi_n : \mathbb{P}^1 \to B_n$ can be also calculated. Namely, the order-sequence for Φ_n is $0 < 1 < q < q^2 < \cdots < q^{n-1}$, the q-Frobenius index is 1, and hence the q-Frobenius order-sequence is $0 < q < q^2 < \cdots < q^{n-1}$. At present, our proof of this fact is not concise. So we will discuss it in a separate paper.

ACKNOWLEDGMENT. We express our thanks to the referee for giving us candid comments.

References

1. Y. Aubry and M. Perret, *A Weil theorem for singular curves*, in: R. Pellikaan, M. Perret and S. Vlăduţ (Eds.), Arithmetic geometry and coding theory (Luminy, 1993), de Gruyter, Berlin, 1996, 1–7. MR1394921 (97g:11061)
2. E. Ballico and A. Hefez, *Non-reflexive projective curves of low degree*, Manuscripta Math. 70 (1991), 385–396. MR1092144 (92g:14026)
3. S. Fukasawa, *Galois points for a non-reflexive plane curve of low degree*, preprint 2010.
4. A. Garcia and M. Homma, *Frobenius order-sequences of curves*, in: G. Frey and J. Ritter (Eds), Algebra and number theory (Essen, 1992), de Gruyter, Berlin, 1994, 27–41. MR1285362 (95i:14023)
5. J. W. P. Hirschfeld, Projective geometries over finite fields (second edition), Oxford University Press, Oxford, 1998. MR1612570 (99b:51006)
6. J. W. P. Hirschfeld, G. Korchmáros and F. Torres, Algebraic curves over a finite field, Princeton Univ. Press, Princeton and Oxford, 2008. MR2386879 (2008m:14040)
7. M. Homma, *Galois points for a Hermitian curve*, Comm. Algebra 34 (2006) 4503–4511. MR2273720 (2007i:14028)
8. M. Homma and S. J. Kim, *Nonsingular plane filling curves of minimum degree over a finite field and their automorphism groups: Supplements to a work of Tallini*, `arXiv:0903.1918`, 2009.
9. M. Homma and S. J. Kim, *Sziklai's conjecture on the number of points of a plane curve over a finite field* III, Finite Fields Appl. 16 (2010) 315–319. MR2678619 (2011h:14028)
10. M. Homma and S. J. Kim, *Toward determination of optimal plane curves with a fixed degree over a finite field*, Finite Fields Appl. 17(2011) 240–253. MR2793248 (2012c:14052)
11. Y. Ihara, *Some remarks on the number of rational points of algebraic curves over finite fields*, J. Fac. Sci. Univ. Tokyo Sect. IA Math. 28 (1981) 721–724. MR656048 (84c:14016)
12. H.-G. Rück and H. Stichtenoth, *A characterization of Hermitian function fields over finite fields*, J. Reine Angew. Math. 457 (1994) 185–188. MR1305281 (95h:11059)
13. B. Segre, *Ovals in a finite projective plane*, Canadian J. Math. 7 (1955) 414–416. MR0071034 (17:72g)
14. K.-O. Stöhr and J. F. Voloch, *Weierstrass points and curves over finite fields*, Proc. London Math. Soc. (3) 52 (1986) 1–19. MR812443 (87b:14010)
15. P. Sziklai, *A bound on the number of points of a plane curve*, Finite Fields Appl. 14 (2008) 41–43. MR2381474 (2008k:14054)

DEPARTMENT OF MATHEMATICAL SCIENCES, YAMAGATA UNIVERSITY, YAMAGATA 990-8560, JAPAN
E-mail address: `s.fukasawa@sci.kj.yamagata-u.ac.jp`

DEPARTMENT OF MATHEMATICS, KANAGAWA UNIVERSITY, YOKOHAMA 221-8686, JAPAN
E-mail address: `homma@kanagawa-u.ac.jp`

DEPARTMENT OF MATHEMATICS AND RINS, GYEONGSANG NATIONAL UNIVERSITY, JINJU 660-701, KOREA
E-mail address: `skim@gnu.kr`

Contemporary Mathematics
Volume **574**, 2012
http://dx.doi.org/10.1090/conm/574/11430

Enumeration of Splitting Subspaces over Finite Fields

Sudhir R. Ghorpade and Samrith Ram

ABSTRACT. We discuss an elementary, yet unsolved, problem of Niederreiter concerning the enumeration of a class of subspaces of finite dimensional vector spaces over finite fields. A short and self-contained account of some recent progress on this problem is included and some related problems are discussed.

1. Introduction

Finite fields have a remarkable property that finite dimensional vector spaces over them are naturally endowed with a canonical and compatible field structure. Indeed, we can simply "move the d" so as to write $\mathbb{F}_q^d \simeq \mathbb{F}_{q^d}$, where d is any positive integer and as usual, $\mathbb{F}_q$ denotes the finite field with q elements. This leads to some interesting notions where the field structure and the linear structure are intertwined. One such notion is that of a splitting subspace, which appears to go back at least to Niederreiter (1995) in connection with his work on pseudorandom number generation. Here is the definition:

Let m, n be positive integers, q a prime power, and $\alpha \in \mathbb{F}_{q^{mn}}$. An m-dimensional $\mathbb{F}_q$-linear subspace W of $\mathbb{F}_{q^{mn}}$ is said to be *α-splitting* if

$$\mathbb{F}_{q^{mn}} = W \oplus \alpha W \oplus \cdots \oplus \alpha^{n-1} W.$$

Concerning these, Niederreiter [**12**] asked the following: given $\alpha \in \mathbb{F}_{q^{mn}}$ such that $\mathbb{F}_{q^{mn}} = \mathbb{F}_q(\alpha)$, what is the number of m-dimensional α-splitting subspaces of $\mathbb{F}_{q^{mn}}$?

Actually, the above question is a slightly more general version of the original question stated as an open problem in [**12**, p. 11] where it is assumed that q is prime and α is a primitive element of $\mathbb{F}_{q^{mn}}$ in the sense that it is a generator of the cyclic group $\mathbb{F}_{q^{mn}}^*$ of nonzero elements of $\mathbb{F}_{q^{mn}}$. But this general version seems quite natural and we will always consider Niederreiter's question in this setting.

The main aim of this article is to make Niederreiter's question better known and to facilitate further research on it. We were motivated by the fact that till recently there had not been any significant progress on this question since it was posed more than 15 years ago. Recent progress came about primarily by relating this question to seemingly different questions in cryptography. This brought to the fore exciting connections not only with cryptography but also matrix theory and finite projective geometry via the so called block companion Singer cycles. As a result, a quantitative formulation of Niederreiter's question suggested itself and a small breakthrough was obtained in the form of a solution in the case of splitting

2010 *Mathematics Subject Classification.* Primary 15A03, 11T06 05E99 Secondary 11T71.

planes, i.e., when $m = 2$. We refer to Appendix A and to [**14, 6, 7, 9**] for these developments. While the myriad connections are no doubt interesting, we wish to underline the fact that Niederreiter's question is a beautiful problem that is easy to state and is of interest in itself. With this in view, we give here an account of the recent progress on this question by focusing mainly on splitting subspaces *per se* and relegating its connections to cryptography and such to an appendix at the end. In particular, we include a short and self-contained proof of the solution to Niederreiter's question in the case of splitting planes. Our original proof (cf. [**7**]) in the case $m = 2$ used a result of Benjamin and Bennett [**1**], which in turn was motivated by a question of Corteel, Savage, Wilf, and Zeilberger [**2**] (see [**5**, Rem. 4.2] for more historical information). Here we have removed the dependence on Benjamin and Bennett [**1**] by means of an auxiliary result (Lemma 3.3) and given a quick and independent proof of it by modifying an argument in [**5**, Thm. 4.1]. We have also used this opportunity to include certain variants of Niederreiter's question and some preliminary results concerning them. Finally, for the convenience of the reader (and at the suggestion of a referee), we include a brief appendix where interconnections with cryptography, Singer cycles, etc., have been outlined.

2. Easy cases and guesses

Fix, throughout this paper, positive integers m, n and a prime power q. Let us first note that for an arbitrary $\alpha \in \mathbb{F}_{q^{mn}}$, there may not be any α-splitting subspace; for example, if $\alpha \in \mathbb{F}_q$, then $\alpha^i W = W$ for every m-dimensional subspace W and every $i \geq 0$, and so W cannot be α-splitting if $n > 1$. To avoid such situations, we will always assume that $\alpha \in \mathbb{F}_{q^{mn}}$ satisfies $\mathbb{F}_{q^{mn}} = \mathbb{F}_q(\alpha)$. In this case, $\{1, \alpha, \alpha^2, \dots, \alpha^{mn-1}\}$ forms a $\mathbb{F}_q$-basis of $\mathbb{F}_{q^{mn}}$ and hence $\{1, \alpha^n, \alpha^{2n}, \dots, \alpha^{(m-1)n}\}$ spans an m-dimensional α-splitting subspace of $\mathbb{F}_{q^{mn}}$, say Λ. Let us define

$$S(\alpha, m, n; q) := \text{the number of } \alpha\text{-splitting subspaces of } \mathbb{F}_{q^{mn}} \text{ of dimension } m.$$

Niederreiter's question is to determine (a nice formula for) $S(\alpha, m, n; q)$. The case when m or n is equal to 1 is quite trivial. Indeed, if $n = 1$, then the only m-dimensional subspace, viz., $W = \mathbb{F}_{q^{mn}}$, is α-splitting for every $\alpha \in \mathbb{F}_{q^{mn}}$. On the other hand, if $m = 1$ and if $\alpha \in \mathbb{F}_{q^{mn}} = \mathbb{F}_{q^n}$ is such that $\mathbb{F}_{q^{mn}} = \mathbb{F}_q(\alpha)$, then every 1-dimensional subspace is α-splitting. Thus

$$S(\alpha, m, n; q) = \frac{q^{mn} - 1}{q^m - 1} \quad \text{if } \min\{m, n\} = 1.$$

In fact, the fraction on the right is always a lower bound for $S(\alpha, m, n; q)$. To see this, it suffices to note two things: (i) if Λ is as above, then $\beta\Lambda$ is an m-dimensional α-splitting subspace for every $\beta \in \mathbb{F}_{q^{mn}}^*$, since $\alpha^j \beta = \beta\alpha^j$ for $0 \leq j \leq n-1$, and (ii) $\Lambda = \mathbb{F}_q(\alpha^n)$ and Λ^* is a subgroup of the cyclic group $\mathbb{F}_{q^{mn}}^*$ of index $(q^{mn}-1)/(q^m-1)$ so that if we let $\beta \in \mathbb{F}_{q^{mn}}^*$ vary over representatives of distinct cosets of Λ^* in $\mathbb{F}_{q^{mn}}^*$, then the corresponding subspaces $\beta\Lambda$ are distinct (in fact, essentially disjoint). To work out a slightly nontrivial example, let us suppose $m = 2$ and $n = 2$. Recall that for any integers a, b with $a \geq b \geq 0$, the number of b-dimensional subspaces of an a-dimensional vector space over $\mathbb{F}_q$ is given by the Gaussian binomial coefficient

$$\begin{bmatrix} a \\ b \end{bmatrix}_q := \frac{(q^a - 1)(q^a - q) \cdots (q^a - q^{b-1})}{(q^b - 1)(q^b - q) \cdots (q^b - q^{b-1})}.$$

Let W be subspace of $\mathbb{F}_{q^4}$. Note that $W = \alpha W$ if and only if $W = 0$ or $W = \mathbb{F}_{q^4}$ (indeed, if $0 \neq x \in W = \alpha W$, then W contains the linearly independent elements x, αx, $\alpha^2 x$, $\alpha^3 x$ and so $W = \mathbb{F}_{q^4}$). Now suppose $\dim W = 2$ and W is not α-splitting. Then $L = W \cap \alpha W$ is a 1-dimensional and $W = L + \alpha^{-1}L$. Conversely, if L is a 1-dimensional subspace of $\mathbb{F}_{q^4}$, then $L + \alpha^{-1}L$ is a 2-dimensional subspace of $\mathbb{F}_{q^4}$ that is not α-splitting. It follows that

$$S(\alpha, 2, 4; q) = \begin{bmatrix} 4 \\ 2 \end{bmatrix}_q - \begin{bmatrix} 4 \\ 1 \end{bmatrix}_q = \frac{q^4 - 1}{q^2 - 1}\, q^2.$$

We now take an inspired leap and propose the following quantitative formulation of Niederreiter's question.

Splitting Subspace Conjecture: Let $\alpha \in \mathbb{F}_{q^{mn}}$ satisfy $\mathbb{F}_{q^{mn}} = \mathbb{F}_q(\alpha)$. Then

$$S(\alpha, m, n; q) = \frac{q^{mn} - 1}{q^m - 1}\, q^{m(m-1)(n-1)}.$$

To be sure, the conjectural formula fits well with the examples considered above as well as the general lower bound for $S(\alpha, m, n; q)$. Still to arrive at it based only on a few examples is indeed quite a leap. As alluded to in the Introduction and explained in the appendix, the true inspiration, in fact, comes from a recent conjecture of Zeng, Han and He [**14**] and the subsequent work in [**6**] and [**7**]. But at any rate, we have a nice specific problem, which seems to be open, in general. Its solution in the only nontrivial case known so far will be considered next.

3. Splitting planes

We begin with a simple but useful observation that goes back to Niederreiter [**12**, Lem. 3] and says the enumeration of splitting subspaces of $\mathbb{F}_{q^{mn}}$ is equivalent to the enumeration of certain ordered bases of $\mathbb{F}_{q^{mn}}$. To make this more precise, let us introduce some notation.

Given any $\alpha, v_1, \ldots, v_m \in \mathbb{F}_{q^{mn}}$, we let

$$\mathcal{B}^{\alpha}_{(v_1,\ldots,v_m)} := \left\{v_1, \ldots, v_m, \alpha v_1, \ldots, \alpha v_m, \ldots, \alpha^{n-1} v_1, \ldots, \alpha^{n-1} v_m\right\},$$

with the understanding that $\mathcal{B}^{\alpha}_{(v_1,\ldots,v_m)}$ is to be regarded as an ordered set with mn elements. In case $\mathcal{B}^{\alpha}_{(v_1,\ldots,v_m)}$ is an ordered basis of $\mathbb{F}_{q^{mn}}$, the set $\{v_1, \ldots, v_m\}$ is necessarily a $\mathbb{F}_q$-basis of an m-dimensional subspace of $\mathbb{F}_{q^{mn}}$ and we will refer to $\mathcal{B}^{\alpha}_{(v_1,\ldots,v_m)}$ as an *α-splitting ordered basis* of $\mathbb{F}_{q^{mn}}$. The number of α-splitting ordered bases of $\mathbb{F}_{q^{mn}}$ will be denoted by $N(\alpha, m, n; q)$.

LEMMA 3.1. *Let $\alpha \in \mathbb{F}_{q^{mn}}$ satisfy $\mathbb{F}_{q^{mn}} = \mathbb{F}_q(\alpha)$, and let $v_1, \ldots, v_m \in \mathbb{F}_{q^{mn}}$. Then $\mathcal{B}^{\alpha}_{(v_1,\ldots,v_m)}$ is an ordered basis of $\mathbb{F}_{q^{mn}}$ if and only if $\{v_1, \ldots, v_m\}$ span an m-dimensional α-splitting subspace of $\mathbb{F}_{q^{mn}}$. Consequently,*

$$S(\alpha, m, n; q) = \frac{N(\alpha, m, n; q)}{|\mathrm{GL}_m(\mathbb{F}_q)|}, \quad \text{that is,} \quad N(\alpha, m, n; q) = S(\alpha, m, n; q) \prod_{i=0}^{m-1} (q^m - q^i).$$

PROOF. The first assertion is obvious. The second follows from the first by noting that the number of distinct ordered bases of an m-dimensional vector space over $\mathbb{F}_q$ is $|\mathrm{GL}_m(\mathbb{F}_q)| = \prod_{i=0}^{m-1}(q^m - q^i)$. □

From now on, we will focus on the case of splitting planes, i.e., the case $m = 2$.

LEMMA 3.2. *Let $\alpha \in \mathbb{F}_{q^{2n}}$ be such that $\mathbb{F}_{q^{2n}} = \mathbb{F}_q(\alpha)$. Then*

$$N(\alpha, 2, n; q) = \left(q^{2n} - \nu - 1\right)\left(q^{2n} - 1\right),$$

where ν denotes the cardinality of the set Σ of pairs (f_1, f_2) of nonzero polynomials in $\mathbb{F}_q[X]$ of degree $< n$ with f_2 monic and f_1, f_2 relatively prime.

PROOF. Fix $v_1 \in \mathbb{F}_{q^{2n}}$ with $v_1 \neq 0$. Then for any $v_2 \in \mathbb{F}_{q^{2n}}$, the ordered set

$$\mathcal{B}^{\alpha}_{(v_1,v_2)} = \left\{v_1,\, v_2,\, \alpha v_1,\, \alpha v_2,\, \dots,\, \alpha^{n-1}v_1,\, \alpha^{n-1}v_2\right\}$$

is a $\mathbb{F}_q$-basis of $\mathbb{F}_{q^{2n}}$ if and only if the ordered set

$$\mathcal{S}_\beta := \left\{1,\, \beta,\, \alpha,\, \alpha\beta,\, \dots,\, \alpha^{n-1},\, \alpha^{n-1}\beta\right\}$$

is linearly independent over $\mathbb{F}_q$, where $\beta := v_2/v_1$. Now, $1, \alpha, \dots, \alpha^{2n-1}$ are linearly independent over $\mathbb{F}_q$ and in particular, so are $1, \alpha, \dots, \alpha^{n-1}$. Thus for any $\beta \in \mathbb{F}^*_{q^{2n}}$, the ordered set $\mathcal{S}_\beta$ is $\mathbb{F}_q$-independent if and only if β cannot be expressed as

$$\frac{a_0 + a_1\alpha + \cdots + a_{n-1}\alpha^{n-1}}{b_0 + b_1\alpha + \cdots + b_{n-1}\alpha^{n-1}}$$

for some $a_i, b_i \in \mathbb{F}_q$ such that not all a_i are zero and not all b_i are zero $(0 \le i \le n-1)$. It follows that $\left\{\beta \in \mathbb{F}^*_{q^{2n}} : \mathcal{S}_\beta \text{ is linearly independent}\right\} = \mathbb{F}^*_{q^{2n}} \setminus \Sigma_\alpha$, where

$$\Sigma_\alpha := \left\{\frac{p_1(\alpha)}{p_2(\alpha)} : p_i \in \mathbb{F}_q[X]^* \text{ with } \deg(p_i) < n \text{ for } i = 1, 2\right\}.$$

Now consider

$$\Sigma := \left\{\frac{f_1}{f_2} : f_i \in \mathbb{F}_q[X]^* \text{ with } \deg(f_i) < n \text{ for } i = 1, 2 \text{ and } \mathrm{GCD}(f_1, f_2) = 1\right\}.$$

The map $\Sigma \to \Sigma_\alpha$ given by $(f_1, f_2) \mapsto f_1(\alpha)/f_2(\alpha)$ is clearly well-defined and surjective. Moreover, if $(f_1, f_2), (g_1, g_2) \in \Sigma$ are such that $f_1(\alpha)g_2(\alpha) = g_1(\alpha)f_2(\alpha)$, then $f_1g_2 = g_1f_2$ because the minimal polynomial of α over $\mathbb{F}_q$ has degree $2n$. Further since $\mathrm{GCD}(f_1, f_2) = 1 = \mathrm{GCD}(g_1, g_2)$ and since f_2, g_2 are monic, it follows that $f_2 = g_2$ and therefore $f_1 = g_1$. Thus Σ_α is in bijection with Σ, and hence upon letting $\nu = |\Sigma|$, we find

$$\left|\left\{\beta \in \mathbb{F}^*_{q^{2n}} : \mathcal{S}_\beta \text{ is linearly independent }\right\}\right| = (q^{2n} - 1 - \nu).$$

Finally, if we vary v_1 over the $(q^{2n} - 1)$ elements of $\mathbb{F}^*_{q^{2n}}$, then we readily see that the number of ordered bases of the form $\mathcal{B}^{\alpha}_{(v_1,v_2)}$ is equal to $\left(q^{2n} - \nu - 1\right)\left(q^{2n} - 1\right)$. □

The cardinality ν of the set Σ appearing in Lemma 3.2 will be determined using the following more general result concerning pairs of relatively prime polynomials.

LEMMA 3.3. *Let N_1, N_2 be positive integers with $N_1 \ge N_2$ and let $\nu(N_1, N_2)$ denote the number of ordered pairs (f_1, f_2) of coprime nonzero polynomials in $\mathbb{F}_q[X]$ with f_2 monic and $\deg f_i < N_i$ for $i = 1, 2$. Then $\nu(N_1, N_2) = q^{N_1+N_2-1} - 1$.*

PROOF. We can partition the set of ordered pairs (f_1, f_2) of nonzero polynomials in $\mathbb{F}_q[X]$ with f_2 monic and $\deg f_i < N_i$ for $i = 1, 2$ into disjoint subsets S_d $(0 \le d < N_2)$, where S_d consists of pairs whose GCD is of degree d. Given any monic polynomial $h \in \mathbb{F}_q[X]$ of degree d and any coprime pair (g_1, g_2) of nonzero polynomials with g_2 monic and $\deg g_i < N_i - d$ for $i = 1, 2$, it is easy to see that $(hg_1, hg_2) \in S_d$. Conversely, if $(f_1, f_2) \in S_d$, then the polynomial $h = \mathrm{GCD}(f_1, f_2)$ is monic of degree d and $(f_1/h, f_2/h)$ is a coprime pair comprising of a nonzero polynomial of degree $N_1 - d$ and a monic polynomial of degree $N_2 - d$. This shows that $|S_d| = q^d \nu(N_1 - d, N_2 - d)$ for $0 \le d < N_2$. Since for any positive integer N, there are $(q^N - 1)$ nonzero polynomials in $\mathbb{F}_q[X]$ of degree $< N$ and of these exactly $(q^N - 1)/(q-1)$ are monic, it follows that

$$(q^{N_1} - 1)\,\frac{(q^{N_2} - 1)}{q-1} = \sum_{0 \le d < N_2} |S_d| = \sum_{0 \le d < N_2} q^d \nu\,(N_1 - d, N_2 - d)\,. \tag{3.1}$$

If $N_2 = 1$, we immediately obtain $\nu(N_1, N_2) = q^{N_1} - 1$. On the other hand, if $N_2 > 1$, then substituting $N_i - 1$ for N_i $(i = 1, 2)$ in the above relation yields

$$(q^{N_1 - 1} - 1)\,\frac{(q^{N_2 - 1} - 1)}{q-1} = \sum_{1 \le d < N_2} q^{d-1} \nu\,(N_1 - d, N_2 - d)\,. \tag{3.2}$$

Multiplying equation (3.2) by q and subtracting the result from (3.1), and then making an elementary calculation, we see that $\nu\,(N_1, N_2) = q^{N_1 + N_2 - 1} - 1$. $\square$

It is now a simple matter to show that the Splitting Subspace Conjecture holds in the affirmative when $m = 2$ (and n is arbitrary).

THEOREM 3.4. *Let $\alpha \in \mathbb{F}_{q^{2n}}$ be such that $\mathbb{F}_{q^{2n}} = \mathbb{F}_q(\alpha)$. Then*

$$S(\alpha, 2, n; q) = \frac{q^{2n} - 1}{q^2 - 1}\, q^{2(n-1)}.$$

PROOF. Follows from Lemmas 3.1, 3.2, and 3.4. $\square$

4. Refinements and Extensions

For $\alpha \in \mathbb{F}_{q^{mn}}$, let $\mathfrak{S}_\alpha$ denote the set of all m-dimensional α-splitting subspaces of $\mathbb{F}_{q^{mn}}$. By a *pointed α-splitting subspace* of dimension m we shall mean a pair (W, x) where $W \in \mathfrak{S}_\alpha$ and $x \in W$. The element x may be referred to as the *base point* of (W, x). Given any $x \in \mathbb{F}_{q^{mn}}$, we let $\mathfrak{S}_\alpha^x := \{W \in \mathfrak{S}_\alpha : x \in W\}$.

PROPOSITION 4.1. *Let $\alpha \in \mathbb{F}_{q^{mn}}$ be such that $\mathbb{F}_{q^{mn}} = \mathbb{F}_q(\alpha)$. Then*

$$|\mathfrak{S}_\alpha^x| = |\mathfrak{S}_\alpha^y| \quad \text{for any } x, y \in \mathbb{F}_{q^{mn}}^*.$$

Moreover, for any $x \in \mathbb{F}_{q^{mn}}^*$, the set $\mathfrak{S}_\alpha^x$ is nonempty and

$$S(\alpha, m, n; q) = |\mathfrak{S}_\alpha^x|\,\frac{q^{mn} - 1}{q^m - 1}.$$

PROOF. If $x, y \in \mathbb{F}_{q^{mn}}^*$ and $\beta = y/x$, then $W \mapsto \beta W$ gives a bijection of $\mathfrak{S}_\alpha^x$ onto $\mathfrak{S}_\alpha^y$. Moreover, for any $x \in \mathbb{F}_{q^{mn}}^*$, the $\mathbb{F}_q$-linear span of $\{x\alpha^{in} : 0 \le i < m\}$ is clearly in $\mathfrak{S}_\alpha^x$ and thus $\mathfrak{S}_\alpha^x$ is nonempty. Finally, by counting in two different ways the set $\{(W, x) : W \in \mathfrak{S}_\alpha \text{ and } x \in W\}$ of all pointed α-splitting subspaces, we find $|\mathfrak{S}_\alpha|\,(q^m - 1) = |\mathfrak{S}_\alpha^x|\,(q^{mn} - 1)$, as desired. $\square$

It may be remarked that the lower bound for $S(\alpha, m, n; q)$ discussed in Section 2 is an immediate consequence of Proposition 4.1. In light of Proposition 4.1, we see that the Splitting Subspace Conjecture is equivalent to the following simpler looking conjecture.

CONJECTURE 4.2 (Pointed Splitting Subspace Conjecture). *Let $\alpha \in \mathbb{F}_{q^{mn}}$ be such that $\mathbb{F}_{q^{mn}} = \mathbb{F}_q(\alpha)$ and let $x \in \mathbb{F}_{q^{mn}}^*$. Then the number of m-dimensional pointed α-splitting subspaces of $\mathbb{F}_{q^{mn}}$ with base point x is equal to $q^{m(m-1)(n-1)}$.*

We remark that $q^{m(m-1)}$ is the number of nilpotent $m \times m$ matrices over $\mathbb{F}_q$, thanks to an old result of Fine and Herstein [**4**]. Thus a particularly nice way to prove the Pointed Splitting Subspace Conjecture could be to set up a natural bijection between $\mathfrak{S}_\alpha^x$ and the set of $(n-1)$-tuples (or if one prefers, pointed n-tuples) of nilpotent $m \times m$ matrices over $\mathbb{F}_q$.

If the Splitting Subspace Conjecture were to hold in the affirmative, then an obvious consequence would be that the number of m-dimensional pointed α-splitting subspaces of $\mathbb{F}_{q^{mn}}$ is independent of the choice of $\alpha \in \mathbb{F}_{q^{mn}}$ as long as it satisfies $\mathbb{F}_{q^{mn}} = \mathbb{F}_q(\alpha)$. In other words, for any $\alpha, \beta \in \mathbb{F}_{q^{mn}}$,

$$S(\alpha, m, n; q) = S(\beta, m, n; q) \quad \text{provided} \quad \mathbb{F}_{q^{mn}} = \mathbb{F}_q(\alpha) = \mathbb{F}_q(\beta).$$

In general we do not know if this weaker statement is true. The following result summarizes the cases where the answer is known.

PROPOSITION 4.3. *Let $\alpha \in \mathbb{F}_{q^{mn}}$ be such that $\mathbb{F}_{q^{mn}} = \mathbb{F}_q(\alpha)$. If*

$$\beta = \frac{a\alpha^{q^r} + b}{c\alpha^{q^r} + d} \quad \text{for some nonnegative integer } r \text{ and } \begin{pmatrix} a & b \\ c & d \end{pmatrix} \in \mathrm{GL}_2(\mathbb{F}_q),$$

then $\mathbb{F}_{q^{mn}} = \mathbb{F}_q(\beta)$ and $S(\alpha, m, n; q) = S(\beta, m, n; q)$.

PROOF. First, note that if $\beta = c\alpha$ for some $c \in \mathbb{F}_q^*$ or $\beta = \alpha + d$ for some $d \in \mathbb{F}_q$, then $\mathbb{F}_{q^{mn}} = \mathbb{F}_q(\beta)$ and in view of Lemma 3.1, we see that $\mathfrak{S}_\alpha = \mathfrak{S}_\beta$ and so $S(\alpha, m, n; q) = S(\beta, m, n; q)$. Further, if $\alpha \neq 0$ (which is necessarily the case if $mn > 1$), then $\mathbb{F}_{q^{mn}} = \mathbb{F}_q(1/\alpha)$ and again in view of Lemma 3.1 and the fact that multiplication by the nonzero element $\alpha^{-(n-1)}$ preserves linear independence, it follows that $\mathfrak{S}_{1/\alpha} = \mathfrak{S}_\alpha$ and so $S(\alpha, m, n; q) = S(1/\alpha, m, n; q)$. Finally, note that for any nonnegative integer r, the elements α and α^{q^r} are Galois conjugate, i.e., they have the same minimal polynomial over $\mathbb{F}_q$, and therefore $\mathbb{F}_{q^{mn}} = \mathbb{F}_q\left(\alpha^{q^r}\right)$ and there is a one-to-one correspondence between α-splitting and α^{q^r}-splitting subspaces, induced by the corresponding element of the Galois group of $\mathbb{F}_{q^{mn}}$ over $\mathbb{F}_q$. Combining these, we obtain the desired result. □

It may be remarked that in view of Proposition 4.3 and the normal basis theorem [**10**, p. 60], we see that there is a $\mathbb{F}_q$-basis $\mathcal{B}$ of $\mathbb{F}_{q^{mn}}$ such that each element of $\mathcal{B}$ generates $\mathbb{F}_{q^{mn}}$ over $\mathbb{F}_q$ and $S(\alpha, m, n; q) = S(\beta, m, n; q)$ for all $\alpha, \beta \in \mathcal{B}$.

Finally, we note that Niederreiter's question can also be posed in a more general situation where instead of considering multiples of an m-dimensional subspace by powers of α, we consider its transforms by an endomorphism of $\mathbb{F}_{q^{mn}}$. More precisely, given any $\mathbb{F}_q$-linear endomorphism $T : \mathbb{F}_{q^{mn}} \to \mathbb{F}_{q^{mn}}$, we say that an m-dimensional subspace W of $\mathbb{F}_{q^{mn}}$ is *T-splitting* if

$$\mathbb{F}_{q^{mn}} = W \oplus T(W) \oplus T^2(W) \oplus \cdots \oplus T^{n-1}(W),$$

where T^j denotes the j-fold composite of T with itself $(0 \leq j < n)$. We let

$$S_T(m, n; q) = \text{the number of } m\text{-dimensional } T\text{-splitting subspaces of } \mathbb{F}_{q^{mn}}.$$

Evidently, if T is the $\mathbb{F}_q$-linear endomorphism of $\mathbb{F}_{q^{mn}}$ given by $x \mapsto \alpha x$, then $S_T(m, n; q) = S(\alpha, m, n; q)$. A more general variant of Niederreiter's question is to determine $S_T(m, n; q)$ for every $\mathbb{F}_q$-linear endomorphism T of $\mathbb{F}_{q^{mn}}$. An answer to this question does not seem to be known, even conjecturally. It should be noted, however, that certain restrictions on the structure of T will be needed in order that $S_T(m, n; q)$ is nonzero and independent of the choice of T in a suitable class.

For example, if $m = 1$, then the existence of m-dimensional T-splitting subspaces of $\mathbb{F}_{q^{mn}}$ evidently forces T to be cyclic and the minimal polynomial of T to be the characteristic polynomial of T. A complete answer to the above variant of Niederreiter's question in this case is given below. On the other hand, if $n = 1$, then $W = \mathbb{F}_{q^{mn}}$ is obviously the only m-dimensional T-splitting subspace, for any $T : \mathbb{F}_{q^{mn}} \to \mathbb{F}_{q^{mn}}$ and thus $S_T(m, 1; q) = 1$.

PROPOSITION 4.4. *Let $T : \mathbb{F}_{q^n} \to \mathbb{F}_{q^n}$ be a cyclic $\mathbb{F}_q$-linear endomorphism and let $p_T \in \mathbb{F}_q[X]$ be the minimal polynomial of T. Suppose $p_T = f_1^{e_1} \cdots f_k^{e_k}$ is the factorization of p_T into positive powers of distinct monic irreducible polynomials $f_i \in \mathbb{F}_q[X]$ with $\deg(f_i) = n_i$ for $i = 1, \ldots, k$. Then*

$$S_T(1, n; q) = \frac{q^n}{q-1} \prod_{i=1}^{k} \left(1 - \frac{1}{q^{n_i}}\right).$$

PROOF. Clearly any 1-dimensional T-splitting subspace W of $\mathbb{F}_{q^n}$ is spanned by a cyclic vector for T. So it suffices to count the number of cyclic vectors for T. Let $v \in \mathbb{F}_{q^n}$ be a cyclic vector for T. Then any other cyclic vector $w \in \mathbb{F}_{q^n}$ of T is necessarily of the form $f(T)v$ where $f \in \mathbb{F}_q[X]$ is such that $\deg f \leq n - 1$. Hence w is cyclic only if the T-annihilator of w is precisely p_T. Now the annihilator of $f(T)v$ is p_T if and only if $\gcd(f, p_T) = 1$. Thus the number of $f \in \mathbb{F}_q[X]$ with $\deg f \leq n - 1$ such that $w = f(T)v$ is cyclic is equal to the number of polynomials in $\mathbb{F}_q[X]$ of degree $\leq n - 1$ that are coprime to p_T. This is given by the q-analogue of the Euler totient function (cf. [**10**, p. 122]) evaluated at the minimal polynomial of T, namely,

$$\Phi_q(p_T) = q^n \prod_{i=1}^{k} \left(1 - \frac{1}{q^{n_i}}\right).$$

Note that $f_1(T)v \neq f_2(T)v$ for distinct polynomials f_1, f_2 of degree at most $n - 1$, for otherwise $p_T \mid (f_1 - f_2)$, which is a contradiction. Thus there are $\Phi_q(p_T)$ distinct cyclic vectors for T. Since each 1-dimensional T-splitting subspace of $\mathbb{F}_{q^n}$ is spanned by precisely $q - 1$ distinct cyclic vectors, it follows that the number of 1-dimensional T-splitting subspaces of $\mathbb{F}_{q^n}$ is $\Phi_q(p_T)/(q-1)$, as desired. □

It may be noted that if $\alpha \in \mathbb{F}_{q^{mn}}$ is such that $\mathbb{F}_{q^{mn}} = \mathbb{F}_q(\alpha)$ and if T is the $\mathbb{F}_q$-linear endomorphism of $\mathbb{F}_{q^{mn}}$ given by $x \mapsto \alpha x$, then the minimal polynomial p_T of T is precisely the minimal polynomial of α over $\mathbb{F}_q$. Hence p_T is irreducible and therefore the formula in Proposition 4.4 reduces to $S(\alpha, 1, n; q) = (q^n - 1)/(q - 1)$, exactly as observed in the beginning of Section 3.

Appendix A. Vector Recurrences and Singer Cycles

As before, we fix positive integers m, n and a prime power q. For any positive integer d, we denote, as usual, by $\mathrm{M}_d(\mathbb{F}_q)$ the set of all $d \times d$ matrices with entries in $\mathbb{F}_q$, and by $\mathrm{GL}_d(\mathbb{F}_q)$ the group of all nonsingular matrices in $\mathrm{M}_d(\mathbb{F}_q)$.

Let $C_0, C_1, \ldots, C_{n-1} \in M_m(\mathbb{F}_q)$. Given any *initial state* in $(\mathbb{F}_q^m)^n$, i.e., an n-tuple $(\mathbf{s}_0, \ldots, \mathbf{s}_{n-1})$ of (row) vectors in $\mathbb{F}_q^m$, the vector recurrence (of order n over $\mathbb{F}_q^m$)

$$\text{(A.1)} \qquad \mathbf{s}_{i+n} = \mathbf{s}_i C_0 + \mathbf{s}_{i+1} C_1 + \cdots + \mathbf{s}_{i+n-1} C_{n-1} \quad \text{for} \quad i = 0, 1, \ldots$$

generates an infinite sequence $\mathbf{s}^\infty = (\mathbf{s}_0, \mathbf{s}_1, \dots)$ of vectors in $\mathbb{F}_q^m$. It is easy to see that there are integers r, n_0 with $1 \le r \le q^{mn} - 1$ and $n_0 \ge 0$ such that $\mathbf{s}_{j+r} = \mathbf{s}_j$ for all $j \ge n_0$. The least positive integer r with this property is called the *period* of $\mathbf{s}^\infty$ and the corresponding least nonnegative integer n_0 is called the *preperiod* of $\mathbf{s}^\infty$. The sequence $\mathbf{s}^\infty$ is said to be *periodic* if its preperiod is 0. The vector recurrence (A.1) is said to be *primitive* if for any choice of nonzero initial state, the infinite sequence generated by it is periodic of period $q^{mn} - 1$. Vector recurrences are also known as word oriented linear feedback shift registers or σ-LFSRs, and they reduce to classical LFSRs or homogeneous linear recurrences of order n (with coefficients in $\mathbb{F}_q$) when $m = 1$. Primitive vector recurrences are of interest in cryptography since they are useful in pseudorandom number generation, or alternatively, for designing fast, secure, and efficient stream ciphers. While the study of (ordinary) LFSRs is classical (see, e.g., [**10**, Chap. 8]), vector recurrences and the corresponding multiple recursive method appears to have been first studied by Niederreiter [**11, 12**]. This method seems to have been rediscovered by Zeng, Han and He [**14**] in the guise of σ-LFSRs. Prior to that, generalizations of LFSRs (that turn out to be special cases of vector recurrences of Niederreiter) were studied by Tsaban and Vishne [**13**] and later by Dewar and Panario [**3**], and these are called transformation shift registers or TSRs. We refer to the recent paper of Hasan, Panario and Wang [**8**] for more on TSRs and related developments.

Enumerating primitive LFSRs of a given order is easy and well-known, whereas it is an open question in the case of σ-LFSRs. The following conjectural formula was proposed in [**6**] as a q-ary version of a conjecture of Zeng, Han and He [**14**].

Primitive Vector Recurrence Conjecture (PVRC): The number of primitive vector recurrences of order n over $\mathbb{F}_q^m$ is

$$\frac{\phi(q^{mn} - 1)}{mn} q^{m(m-1)(n-1)} \prod_{i=1}^{m-1} (q^m - q^i). \tag{A.2}$$

To relate the above to matrices, note that the maximum possible order of an element of the finite group $\mathrm{GL}_d(\mathbb{F}_q)$ is $q^d - 1$ (see, e.g., [**6**, Prop. 3.1]) and elements of order $q^d - 1$ are called *Singer cycles* in $\mathrm{GL}_d(\mathbb{F}_q)$. By an (m,n)*-block companion Singer cycle* over $\mathbb{F}_q$ we shall mean a Singer cycle T in $\mathrm{GL}_{mn}(\mathbb{F}_q)$ of the form

$$T = \begin{pmatrix} \mathbf{0} & \mathbf{0} & \mathbf{0} & . & . & \mathbf{0} & \mathbf{0} & C_0 \\ I_m & \mathbf{0} & \mathbf{0} & . & . & \mathbf{0} & \mathbf{0} & C_1 \\ . & . & . & . & . & . & . & . \\ . & . & . & . & . & . & . & . \\ \mathbf{0} & \mathbf{0} & \mathbf{0} & . & . & I_m & \mathbf{0} & C_{n-2} \\ \mathbf{0} & \mathbf{0} & \mathbf{0} & . & . & \mathbf{0} & I_m & C_{n-1} \end{pmatrix}, \tag{A.3}$$

where $C_0, C_1, \dots, C_{n-1} \in \mathrm{M}_m(\mathbb{F}_q)$ and I_m denotes the $m \times m$ identity matrix over $\mathbb{F}_q$, while $\mathbf{0}$ indicates the zero matrix in $\mathrm{M}_m(\mathbb{F}_q)$. The relation between (m,n)-block companion matrices such as T above and vector recurrences will be clearer if one observes that (A.1) is equivalent to the relation $S_{i+1} = S_i T$ for $i \ge 0$, where S_i is the ith state vector $(\mathbf{s}_i, \dots, \mathbf{s}_{i+n-1})$ and T is as in (A.3). Primitive vector recurrences correspond precisely to (m,n)-block companion Singer cycles [**6**, Thm. 5.2] and thus the PVRC is equivalent to

Block Companion Singer Cycle Conjecture (BCSCC): The number of (m,n)-block companion Singer cycles over $\mathbb{F}_q$ is given by (A.2).

It turns out that the map that sends a matrix in $M_{mn}(\mathbb{F}_q)$ to its characteristic polynomial maps the set of (m,n)-block companion Singer cycles over $\mathbb{F}_q$ onto the set of primitive polynomials in $\mathbb{F}_q[X]$ of degree mn (cf. [**6**, Thm. 6.1]). Recall that a polynomial in $\mathbb{F}_q[X]$ of degree d is said to be *primitive* if it is the minimal polynomial over $\mathbb{F}_q$ of a generator of the cyclic group $\mathbb{F}_{q^d}^*$ of nonzero elements of $\mathbb{F}_{q^d}$. Evidently, the number of primitive polynomials in $\mathbb{F}_q[X]$ of degree d is $\varphi(q^d-1)/d$. With this in view, BCSCC is implied by the following stronger conjecture.

Primitive Fiber Conjecture (PFC): For any primitive polynomial f in $\mathbb{F}_q[X]$ of degree mn, the number of (m,n)-block companion Singer cycles over $\mathbb{F}_q$ having f as its characteristic polynomial is

$$q^{m(m-1)(n-1)} \prod_{i=1}^{m-1} (q^m - q^i). \tag{A.4}$$

The above conjecture, proposed first in [**6**], was further strengthened in [**7**] as follows.

Irreducible Fiber Conjecture (IFC): For any irreducible polynomial f in $\mathbb{F}_q[X]$ of degree mn, the number of (m,n)-block companion Singer cycles over $\mathbb{F}_q$ having f as its characteristic polynomial is given by (A.4).

To relate irreducible fibers to splitting subspaces, it suffices to observe (see, e.g., Lemmas 5.1 and 5.2 of [**7**]) that if $f \in \mathbb{F}_q[X]$ is an irreducible polynomial of degree mn and if $\alpha \in \mathbb{F}_{q^{mn}}$ is any root of f, then the the number of (m,n)-block companion Singer cycles over $\mathbb{F}_q$ having f as its characteristic polynomial is precisely $N(\alpha,m,n;q)/(q^{mn}-1)$, where $N(\alpha,m,n;q)$ is as in Section 3 above. With this in view, the relation between the conjectures stated above and the Splitting Subspace Conjecture (SSC) as well as the Pointed Splitting Subspace Conjecture (PSSC) stated earlier in this paper can be summarized as follows.

$$\text{PSSC} \iff \text{SSC} \iff \text{IFC} \implies \text{PFC} \implies \text{BCSCC} \iff \text{PVRC}.$$

In view of the results proved in the previous sections, it is seen that each of these conjectures holds in the affirmative when $m \le 2$ or $n \le 1$. It may also be remarked that in order to prove the PFC, one may fix a primitive polynomial $f \in \mathbb{F}_q[X]$ of degree mn and a matrix $T \in \mathrm{M}_{mn}(\mathbb{F}_q)$ with f as its characteristic polynomial. Then matrices in $M_{mn}(\mathbb{F}_q)$ having f as its characteristic polynomial are necessarily similar to T. With this in view, Lachaud [**9**] has made a fine analysis of the similarity class of T and the collection of (m,n)-block companion matrices in it. He shows that T may be chosen to be a block diagonal matrix and the (m,n)-block companion matrices in the similarity class of T correspond to $V^{-1}TV$, where V is a so called block Vandermonde matrix. This analysis and the results of Lachaud [**9**] lend further insight into the above conjectures. Nonetheless, the general case remains open.

References

1. A. T. Benjamin and C.D. Bennett, The probability of relatively prime polynomials, *Math. Mag.* **80** (2007), 196–202. MR2322084 (2008b:11036)
2. S. Corteel, C. Savage, H. Wilf, and D. Zeilberger, A Pentagonal Number Sieve, *J. Combin. Theory Ser. A* **82** (1998), 186–192. MR1620873 (99d:11111)
3. M. Dewar and D. Panario, Linear transformation shift registers, *IEEE Trans. Inform. Theory*, **49** (2003), 2047–2052. MR2004712 (2004j:94020)

4. N. J. Fine and I. N. Herstein, The probability that a matrix be nilpotent, *Illinois J. Math.* **2** (1958), 499–504. MR0096677 (20:3160)
5. M. García-Armas, S. R. Ghorpade, and S. Ram, Relatively prime polynomials and nonsingular Hankel matrices over finite fields, *J. Combin. Theory Ser. A* **118** (2011), 819–828. MR2745427 (2011i:15038)
6. S. R. Ghorpade, S. U. Hasan, and M. Kumari, Primitive polynomials, Singer cycles, and word-oriented linear feedback shift registers, *Des. Codes Cryptogr.* **58** (2011), 123–134. MR2770307 (2012e:11200)
7. S. R. Ghorpade and S. Ram, Block companion Singer cycles, primitive recursive vector sequences, and coprime polynomial pairs over finite fields, *Finite Fields Appl.* **17** (2011), 461–472. MR2831705
8. S. U. Hasan, D. Panario, and Q. Wang, Word-oriented transformation shift registers and their linear complexity, *Sequences and Their Applications – SETA 2012*, Lecture Notes in Comput. Sci., Springer, Berlin (2012), to appear.
9. G. Lachaud, Construction of block companion matrices in a conjugacy class, preprint, 2011.
10. R. Lidl and H. Niederreiter, *Finite Fields*, Enc. of Math. and its Appl., Vol. 20, Cambridge University Press, Cambridge, 1983. MR746963 (86c:11106)
11. H. Niederreiter, Factorization of polynomials and some linear-algebra problems over finite fields, *Linear Algebra Appl.* **192** (1993), 301–328. MR1236747 (95b:11114)
12. H. Niederreiter, The multiple-recursive matrix method for pseudorandom number generation, *Finite Fields Appl.* **1** (1995), 3–30. MR1334623 (96k:11103)
13. B. Tsaban and U. Vishne, Efficient linear feedback shift registers with maximal period, *Finite Fields Appl.* **8** (2002), 256–267. MR1894517 (2003i:94040)
14. G. Zeng, W. Han and K. He, High efficiency feedback shift register: σ-LFSR, *Cryptology e-Print Archive: Report 2007/114* (available: `http://eprint.iacr.org/2007/11`

Department of Mathematics, Indian Institute of Technology Bombay, Powai, Mumbai 400076, India.

E-mail address: `srg@math.iitb.ac.in`

Department of Mathematics, Indian Institute of Technology Bombay, Powai, Mumbai 400076, India.

E-mail address: `samrith@gmail.com`

Contemporary Mathematics
Volume **574**, 2012
http://dx.doi.org/10.1090/conm/574/11446

The characteristic polynomials of abelian varieties of dimension 4 over finite fields

Safia Haloui and Vijaykumar Singh

ABSTRACT. We describe the set of characteristic polynomials of abelian varieties of dimension 4 over finite fields.

1. Introduction and results

The aim of this paper is to give a description of the set of characteristic polynomials of abelian varieties of dimension 4 as it was done in [**1**] for dimension 3.

It is well known that the characteristic polynomial of an abelian variety of dimension g over $\mathbb{F}_q$ (with $q = p^n$) is monic, with integer coefficients, of degree $2g$ and that the sets of its roots consist of couples of complex conjugated numbers of modulus $\sqrt{q}$. Any polynomial having those properties is called a *Weil polynomial.* Obviously, every Weil polynomial of degree 8 has the form

$$p(t) = t^8 + a_1t^7 + a_2t^6 + a_3t^5 + a_4t^4 + qa_3t^3 + q^2a_2t^2 + q^3a_1t + q^4$$

for certain integers a_1, a_2, a_3 and a_4. In Section 2 we prove the following proposition which gives a characterization of the quadruples (a_1, a_2, a_3, a_4) corresponding to Weil polynomials of degree 8 (see [**9, 4, 1**] for a characterization of Weil polynomials of lower degrees):

THEOREM 1.1. *Let* $p(t) = t^8+a_1t^7+a_2t^6+a_3t^5+a_4t^4+qa_3t^3+q^2a_2t^2+q^3a_1t+q^4$ *be a polynomial with integer coefficients. Then* $p(t)$ *is a Weil polynomial if and only if the following conditions hold:*

(1) $|a_1| \leq 8\sqrt{q}$,
(2) $6\sqrt{q}|a_1| - 20q \leq a_2 \leq \frac{3a_1^2}{8} + 4q$,
(3) $-9qa_1 - 4\sqrt{q}a_2 - 16q\sqrt{q} \leq a_3 \leq -9qa_1 + 4\sqrt{q}a_2 + 16q\sqrt{q}$,
(4) $-\frac{a_1^3}{8} + \frac{a_1a_2}{2} + qa_1 - (\frac{2}{3}(\frac{3a_1^2}{8} - a_2 + 4q))^{3/2} \leq a_3 \leq -\frac{a_1^3}{8} + \frac{a_1a_2}{2} + qa_1 + (\frac{2}{3}(\frac{3a_1^2}{8} - a_2 + 4q))^{3/2}$,
(5) $2\sqrt{q}|qa_1 + a_3| - 2qa_2 - 2q^2 \leq a_4$,
(6) $\frac{9a_1^4}{256} - \frac{3a_1^2a_2}{16} + \frac{a_1a_3}{4} + \frac{a_2^2}{6} + \frac{2qa_2}{3} + \frac{2q^2}{3} + \omega + \overline{\omega} \leq a_4 \leq \frac{9a_1^4}{256} - \frac{3a_1^2a_2}{16} + \frac{a_1a_3}{4} + \frac{a_2^2}{6} + \frac{2qa_2}{3} + \frac{2q^2}{3} + j\omega + j^2\overline{\omega}$

2010 *Mathematics Subject Classification.* Primary 14G15, 11C08, 11G10, 11G25.
Key words and phrases. Abelian varieties over finite fields, Weil polynomials.

where

$$\omega = \frac{1}{24}\Big(8(-\tfrac{3a_1^2}{8} + a_2 - 4q)^6 + 540(-\tfrac{3a_1^2}{8} + a_2 - 4q)^3(\tfrac{a_1^3}{8} - qa_1 - \tfrac{a_1a_2}{2} + a_3)^2$$
$$-729(\tfrac{a_1^3}{8} - qa_1 - \tfrac{a_1a_2}{2} + a_3)^4$$
$$+i9|\tfrac{a_1^3}{8} - qa_1 - \tfrac{a_1a_2}{2} + a_3|(-(\tfrac{a_1^3}{8} - qa_1 - \tfrac{a_1a_2}{2} + a_3)^2 - \tfrac{8}{27}(-\tfrac{3a_1^2}{8} + a_2 - 4q)^3)^{3/2}\Big)^{1/3},$$
$\omega^{1/3} = |\omega|^{1/3} e^{\frac{arg(\omega)i}{3}}$ *and* $j = e^{\frac{2i\pi}{3}}$.

Now it remains to give criterions to determine when a Weil polynomial is the characteristic polynomial of an abelian variety of dimension 4. Since the characteristic polynomial of a non-simple abelian variety is a product of characteristic polynomials of abelian varieties of smaller dimension and our problem is already solved for smaller dimensions [**9, 4, 1**], it is sufficient to consider the simple case.

By results of Honda and Tate, the characteristic polynomial of a simple abelian variety of dimension 4 over $\mathbb{F}_q$ has the form $p(t) = h(t)^e$ where $h(t)$ is an irreducible Weil polynomial and e is an integer. Xing [**11**] and Maisner and Nart [**2**] gave independently a description of characteristic polynomials of abelian varieties of dimension 4 with $e > 1$. Therefore, we can restrict our attention to the case $e = 1$, that is, $p(t)$ is irreducible.

If $p(t)$ is irreducible, the determination of the possible Newton polygons for $p(t)$ (Section 3) gives us the following proposition:

THEOREM 1.2. *Let* $p(t) = t^8+a_1t^7+a_2t^6+a_3t^5+a_4t^4+qa_3t^3+q^2a_2t^2+q^3a_1t+q^4$ *be an irreducible Weil polynomial. Then $p(t)$ is the characteristic polynomial of an abelian variety of dimension 4 if and only if one of the following conditions holds:*

(1) $v_p(a_4) = 0$,
(2) $v_p(a_3) = 0$, $v_p(a_4) \geq n/2$ *and* $p(t)$ *has no root of valuation* $n/2$ *in* $\mathbb{Q}_p$,
(3) $v_p(a_2) = 0$, $v_p(a_3) \geq n/2$, $v_p(a_4) \geq n$ *and* $p(t)$ *has no root of valuation* $n/2$ *in* $\mathbb{Q}_p$,
(4) $v_p(a_1) = 0$, $v_p(a_2) \geq n/2$, $v_p(a_3) \geq n$, $v_p(a_4) \geq 3n/2$ *and* $p(t)$ *has no root of valuation* $n/2$ *nor factor of degree* 3 *in* $\mathbb{Q}_p$,
(5) $v_p(a_1) = 0$, $v_p(a_2) \geq n/3$, $v_p(a_3) \geq 2n/3$, $v_p(a_4) = n$ *and* $p(t)$ *has no root of valuation* $n/3$ *and* $2n/3$ *in* $\mathbb{Q}_p$,
(6) $v_p(a_1) \geq n/3$, $v_p(a_2) \geq 2n/3$, $v_p(a_3) = n$, $v_p(a_4) \geq 3n/2$ *and* $p(t)$ *has no root in* $\mathbb{Q}_p$,
(7) $v_p(a_1) \geq n/4$, $v_p(a_2) \geq n/2$, $v_p(a_3) = 3n/4$, $v_p(a_4) = n$ *and* $p(t)$ *has no root nor factor of degree* 2 *and* 3 *in* $\mathbb{Q}_p$,
(8) $v_p(a_1) \geq n/2$, $v_p(a_2) \geq n$, $v_p(a_3) = 3n/2$, $v_p(a_4) \geq 2n$ *and* $p(t)$ *has no root nor factor of degree* 3 *in* $\mathbb{Q}_p$.

The p-ranks of abelian varieties in cases (1), (2), (3), (4), (5), (6), (7) and (8) are respectively 4, 3, 2, 1, 1, 0, 0 *and* 0. *The abelian varieties in case (8) are supersingular.*

It is possible to make condition (8) of Theorem 1.2 more explicit. Indeed, in [**6**], Singh, McGuire and Zaytsev gave the list of irreducible characteristic polynomials of supersingular abelian varieties of dimension 4 , where q is not a square. We complete the classification by finding the list in the case q is a square in the following proposition (see Section 4).

THEOREM 1.3. *The polynomial $p(t)$ is the irreducible characteristic polynomial of a supersingular abelian variety of dimension 4 if and only if one of the following conditions holds*

- *q is a square and (a_1, a_2, a_3, a_4) belongs to the following list:*
  (1) $(-q^{1/2}, 0, q^{3/2}, -q^2)$, $p \not\equiv 1 \mod 15$,
  (2) $(q^{1/2}, 0, -q^{3/2}, -q^2)$, $p \not\equiv 1 \mod 30$,
  (3) $(0, 0, 0, 0)$, $p \not\equiv 1 \mod 16$,
  (4) $(0, -q, 0, q^2)$, $p \not\equiv 1 \mod 20$,
  (5) $(0, 0, 0, -q^2)$, $p \not\equiv 1 \mod 24$,
- *q is not a square and (a_1, a_2, a_3, a_4) belongs to the following list:*
  (1) $(\pm\sqrt{pq}, q, 0, -q^2)$, $p = 2$,
  (2) $(\pm\sqrt{pq}, 2q, \pm q\sqrt{pq}, q^2)$, $p = 3$,
  (3) $(0, 0, 0, 0)$,
  (4) $(0, -q, 0, q^2)$,
  (5) $(0, q, 0, q^2)$, $p \neq 5$,
  (6) $(0, 0, 0, -q^2)$, $p \neq 2$,
  (7) $(\pm\sqrt{pq}, 2q, \pm q\sqrt{pq}, 3q^2)$, $p = 5$.

2. The coefficients of Weil polynomials of degree 8

In this section, we prove Theorem 1.1. Let $p(t) = t^8 + a_1t^7 + a_2t^6 + a_3t^5 + a_4t^4 + qa_3t^3 + q^2a_2t^2 + q^3a_1t + q^4$ be a polynomial with integer coefficients. We clearly have $p(t) = t^{2g}/q^g p(q/t)$. Arguing as in the proof of [**8**, Theorem 5.1.15. (e)], we can arrange the roots of $p(t)$ as $\omega_1, q/\omega_1, \ldots, \omega_4, q/\omega_4$. Letting $x_i = -(\omega_i + q/\omega_i)$ we have $p(t) = \prod_{i=1}^4 (t^2 + x_it + q)$.

If $p(t)$ is a Weil polynomial then $|\omega_i| = \sqrt{q}$ and thus $q/\omega_i = \overline{\omega_i}$; we deduce that the x_i's are real numbers of absolute value less than or equal to $2\sqrt{q}$. Conversely, if x_i is real and $|x_i| \leq 2\sqrt{q}$ then the polynomial $t^2 + x_it + q \in \mathbb{R}[t]$ has a negative discriminant and thus its roots are complex conjugated and of modulus $\sqrt{q}$; if this true for $i = 1, \ldots, 4$ then $p(t)$ is a Weil polynomial.

Therefore $p(t)$ is a Weil polynomial if and only if the polynomials $f^+(t) = \prod_{i=1}^4 (t - (2\sqrt{q} + x_i))$ and $f^-(t) = \prod_{i=1}^4 (t - (2\sqrt{q} - x_i))$ have only real and positive roots.

First, we determine a necessary and sufficient condition of some polynomial of degree 4 having all real roots.

Let $f(t) = t^4 + r_1t^3 + r_2t^2 + r_3t + r_4$ be a monic polynomial of degree 4 with real coefficients. Looking at the table of variation of $f(t)$, we see that there exists some r_4 for which $f(t)$ has all real roots if and only if $f'(t)$ has all real roots. This condition is equivalent to

$$\Delta_{f'} \geq 0 \tag{1}$$

where $\Delta_{f'}$ is the discriminant of $f'(t)$.

The discriminant of $f(t)$ is a polynomial of degree 3 in r_4 which we will denote $\Delta_f(t)$ (that is, $\Delta_f(r_4)$ is the discriminant of $f(t)$). It is well known that if $f(t)$ has all real roots then $\Delta_f(r_4) \geq 0$. Moreover, the function which associate to r_4 the number of roots of $f(t)$ is constant on the intervals delimited by the roots of $\Delta_f(t)$ (because $\Delta_f(r_4) = 0$ when $f(t)$ has a multiple root).

When r_4 is very big, the graph of $f(t)$ doesn't touch the x-axis and therefore $f(t)$ has no real roots. Thus if γ_3 is the biggest root of $\Delta_f(t)$, by the previous discussion, $f(t)$ has no real root for $r_4 \in]\gamma_3; +\infty[$.

We deduce that if (1) is satisfied then $\Delta_f(t)$ must have 3 real roots $\gamma_1 \leq \gamma_2 \leq \gamma_3$ and $f(t)$ has all real roots if and only if

$$\gamma_1 \leq r_4 \leq \gamma_2. \tag{2}$$

The roots of $\Delta_f(t)$ can be found using Cardan's method. Let us recall quickly what it is.

Given a polynomial $h(t) = t^3 + u_2 t + u_3$, we set $\delta = -u_3^2 - \frac{4}{27}u_2^3$. Then $h(t)$ has all real roots if and only if $\delta \geq 0$. If this is the case, the roots of $h(t)$ are $\gamma_1 = \omega + \overline{\omega}$, $\gamma_2 = j\omega + j^2\overline{\omega}$ and $\gamma_3 = j^2\omega + j\overline{\omega}$ where $j = e^{\frac{2i\pi}{3}}$ and $\omega = (\frac{-u_3 + i\sqrt{\delta}}{2})^{1/3}$. Moreover, with the convention $\omega^{1/3} = |\omega|^{1/3} e^{\frac{arg(\omega)i}{3}}$, we have $\gamma_1 \leq \gamma_2 \leq \gamma_3$.

In the general case, we have $h(t) = v_0 t^3 + v_1 t^2 + v_2 t + v_3$ and we conclude using the fact that $\frac{1}{v_0} h(t - \frac{v_1}{3v_0}) = t^3 + u_2 t + u_3$ with $u_2 = -\frac{v_1^2}{3v_0^2} + \frac{v_2}{v_0}$ and $u_3 = \frac{2v_1^3}{27v_0^3} - \frac{v_1 v_2}{3v_0^2} + \frac{v_3}{v_0}$.

For $i = 1, 2, 3, 4$, let s_i denote the ith symmetric function of the x_i's (that is, $\prod_{i=1}^4 (t + x_i) = t^4 + \sum_{i=1}^4 s_i t^{4-i}$). Expanding the expression $p(t) = \prod_{i=1}^4 (t^2 + x_i t + q)$, we find:

$$\begin{aligned}
s_1 &= a_1 \\
s_2 &= a_2 - 4q \\
s_3 &= a_3 - 3qa_1 \\
s_4 &= a_4 - 2qa_2 + 2q^2.
\end{aligned}$$

Now, in order to simplify the calculation, we remark that $f^+(t)$ and $f^-(t)$ have all real roots if and only if the polynomial $f(t) = \prod_{i=1}^4 (t + x_i - \frac{a_1}{4})$ has. Therefore, it is equivalent to apply our results to $f(t)$.

Expanding the expression of $f(t)$, we find that $f(t) = t^4 + r_2 t^2 + r_3 t + r_4$, where

$$\begin{aligned}
r_2 &= -\frac{3s_1^2}{8} + s_2 \\
r_3 &= \frac{s_1^3}{8} - \frac{s_1 s_2}{2} + s_3 \\
r_4 &= -\frac{3s_1^4}{256} + \frac{s_1^2 s_2}{16} - \frac{s_1 s_3}{4} + s_4.
\end{aligned}$$

Substituting s_1, s_2, s_3 and s_4 with their expressions in a_1, a_2, a_3 and a_4 we obtain

$$\begin{aligned}
r_2 &= -\frac{3a_1^2}{8} + a_2 - 4q \\
r_3 &= \frac{a_1^3}{8} - qa_1 - \frac{a_1 a_2}{2} + a_3 \\
r_4 &= -\frac{3a_1^4}{256} + \frac{qa_1^2}{2} + \frac{a_1^2 a_2}{16} - \frac{a_1 a_3}{4} - 2qa_2 + 2q^2 + a_4.
\end{aligned}$$

We have

$$\Delta_f(t) = 256t^3 - 128r_2^2 t^2 + 16r_2(r_2^3 + 9r_3^2)t - r_3^2(4r_2^3 + 27r_3^2).$$

Now, we use Cardan's method. Set

$$\begin{aligned} u_2 &= -\frac{r_2^4}{48}+\frac{9r_2r_3^2}{16} \\ u_3 &= \frac{r_2^6}{864}+\frac{5r_2^3r_3^2}{64}-\frac{27r_3^4}{256} \\ \delta &= -u_3^2-\frac{4}{27}u_2^3 = \frac{r_3^2(-8r_2^3-27r_3^2)^3}{1769472}. \end{aligned}$$

Suppose that (1) is satisfied. Then $\delta \geq 0$ and the roots of $\Delta_f(t)$ are

$$\begin{aligned} \gamma_1 &= \omega+\overline{\omega}+\frac{r_2^2}{6} \\ \gamma_2 &= j\omega+j^2\overline{\omega}+\frac{r_2^2}{6} \\ \gamma_3 &= j^2\omega+j\overline{\omega}+\frac{r_2^2}{6}. \end{aligned}$$

where

$$\omega = \frac{1}{24}\left(8r_2^6+540r_2^3r_3^2-729r_3^4+i9|r_3|(-r_3^2-\frac{8}{27}r_2^3)^{3/2}\right)^{1/3}.$$

Substituting r_2, r_3 and r_4 with their expressions in a_1, a_2, a_3 and a_4 we obtain condition (6) of Theorem 1.1.

Next we have to determine when (1) is satisfied. We have:

$$\Delta_{f'} = -16(8r_2^3+27r_3^2).$$

Therefore, (1) is equivalent to

$$r_2 \leq 0 \text{ and } -(\frac{-2r_2}{3})^{3/2} \leq r_3 \leq (\frac{-2r_2}{3})^{3/2}.$$

This gives us the second inequality of condition (2) and condition (4) of Theorem 1.1.

Finally, we determine when the polynomials $f^+(t)$ and $f^-(t)$ have only positive roots.

For $i=1,2,3$, let r_i^+ and r_i^- denote the respective ith coefficients of $f^+(t)$ and $f^-(t)$. Expending the expressions $f^+(t)=\prod_{i=1}^4(t-(2\sqrt{q}+x_i))$ and $f^-(t)=\prod_{i=1}^4(t-(2\sqrt{q}-x_i))$, we find:

$$\begin{aligned} r_1^+ &= -8\sqrt{q}-s_1 \\ r_2^+ &= 24q+6\sqrt{q}s_1+s_2 \\ r_3^+ &= -32q\sqrt{q}-12qs_1-4\sqrt{q}s_2-s_3 \\ r_4^+ &= 16q^2+8q\sqrt{q}s_1+4qs_2+2\sqrt{q}s_3+s_4 \end{aligned}$$

and

$$\begin{aligned} r_1^- &= -8\sqrt{q}+s_1 \\ r_2^- &= 24q-6\sqrt{q}s_1+s_2 \\ r_3^- &= -32q\sqrt{q}+12qs_1-4\sqrt{q}s_2+s_3 \\ r_4^- &= 16q^2-8q\sqrt{q}s_1+4qs_2-2\sqrt{q}s_3+s_4. \end{aligned}$$

Substituting s_1, s_2, s_3 and s_4 with their expressions in a_1, a_2, a_3 and a_4 we obtain

$$\begin{aligned} r_1^+ &= -8\sqrt{q} - a_1 \\ r_2^+ &= 20q + 6\sqrt{q}a_1 + a_2 \\ r_3^+ &= -16q\sqrt{q} - 9qa_1 - 4\sqrt{q}a_2 - a_3 \\ r_4^+ &= 2q^2 + 2q\sqrt{q}a_1 + 2qa_2 + 2\sqrt{q}a_3 + a_4 \end{aligned}$$

and

$$\begin{aligned} r_1^- &= -8\sqrt{q} + a_1 \\ r_2^- &= 20q - 6\sqrt{q}a_1 + a_2 \\ r_3^- &= -16q\sqrt{q} + 9qa_1 - 4\sqrt{q}a_2 + a_3 \\ r_4^- &= 2q^2 - 2q\sqrt{q}a_1 + 2qa_2 - 2\sqrt{q}a_3 + a_4. \end{aligned}$$

Suppose that $f^+(t)$ and $f^-(t)$ have all real roots. Then by [**7**, §2, Lemma], $f^+(t)$ and $f^-(t)$ have only positive roots if and only if $(-1)^i r_i^+ \geq 0$ and $(-1)^i r_i^- \geq 0$ for $i = 1, 2, 3, 4$. This gives us the remaining conditions of Theorem 1.1 and concludes the proof.

Remark. We could have used [**7**, §2, Lemma] to determine when a polynomial of degree 4 has only real roots but the computation and the results would have been longer.

3. Newton polygons

Let $p(t)$ be an irreducible Weil polynomial. By [**3**], $p(t)^e$ is the characteristic polynomial of a simple abelian variety, where e the least common denominator of $v_p(f(0))/n$ where $f(t)$ runs through the irreducible factors of $p(t)$ over $\mathbb{Q}_p$. Thus $p(t)$ is the characteristic polynomial of an abelian variety of dimension 4 if and only if e is equal to 1 that is, $v_p(f(0))/n$ are integers.

In order to determine when this condition is satisfied, we consider the Newton polygon of $p(t)$ (see [**10**]). Each of its edges define a factor of $p(t)$ over $\mathbb{Q}_p$. The degree of this factor is the length of the projection onto the x-axis of the edge and all the roots of this factor have a valuation equal to the slope of the edge. Therefore $e = 1$ implies that the length of the projection onto the x-axis of any edge times its slope is a multiple of n.

We graph the Newton polygons satisfying this condition and in each case, we give a necessary and sufficient condition to have $e = 1$. The obtained results are summarized in Theorem 1.2.

Ordinary case: $\mathbf{v_p(a_4) = 0}$

The Newton polygon of $p(t)$ is represented in Figure 1 and we always have $e = 1$.

p-rank 3 case: $\mathbf{v_p(a_4) > 0}$ and $\mathbf{v_p(a_3) = 0}$

The only Newton polygon for which $e = 1$ is represented in Figure 2.

This is the Newton polygon of $p(t)$ if and only if $v_p(a_4) \geq n/2$. If this condition holds, $p(t)$ has a factor in $\mathbb{Q}_p$ of degree 2 with roots of valuation $n/2$ and $e = 1$ if and only if this factor is irreducible, that is, if and only if $p(t)$ has no root of valuation $n/2$ in $\mathbb{Q}_p$.

p-rank 2 case: $\mathbf{v_p(a_4) > 0}$, $\mathbf{v_p(a_3) > 0}$ and $\mathbf{v_p(a_2) = 0}$

The only Newton polygon for which $e = 1$ is represented in Figure 3.

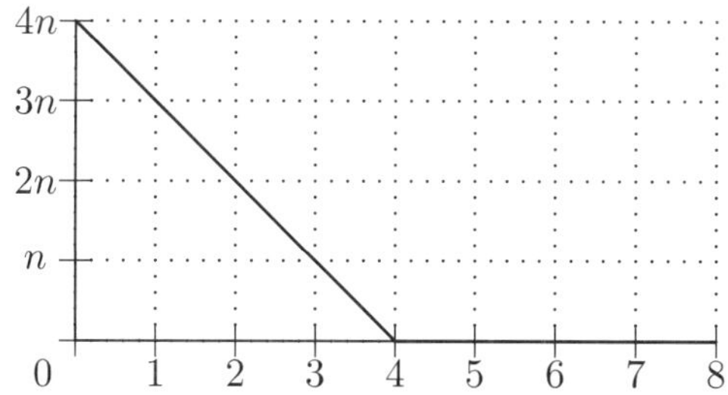

FIGURE 1. Ordinary case

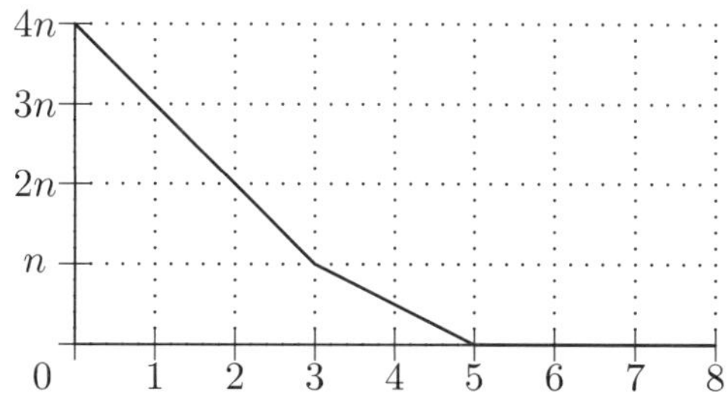

FIGURE 2. p-rank 3 case

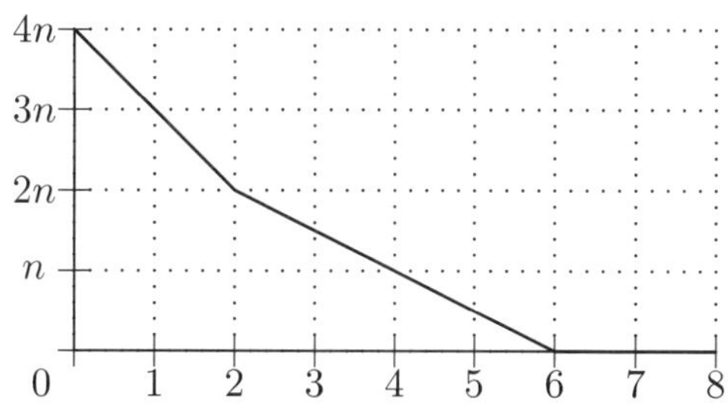

FIGURE 3. p-rank 2 case

This is the Newton polygon of $p(t)$ if and only if $v_p(a_3) \geq n/2$ and $v_p(a_4) \geq n$. If these conditions hold, $p(t)$ has a factor in $\mathbb{Q}_p$ of degree 4 with roots of valuation $n/2$ and $e = 1$ if and only if this factor has no root in $\mathbb{Q}_p$, that is, if and only if $p(t)$ has no root of valuation $n/2$ in $\mathbb{Q}_p$.

p-rank 1 case: $\mathbf{v_p(a_4) > 0}$, $\mathbf{v_p(a_3) > 0}$, $\mathbf{v_p(a_2) > 0}$ and $\mathbf{v_p(a_1) = 0}$

There are two Newton polygons for which $e = 1$. One is represented in Figure 4.

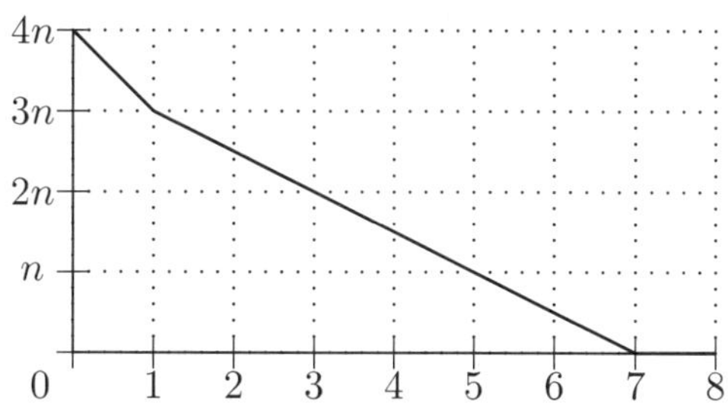

FIGURE 4. p-rank 1 first case

This is the Newton polygon of $p(t)$ if and only if $v_p(a_2) \geq n/2$, $v_p(a_3) \geq n$ and $v_p(a_4) \geq 3n/2$. If these conditions hold, $e = 1$ if and only if $p(t)$ has no root of valuation $n/2$ nor factor of degree 3 in $\mathbb{Q}_p$.

The other Newton polygon is represented in Figure 5.

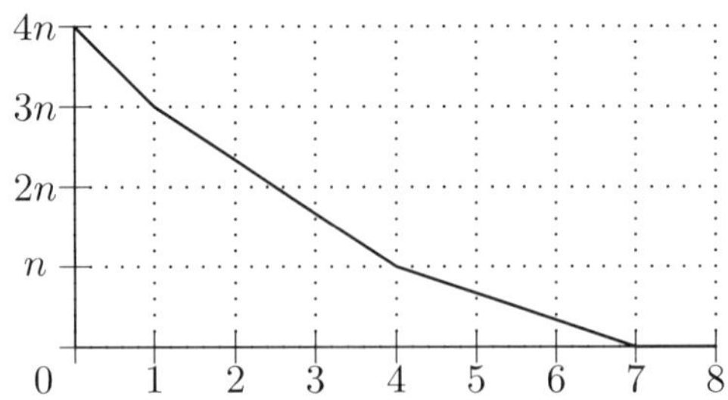

FIGURE 5. p-rank 1 second case

This is the Newton polygon of $p(t)$ if and only if $v_p(a_2) \geq n/3$, $v_p(a_3) \geq 2n/3$ and $v_p(a_4) = n$. If these conditions hold $e = 1$ if and only if $p(t)$ has no root of valuation $n/3$ and $2n/3$ in $\mathbb{Q}_p$.

p-rank 0 case: $\mathbf{v_p(a_4) > 0}$, $\mathbf{v_p(a_3) > 0}$, $\mathbf{v_p(a_2) > 0}$ and $\mathbf{v_p(a_1) > 0}$

There are three Newton polygons for which $e = 1$. One is represented in Figure 6.

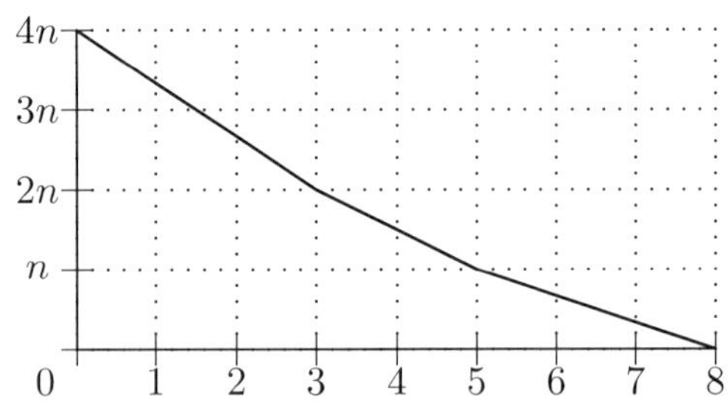

FIGURE 6. p-rank 0 first case

This is the Newton polygon of $p(t)$ if and only if $v_p(a_1) \geq n/3$, $v_p(a_2) \geq 2n/3$, $v_p(a_3) = n$ and $v_p(a_4) \geq 3n/2$. If these conditions hold, $e = 1$ if and only if $p(t)$ has no root in $\mathbb{Q}_p$.

The second Newton polygon is represented in Figure 7.

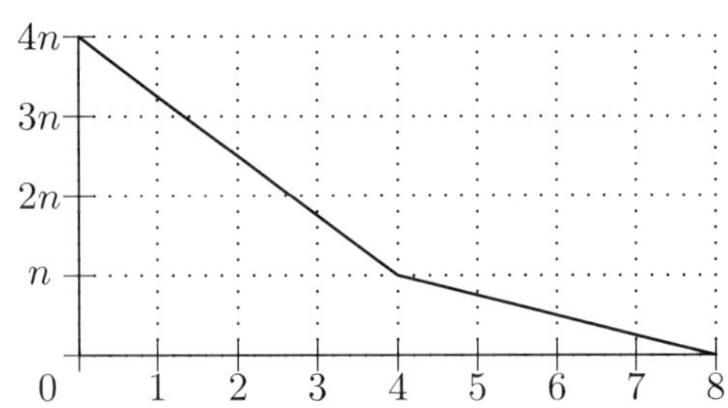

FIGURE 7. p-rank 0 second case

This is the Newton polygon of $p(t)$ if and only if $v_p(a_1) \geq n/4$, $v_p(a_2) \geq n/2$, $v_p(a_3) \geq 3n/4$ and $v_p(a_4) = n$. If these conditions hold, $e = 1$ if and only if $p(t)$ has no factor of degrees 1, 2 and 3 in $\mathbb{Q}_p$.

The last Newton polygon is represented in Figure 8; the corresponding abelian varieties are supersingular.

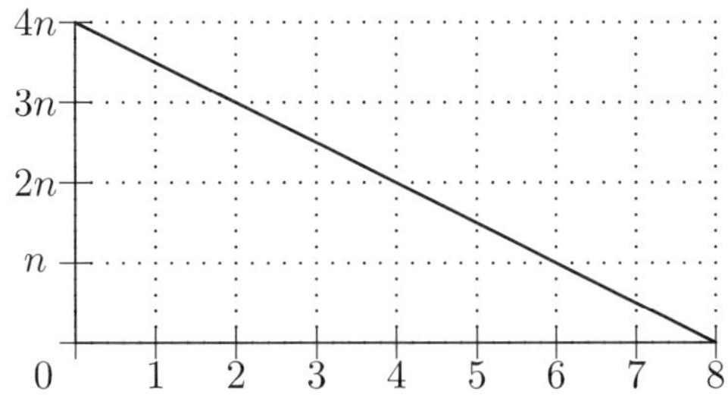

FIGURE 8. Supersingular case

This is the Newton polygon of $p(t)$ if and only if $v_p(a_1) \geq n/2$, $v_p(a_2) \geq n$, $v_p(a_3) \geq 3n/2$ and $v_p(a_3) \geq 2n$. If these conditions hold, $e = 1$ if and only if $p(t)$ has no root nor factor of degree 3 in $\mathbb{Q}_p$.

4. Supersingular case

In [**6**], Singh, McGuire and Zaytsev gave the list of irreducible characteristic polynomials of supersingular abelian varieties of dimension 4 , where q is not a square. Here we derive the list when q is a square.

Let $p(t)$ be an irreducible supersingular Weil polynomial of degree 8, where q is a square. By Honda-Tate Theorem, $\frac{1}{q^4}p(\sqrt{q}t)$ is a cyclotomic polynomial of degree 8 i.e; $\frac{1}{q^4}p(t\sqrt{q}) = \Phi_m(t)$ such that $\phi(m) = 8$ or $m \in \{15, 16, 20, 24, 30\}$. Therefore for each m above, $p(t) = q^g\Phi_m(\frac{t}{\sqrt{q}})$ gives a supersingular Weil polynomial of degree 8. Let $p(t) = \prod_i p_i(t)$ be the decomposition in irreducible factors of $p(t)$ over $\mathbb{Q}_p$ with $\pi = \sqrt{q}\zeta_n$ as a root, where ζ_n is primitive nth root of unity. To determine the dimension of the corresponding abelian variety, recall from [**3**], $p(t)^e$ is a characteristic polynomial of an abelian variety of dimension $4e$, where e is the least common denominator of $\dfrac{v_p(\pi)}{v_p(q)} \deg p_i(t) = \dfrac{\deg p_i(t)}{2}$. Since $p(t) = q^g\Phi_m(\frac{t}{\sqrt{q}})$, $\deg p_i(t) = \deg r_i(t)$ where $\Phi_m(t) = \prod_i r_i(t)$. But from chapter IV.4 in [**5**], we have $\deg r_i(t) = r$ where

$$r = \begin{cases} \text{order of } p \text{ in the multiplicative group } (\frac{\mathbb{Z}}{m\mathbb{Z}})^* \ , & \text{if } (p, m) = 1; \\ f(p^k - p^{k-1}), \text{ where } f \text{ is order of } p \text{ in the multiplicative group } (\frac{\mathbb{Z}}{s\mathbb{Z}})^*, \\ & \text{if } m = p^k s. \end{cases}$$

Hence, $e = 1$ if r is even. Since r divides $\phi(m)$ and in each case of m above, $\phi(m) = 2^3$, implies r is either even or $r = 1$. The later case only happens when $p \equiv 1 \mod m$. Hence, $p(t) = q^g\Phi_m(\frac{t}{\sqrt{q}})$, where $p \not\equiv 1 \mod m$ is an irreducible characteristic polynomial of a supersingular abelian variety of dimension 4, for each $m \in \{15, 16, 20, 24, 30\}$.

Acknowledgements

We would like to thank Christophe Ritzenthaler for suggesting this work.

References

[1] S. Haloui. The characteristic polynomials of abelian varieties of dimensions 3 over finite fields. *J. Number Theory.* no 130, p. 2745-2752, 2010. MR2684495 (2012b:14043)

[2] D. Maisner, E. Nart, appendix of E. W. Howe. Abelian surfaces over finite fields as jacobians. *Experiment. Math.*. Vol 11, p. 321-337, 2002. MR1959745 (2003k:14057)

[3] J. Milne, W. Waterhouse. Abelian varieties over finite fields. *1969 Number Theory Institute*, Proceedings of Symposia in Pure Math., Vol XX, p.53-64, A.M.S., Providence, RI, 1971. Vol 76, p. 351-366, 1990. MR0314847 (47:3397)

[4] H.G. Rück. Abelian surfaces and Jacobian varieties over finite fields. *Compositio Math.*. Vol 76, p. 351-366, 1990. MR1080007 (92e:14016)

[5] J.P.Serre, *Local Fields*, Springer-Verlag, 1979. MR554237 (82e:12016)

[6] V. Singh, A. Zaytsev and G. McGuire. On the characteristic polynomial of frobenius of supersingular abelian varieties of dimension up to 7 over finite fields. Preprint: `http://arxiv.org/abs/1005.3635`.

[7] C. Smyth. Totally positive algebraic integers of small trace. *Ann. Inst. Fourrier.* Vol 33, p. 285-302, 1973.

[8] H. Stichtenoth. *Algebraic function fields and codes.* GTM 254, Springer, 2009. MR2464941 (2010d:14034)

[9] W. Waterhouse. Abelian varieties over finite fields. *Ann. sci. Ecole Norm. Sup.*. (4), t. 2, p. 521-560, 1969. MR0265369 (42:279)

[10] E. Weiss. *Algebraic number theory.* McGraw, New-York, 1963. MR0159805 (28:3021)

[11] C.P. Xing. The characteristic polynomials of abelian varieties of dimension three and four over finite fields. *Science in China.* Vol 37, no 3, p. 147-150, 1994. MR1275799 (95e:11065)

Institut de Mathématiques de Luminy, Marseille, France — and — Claude Shannon Institute, Dublin, Ireland

E-mail address: `haloui@iml.univ-mrs.fr`

Institut de Mathématiques de Luminy, Marseille, France — and — Claude Shannon Institute, Dublin, Ireland

E-mail address: `vijaykumar.singh@ucdconnect.ie`

Contemporary Mathematics
Volume **574**, 2012
http://dx.doi.org/10.1090/conm/574/11431

New bounds on the maximum number of points on genus-4 curves over small finite fields

Everett W. Howe

Dedicated to the memory of my mother, Norma Howe

ABSTRACT. For prime powers $q < 100$, we compute new upper and lower bounds on $N_q(4)$, the maximal number of points on a genus-4 curve over a finite field with q elements. We determine the exact value of $N_q(4)$ for 17 prime powers q for which the value was previously unknown.

1. Introduction

For every prime power q and integer $g \geq 0$, let $N_q(g)$ denote the maximal number of points on a curve of genus g over $\mathbf{F}_q$. In this paper we compute new upper and lower bounds on $N_q(4)$ for prime powers q less than 100, and we find the exact value of $N_q(4)$ for 17 prime powers q for which the value was previously unknown.

Weil's famous 'Riemann Hypothesis' for curves over finite fields [**22**–**25**] gives the fundamental result that

$$N_q(g) \leq q + 1 + 2g\sqrt{q},$$

which generalizes Hasse's theorem for the number of points on an elliptic curve over a finite field [**6**]. Weil's bound was improved significantly in the case where g is large with respect to q by Manin [**14**], Ihara [**11**], and Drinfel′d and Vlăduţ [**20**], and for fixed q we have the Drinfel′d-Vlăduţ bound

$$N_q(g) \leq (\sqrt{q} - 1 + o(1))g \qquad \text{as } g \to \infty.$$

When q is a square this bound is asymptotically optimal, in the sense that

$$\limsup N_q(g)/g = \sqrt{q} - 1.$$

The value of this lim sup is not known for any nonsquare q.

For any particular choice of q and g, it can be a difficult computational problem to determine the actual value of $N_q(g)$ and to find a curve that attains this number of points. When g is less than $(q - \sqrt{q})/2$, the best upper bound known is often Serre's improvement to the Weil bound [**17**]:

$$N_q(g) \leq q + 1 + g\lfloor 2\sqrt{q}\rfloor.$$

2010 *Mathematics Subject Classification.* Primary 11G20; Secondary 14G05, 14G10, 14G15.
Key words and phrases. Curve, rational point, zeta function, finite field.

q	old	new	q	old	new	q	old	new
2	8	8	23	–58	57	59	–118	116
3	12	12	25	66	66	61	–122	118–119
4	15	15	27	64	64	64	129	129
5	18	18	29	–70	67–68	67	–132	129
7	24	24	31	–73	72	71	132–136	134
8	25	25	32	71–72	71–72	73	–139	138
9	30	30	37	–84	82	79	–148	148
11	33–34	33	41	–90	88	81	154	154
13	–39	38	43	–93	92	83	–154	152
16	45	45	47	–100	98	89	–162	160–162
17	–48	46	49	–106	102–106	97	–174	174
19	–52	48–50	53	–110	108			

TABLE 1. Old and new ranges for $N_q(4)$, for $q < 100$.

But for every fixed genus $g > 2$, nothing is known about the proportion of prime powers for which this bound is attained.

For $g = 1$ and $g = 2$, the exact value of $N_q(g)$ can be calculated. For $g = 1$ this is due essentially to Deuring [**4**] (see [**21**, Thm. 4.1, p. 536]), and for $g = 2$ to Serre [**17**–**19**] (see also [**10**]). But even for $g = 3$ difficulties arise, as is explained in [**12**].

The online tables available at `http://manypoints.org` collect the best known upper and lower bounds on $N_q(g)$ for $g \leq 50$ and for various values of q: the primes less than 100, the prime powers p^i for $p < 20$ and $i \leq 5$, and the powers of 2 up to 2^7. Despite the lack of an explicit formula for $N_q(3)$, for all of the prime powers q listed in these tables the value of $N_q(3)$ has been calculated. This leaves $N_q(4)$ as the next challenge. In this paper, we give new upper and lower bounds for $N_q(4)$ for 22 of the 23 prime powers $q < 100$ for which the exact value had not been previously computed; we find the exact value of $N_q(4)$ for 17 of these prime powers.

Our results are given in Table 1. The entries in the 'old' column show the results that were listed on `manypoints.org` on 1 August 2011. Note that the entries in the 'new' column show that there are now only six values of q less than 100 for which the exact value of $N_q(4)$ is not known.

We obtain our new upper bounds on $N_q(4)$ by using the results of [**9**]. For each q, we use the computer programs associated with that paper to obtain restrictions on genus-4 curves over $\mathbf{F}_q$ with many points. Sometimes we learn that a curve with a certain number of points must be a double cover of one of several elliptic curves; in Section 3 we show how to enumerate such double covers. Sometimes we learn that a curve with a certain number of points must correspond to a Hermitian lattice over a quadratic order; we discuss these cases in Sections 4 and 5. In two cases we find that the Jacobian of a curve with a certain number of points must have complex multiplication by $\mathbf{Z}[\zeta_5]$; our analysis of Hermitian forms over this ring in Section 6 shows that such curves do not exist. Our new lower bounds come from explicit examples of curves, which we present in Section 7.

We have implemented all of our calculations in the computer algebra package Magma [**2**]. As we mentioned above, we discover properties of genus-4 curves over $\mathbf{F}_q$ with a given number of points by using the programs associated with the paper [**9**]. These programs are found in the package `IsogenyClasses.magma`, which is available on the author's website: Go to the bibliography page

`http://alumni.caltech.edu/~however/biblio.html`

and follow the link associated with the paper [**9**]. The programs we use to enumerate double covers are also available online, by starting at the URL given above and following the link associated with the present paper.

2. Restrictions on genus-4 curves with many points

In this section we present the results we obtained from running the programs in `IsogenyClasses.magma`.

Table 2 lists the (q, N) pairs that we will have to eliminate in order to prove our new upper bounds; for each q and N, the table shows what `IsogenyClasses.magma` tells us about genus-4 curves over $\mathbf{F}_q$ with N points. An entry of "None exist" means that `IsogenyClasses.magma` shows that no genus-4 curve over $\mathbf{F}_q$ has exactly N points. Entries of the form "Double cover of elliptic curve with trace t" mean that any genus-4 curve over $\mathbf{F}_q$ with N points must be a double cover of an elliptic curve over $\mathbf{F}_q$ with trace t. An entry of "Hermitian module over R", where R is an order in an imaginary quadratic field, means that any genus-4 curve over $\mathbf{F}_q$ with N points must have a Jacobian that is isogenous to the fourth power of an ordinary elliptic curve over $\mathbf{F}_q$ whose Frobenius endomorphism generates the order R. An entry of "Hermitian module over $\mathbf{Z}[\zeta_5]$" means that any genus-4 curve over $\mathbf{F}_q$ with N points must have a Jacobian that is isogenous to the square of an ordinary abelian surface with complex multiplication by the ring of integers of the 5th cyclotomic field.

3. Double covers of elliptic curves

Table 2 shows that there are a number of pairs (q, N) such that a genus-4 curve over $\mathbf{F}_q$ with N points must be a double cover of an elliptic curve with a certain trace t. Thus, one way to show that there are no genus-4 curves over $\mathbf{F}_q$ with N points would be to enumerate all of the elliptic curve over $\mathbf{F}_q$ of trace t, enumerate all of the genus-4 double covers of these curves, count the number of points on the double covers, and verify that none of them has N points. This is the strategy we use for the appropriate entries in Table 2.

In this section we explain more details of this procedure. The actual Magma programs we use can be found at the URL mentioned in the introduction; follow the link associated with this paper, and then download the file `Genus4.magma`.

Our first comment is that if C is a double cover of an elliptic curve E over a finite field k, and if σ is an automorphism of k, then C^σ is a double cover of E^σ. Since C^σ and C have the same number of points, it will suffice for us to enumerate all of the double covers of a set of representatives of the trace-t elliptic curves up to automorphisms of k.

The finite fields we must investigate are small enough that we use a completely naive method of finding representatives for these elliptic curves. We will only be

q	N	Properties of a genus-4 curve over $\mathbf{F}_q$ with N points
11	34	Hermitian module over $\mathbf{Z}[\zeta_5]$
13	39	Double cover of elliptic curve with trace -7
17	48	Double cover of elliptic curve with trace -8
17	47	None exist
19	52	Hermitian module over $\mathbf{Z}[\sqrt{-3}]$
19	51	None exist
23	58	Double cover of elliptic curve with trace -9
29	70	Hermitian module over $\mathbf{Z}[2i]$
29	69	None exist
31	73	Double cover of elliptic curve with trace -11
37	84	Double cover of elliptic curve with trace -12
37	83	None exist
41	90	Hermitian module over $\mathbf{Z}[\sqrt{-5}]$
41	89	None exist
43	93	Double cover of elliptic curve with trace -13
47	100	Hermitian module over $\mathbf{Z}[\alpha]$
47	99	None exist
53	110	Hermitian module over $\mathbf{Z}[2i]$
53	109	None exist
59	118	Double cover of elliptic curve with trace -15
59	117	Double cover of elliptic curve with trace -15
61	122	Hermitian module over $\mathbf{Z}[\alpha]$
61	121	None exist
61	120	Double cover of elliptic curve with trace -13, or Hermitian module over $\mathbf{Z}[\zeta_5]$
67	132	Hermitian module over $\mathbf{Z}[\sqrt{-3}]$
67	131	None exist
67	130	Double cover of elliptic curve with trace -14
71	136	Hermitian module over $\mathbf{Z}[\sqrt{-7}]$
71	135	None exist
73	139	Double cover of elliptic curve with trace -17
83	154	Double cover of elliptic curve with trace -18
83	153	Double cover of elliptic curve with trace -18

TABLE 2. What `IsogenyClasses.magma` tells us about genus-4 curves over $\mathbf{F}_q$ with N points. Here ζ_5 is a primitive 5th root of unity, $i = \sqrt{-1}$, and α satisfies $\alpha^2 + \alpha + 5 = 0$.

working with fields of characteristic larger than 3, so every elliptic curve can be written in the form $y^2 = x^3 + ax + b$, where a and b are elements of k with $4a^3 + 27b^2 \neq 0$. The curve determined by one such pair (a, b) is isomorphic to a Galois conjugate of the curve determined by another pair (a', b') if and only if $(a', b') = (a^\sigma u^4, b^\sigma u^6)$ for some $u \in k^*$ and some automorphism σ of k. The function `ECs` takes as input a finite field k and a trace t, explicitly computes the orbits of the set of (a, b) pairs under the combined action of k^* and $\operatorname{Aut} k$, and computes the trace of one representative elliptic curve from each orbit. It returns those representatives that have the desired trace t.

Next, given an E of trace t, we must enumerate its genus-4 double covers C. We use the same general idea as in [**8**, §6.1]; our description of the method is adapted from the version given there.

The function field of such a C is obtained from that of E by adjoining a root of $z^2 = f$, where f is a function on E. By the Riemann-Hurwitz formula, in order for C to have genus 4 the divisor of f must be of the form

$$P_1 + \cdots + P_6 + 2D,$$

where the P_i are distinct geometric points on E such that the divisor $P_1 + \cdots + P_6$ is k-rational, and where D is a divisor of degree -3. There is a function g on E such that

$$D + \operatorname{div} g = Q - 4\infty,$$

where ∞ is the infinite point on E and where Q is a rational point on E. Replacing f with fg^2 gives an isomorphic double cover of E. Thus, we may assume that C is given by adjoining a root of $z^2 = f$, where f is a function on E whose divisor is of the form

$$P_1 + \cdots + P_6 + 2Q - 8\infty, \quad \text{where the } P_i \text{ are distinct.} \tag{3.1}$$

We can also change the map $C \to E$ by following it with a translation map on E. Translating E by a rational point R has the effect of replacing f with a function whose divisor is

$$(P_1 + R) + \cdots + (P_6 + R) + 2(Q + R) - 8R$$

(where the sums in parentheses take place in the algebraic group E). By modifying this new f by the square of a function we can get the divisor of f to be

$$(P_1 + R) + \cdots + (P_6 + R) + 2(Q - 3R) - 8\infty.$$

We see that we only need to consider Q that represent distinct classes of $E(k)$ modulo $3E(k)$. Note that every class other than the class of the identity element contains a representative that is not a 2-torsion point, so we may assume that Q does not have order 2. And finally, we note that we need only look at one representative from each orbit of $\operatorname{Aut} E$ acting on $E(k)/3E(k)$.

Next, given an E and a $Q \in E(k)$ not of order 2, we must enumerate all functions on E with divisors of the form (3.1). Suppose first that $Q \neq \infty$. By translating coördinates on E, we may assume that E is given by an equation $y^2 = x^3 + rx^2 + sx + t^2$ and that Q is the point $(0, t)$. Since Q is not a 2-torsion point we have $t \neq 0$, and the tangent line to E at Q is given by $y = mx + t$, where $m = s/(2t)$. Then there are two cases to consider: functions for which none of the P_i is equal to ∞, and functions for which one of the P_i is equal to ∞.

If f is a function on E whose divisor has the desired form and for which no P_i is ∞, then f has degree 8 and lies in the Riemann-Roch space $\mathscr{L}(8\infty - 2Q)$. We check that this space is spanned by the functions

$$\{x^4,\ x^2(y-t),\ x^3,\ x(y-t),\ x^2,\ y-mx-t\}.$$

We can then run through all of the f's in the k-span of these functions, considering only those linear combinations where the coefficient of x^4 is nonzero (so that the function actually has degree 8). In fact, since $z^2 = f$ and $z^2 = d^2 f$ give isomorphic extensions of E, we can restrict our attention to linear combinations where the coefficient of x^4 is either 1 or a fixed nonsquare value.

For a given linear combination f, we can easily compute the number of points on the extension $z^2 = f$ under the assumption that the divisor of f is of the form (3.1). (We get either 2 or 0 points over ∞ depending on the leading term of the Laurent expansion of f at ∞, and likewise for the points over Q; for the other points P of $E(k)$, we get either 0, 1, or 2 points over P depending on whether $f(P)$ is a nonsquare, zero, or a nonzero square.) If the number we calculate is larger than our previous best count, we can then check whether the divisor of f is in fact of the form (3.1).

The case where one of the P_i is equal to ∞ is very similar; the difference is that we now look in the Riemann-Roch space $\mathscr{L}(7\infty - 2Q)$, and that now the extension $z^2 = f$ always has exactly one point lying over ∞.

The function `double_covers_genus_4` in the file `Genus4.magma` takes as input an elliptic curve E over a finite field, and runs the algorithm sketched above to find the largest number of points on a genus-4 double cover of E. The function `double_covers_given_trace` takes as input a prime power q and a trace t, and finds the maximal number of points on a genus-4 curve over $\mathbf{F}_q$ that is a double cover of some elliptic curve over $\mathbf{F}_q$ of trace t.

For each pair (q, N) whose associated entry in Table 2 says that a genus-4 curve over $\mathbf{F}_q$ with N points must be a double cover of an elliptic curve of trace t, we ran our program with input (q, t). For each pair, we found that the maximal number of points on a genus-4 double cover of an elliptic curve with trace t is less than N. Thus, for these pairs (q, N), we find that $N_q(4) < N$.

4. Hermitian forms over maximal quadratic orders

The programs in `IsogenyClasses.magma` show that for several of the (q, N) pairs listed in Table 2, every genus-4 curve over $\mathbf{F}_q$ having N points must have Jacobian isogenous to E^4, where E is an ordinary elliptic curve over $\mathbf{F}_q$ whose Frobenius endomorphism generates a specific imaginary quadratic order R. The study of abelian varieties isogenous to E^4 is simplified by the use of Serre's 'Hermitian modules' [**13**, Appendix] or Deligne's equivalence of categories [**3**] between ordinary abelian varieties and modules over a certain ring, combined with the description of polarizations on these modules provided in [**7**].

In this section we analyze the three cases from Table 2 where the quadratic order in question is maximal; these are the cases $(q, N) = (41, 90)$, $(q, N) = (47, 100)$, and $(q, N) = (61, 122)$, where the corresponding quadratic order $\mathscr{O}$ has discriminant -20, -19, and -19, respectively. As is explained in [**13**, Appendix], a principally-polarized abelian variety isogenous to E^4 corresponds to a principally-polarized

Hermitian $\mathcal{O}$-module of rank 4. Schiemann [**16**] has computed all such principally-polarized Hermitian modules (up to isomorphism) for the quadratic orders we are concerned with; the lists can be found at

`http://www.math.uni-sb.de/ag/schulze/Hermitian-lattices/`

as well as on the author's web site, mentioned in the introduction.

Schiemann calculated the automorphism groups of these polarized Hermitian modules; these groups are isomorphic to the automorphism groups of the corresponding polarized abelian four-folds. Schiemann's tables show that for all of the Hermitian modules we must consider, the automorphism groups have order divisible by 4. By Torelli's theorem [**15**, Thm. 12.1, p. 202], if the polarized Jacobian of a curve C has an automorphism group of order divisible by 4, then either

- C has an automorphism ι of order 2 such that the quotient curve $C_0 = C/\langle\iota\rangle$ has positive genus, or
- C is a hyperelliptic curve with an automorphism of order 4 whose square is the hyperelliptic involution.

For the (q, N) pairs we have to consider, we find that if C is a genus-4 curve over $\mathbf{F}_q$ with N points, then either C is a double cover of an elliptic curve isogenous to E, or C is a double cover of a genus-2 curve whose Jacobian is isogenous to E^2, or C is a hyperelliptic curve with an automorphism whose square is the hyperelliptic involution.

We have already seen how to enumerate the genus-4 double covers of an elliptic curve; running `double_covers_given_trace` for the (q, t) pairs coming from our (q, N) pairs, we find no genus-4 curves over $\mathbf{F}_q$ with N points. In Section 4.1 below we will show how to enumerate the genus-4 double covers of genus-2 curves, and in Section 4.2 we will show how to enumerate the genus-4 hyperelliptic curves with an automorphism whose square is the hyperelliptic involution. For our (q, N) pairs, we find no genus-4 curves over $\mathbf{F}_q$ having N points.

4.1. Double covers of genus-2 curves. Suppose $\varphi : C \to C_0$ is a degree-2 map from a genus-4 curve to a genus-2 curve over a finite field k of characteristic greater than 2. The function field of C is obtained from that of C_0 by adjoining a root of $z^2 = f$, where f is a function on C_0. By the Riemann-Hurwitz formula, in order for C to have genus 4 the divisor of f must be of the form

$$P_1 + P_2 + 2D,$$

where the P_i are distinct geometric points on E such that the divisor $P_1 + P_2$ is k-rational, and where D is a divisor of degree -1. Let ∞ be any rational point on C_0. The Riemann-Roch theorem shows that the Riemann-Roch space $\mathscr{L}(D + 4\infty)$ has dimension 2; it follows that it must contain a function g such that $\operatorname{div} g + D + 3\infty$ is effective. Then

$$\operatorname{div} fg^2 = P_1 + P_2 + 2D' - 6\infty$$

for some effective divisor D' of degree 2. Replacing f with fg^2, we find that C can be obtained from C_0 by adjoining a root of $z^2 = f$, with $\operatorname{div} f = P_1 + P_2 + 2D' - 6\infty$ for some effective D' of degree 2.

Thus, to enumerate all genus-4 curves that are double covers of C_0, we can simply enumerate all effective degree-2 divisors D', compute the Riemann-Roch space $\mathscr{L}(6\infty - 2D')$, loop through the function f in this space (up to squares in k) that are not squares in $\overline{k}(C_0)$, and consider the curves $z^2 = f$. It is easy to count the number of points on a curve defined by such an extension. We implement this algorithm in the function `double_covers_genus_4`, the same function we used for doubles covers of elliptic curves; when the input to the function is a curve of genus 2, the function runs through the procedure sketched above, and outputs the largest number of points it finds.

For the case $(q, N) = (41, 90)$ there are three possibilities for the base curve C_0; this can be determined by brute force (by enumerating all genus-2 curves over k) or by theory, by noting that Schiemann's tables show exactly three unimodular rank-2 Hermitian $\mathcal{O}$-modules that are not products of two rank-1 modules. The three curves are

$$\begin{aligned} y^2 &= x^6 + 7x^4 + 8x^2 - 7, \\ y^2 &= x^6 + 7x^4 + 3x^2 + 7, \quad \text{and} \\ y^2 &= x^6 - 3x^4 - 3x^2 + 1. \end{aligned}$$

Running `double_covers_genus_4` on these curves, we find no genus-4 curves over $\mathbf{F}_{41}$ with 90 points. Similarly, we find one possible C_0 for each of the other two (q, N) pairs we must consider. For $\mathbf{F}_{47}$ and $\mathbf{F}_{61}$, these curves are

$$y^2 = x^6 + 7x^4 - 9x^2 - 6 \quad \text{and} \quad y^2 = x^6 + 10x^4 - 11x^2 - 1,$$

respectively. Running `double_covers_genus_4` on these curves, we find no genus-4 curves over $\mathbf{F}_{47}$ with 100 points, and no genus-4 curves over $\mathbf{F}_{61}$ with 122 points.

4.2. Hyperelliptic curves with automorphisms of order 4. A hyperelliptic curve over $\mathbf{F}_q$ can have at most $2(q + 1)$ rational points, so the only one of our three (q, N) pairs that we need be concerned with is $(q, N) = (61, 122)$.

Suppose C is a genus-4 hyperelliptic curve over a finite field k with an automorphism α of order 4 whose square is the hyperelliptic involution. (In fact, [**5**, Thm. 1.1] shows that *every* order-4 automorphism of a genus-4 hyperelliptic curve has square equal to the hyperelliptic involution.) We can choose a parameter for $\mathbf{P}^1$ so that the automorphism of $\mathbf{P}^1$ induced by α is $x \mapsto a/x$ for some $a \in k^*$.

Writing C as $y^2 = f$ for some separable polynomial $f \in k[x]$ of degree 9 or 10, we see that the automorphism α must have the form $(x, y) \mapsto (a/x, by/x^5)$, and since α^2 is the hyperelliptic involution we must have $b^2 = -a^5$. Replacing x with a^2x/b, we compute that α is given by $(x, y) \mapsto (-1/x, y/x^5)$.

We find that f must be a linear combination of the following polynomials:

$$\{x^{10} + 1,\ x^9 - x,\ x^8 + x^2,\ x^7 - x^3,\ x^6 + x^4\}.$$

For the one pair (q, N) we must consider, we can examine all linear combinations of these polynomials (up to squares in $\mathbf{F}_q$), and compute the number of points on the associated hyperelliptic curves. We find no curve over $\mathbf{F}_{61}$ having 122 points.

The function `hyperelliptic` in the file `Genus4.magma` implements this algorithm.

5. Hermitian forms over nonmaximal quadratic orders

For five of the entries (q, N) in Table 2, we see that the Jacobian of a genus-4 curve over $\mathbf{F}_q$ with N points must be isogenous to E^4, where E lies in an isogeny class of ordinary elliptic curves over $\mathbf{F}_q$ all of whose elements have complex multiplication by a non-maximal order R in a quadratic field K; here R is the ring generated over $\mathbf{Z}$ by the Frobenius endomorphism of E. The five entries are $(19, 52)$, $(29, 70)$, $(53, 110)$, $(67, 132)$, and $(71, 136)$, and the rings R are the orders of conductor 2 inside the maximal orders of, respectively, $\mathbf{Q}(\sqrt{-3})$, $\mathbf{Q}(i)$, $\mathbf{Q}(i)$, $\mathbf{Q}(\sqrt{-3})$, and $\mathbf{Q}(\sqrt{-7})$. The discriminants of the rings R are, respectively, $-12, -16, -16, -12$, and -28. We will show that for each of these pairs (q, N), a genus-4 curve over $\mathbf{F}_q$ having N points must be a double cover of a curve of genus 1 or 2.

PROPOSITION 5.1. *Let t be the trace of Frobenius for an isogeny class $\mathscr{C}$ of ordinary elliptic curves over a finite field $\mathbf{F}_q$ such that $\Delta := t^2 - 4q$ lies in $\{-12, -16, -28\}$. Suppose C is a genus-4 curve over $\mathbf{F}_q$ whose Jacobian is isogenous to the fourth power of an elliptic curve in $\mathscr{C}$.*

(1) *If $\Delta = -12$ or $\Delta = -16$ then C is a double cover of a curve in $\mathscr{C}$.*
(2) *If $\Delta = -28$ then either C is a double cover of a curve in $\mathscr{C}$, or C is a double cover of a genus-2 curve whose Jacobian is isogenous to the square of a curve in $\mathscr{C}$.*

For our five (q, N) pairs, the function `double_covers_given_trace` finds no genus-4 curve with N points. For the pair $(q, N) = (71, 136)$ we must also run `double_covers_genus_4` on three genus-2 curves:

$$\begin{aligned} y^2 &= x^6 - 29x^4 - 29x^2 + 1, \\ y^2 &= x^6 - 13x^4 - 13x^2 + 1, \quad \text{and} \\ y^2 &= x^6 - 12x^4 - 15x^2 - 1. \end{aligned}$$

Again, we find no genus-4 curve with N points.

The proof of Proposition 5.1 relies on several lemmas. To begin, we show that for each of our five cases, there are only four abelian varieties to consider.

LEMMA 5.2. *Let t be the trace of Frobenius for an isogeny class $\mathscr{C}$ of ordinary elliptic curves over a finite field $\mathbf{F}_q$ such that $\Delta := t^2 - 4q$ lies in $\{-12, -16, -28\}$. Then $\mathscr{C}$ contains exactly two elliptic curves, one of them with endomorphism ring equal to the maximal order $\mathscr{O}$ in $K := \mathbf{Q}(\sqrt{\Delta})$, and one with endomorphism ring equal to the order R of conductor 2 in $\mathscr{O}$.*

Let E and F be these two elliptic curves, where E has the larger endomorphism ring. Then every abelian variety over $\mathbf{F}_q$ isogenous to E^n is isomorphic to $E^i \times F^{n-i}$ for some i, and the varieties arising from distinct values of i are not isomorphic to one another.

PROOF. By Deligne's equivalence of categories [**3**], the elliptic curves in $\mathscr{C}$ correspond to the isomorphism classes of nonzero finitely-generated R-modules in K. The endomorphism ring of such a module is either R or $\mathscr{O}$, and since both R and $\mathscr{O}$ have class number 1, there is one elliptic curve E with $\operatorname{End} E \cong \mathscr{O}$ and one elliptic curve F with $\operatorname{End} F \cong R$.

Likewise, Deligne's theorem shows that the varieties isogenous to E^n correspond to rank-n modules over R. By using a result of Borevich and Faddeev [**1**], we see that every such module is isomorphic to a sum of copies of $\mathscr{O}$ plus a sum of copies of R, and that two such modules are isomorphic if and only if they have the same number of each type of summand. The second statement of the lemma follows upon applying Deligne's equivalence of categories. □

Next, we show that for each of our cases, any curve with Jacobian isomorphic to $E^i \times F^{4-i}$ with $i > 0$ must be a double cover of E.

LEMMA 5.3. *Let t be the trace of Frobenius for an isogeny class of ordinary elliptic curves over a finite field $\mathbf{F}_q$ such that $\Delta := t^2 - 4q$ lies in $\{-12, -16, -28\}$, and let E and F be as in Lemma 5.2. If C is a genus-4 curve over $\mathbf{F}_q$ whose Jacobian is isomorphic to $E^i \times F^{4-i}$ for some $i > 0$, then C is a double cover of E.*

PROOF. Our proof is computational, and uses the function `pullback_bound` from the package `IsogenyClasses.magma`. Given a fundamental discriminant D with $-11 \le D < 0$, an integer n, and a dimension d, this function will return an integer B such that for every $d\times d$ Hermitian matrix of determinant n with entries in the quadratic order $\mathcal{O}_D$ of discriminant D, the Hermitian form on $\mathcal{O}_D^d$ determined by this matrix will have a nonzero vector of length at most B. As is explained in [**9**, §4], this translates into a statement about polarizations on E^d, where E has complex multiplication by $\mathcal{O}_D$; namely, for every degree-n^2 polarization λ on E^d, there is an embedding $\varphi : E \to E^d$ such that $\varphi^*\lambda$ is a polarization on E of degree at most B^2.

Now suppose C is a curve with Jacobian isomorphic to $E^i \times F^{4-i}$, with $i > 0$, and let λ be the canonical principal polarization on $\operatorname{Jac} C$. There is an isogeny of degree 2^{4-i} from E^4 to $\operatorname{Jac} C$, and pulling back λ via this isogeny gives a polarization μ of degree 4^{4-i} on E^4. The function `pullback_bound` tells us that we can pull back μ to a polarization of degree 1 or 4 on E, and by [**9**, Lem. 4.3] this gives us a map of degree 1 or 2 from C to E. A map of degree 1 is clearly impossible, so we find that C is a double cover of E. □

We see that to prove Proposition 5.1 we need only consider curves whose Jacobians are isomorphic to F^4; in other words, we need only consider principal polarizations on F^4. For $\Delta = -12$ and $\Delta = -16$ we will show that every such polarization can be pulled back to either E or F to get a polarization of degree 1 or 4. For $\Delta = -28$ we will show that if such a polarization cannot be pulled back to a polarization of degree 1 or 4 on E or F, and if the polarized variety is the Jacobian of a curve, then the curve has an involution that makes it a double cover of a genus-2 curve.

Let φ be a degree-2 isogeny from E to F, and let $\Phi : E^4 \to F^4$ be the degree-16 product isogeny $\varphi\times\varphi\times\varphi\times\varphi$. Denote the kernel of Φ by G. Note that the smallest $\mathcal{O}$-stable subgroup of E^4 that contains G is $E^4[2]$.

Suppose λ is a principal polarization on F^4, and let $\mu = \Phi^*\lambda$ be the pullback of λ to E^4. Then μ is a polarization of degree 16^2 on E^4, and since $\ker\mu$ is stable under the action of $\mathcal{O}$ and contains G, we must have $\ker\mu = E^4[2]$. That means that μ must be twice a principal polarization.

Thus, to enumerate the possible principal polarizations λ on F^4, we can enumerate the principal polarizations on E^4 up to isomorphism, multiply each of these by 2 to get a polarization μ of degree 16^2 on E^4, enumerate the subgroups G of $E^4[2]$ that generate all of $E^4[2]$ as an $\mathcal{O}$-module and that are isotropic with respect to the Weil pairing on $E^4[2]$ determined by μ, and then compute the matrix in $M_4(R)$ that represents the polarization we get by pushing μ down through an isogeny with kernel G. The Magma programs we use to do this can be found at the URL mentioned in the introduction; follow the link associated with this paper, and then download the file `NonMaximalOrders.magma`.

5.1. Proof of Proposition 5.1 when $\Delta = -12$. Schiemann's tables [**16**] show that up to isomorphism the only principal polarization on E^4 is the product polarization, so our only μ is represented by the diagonal matrix with 2's on the diagonal. Embedding E into E^4 along one of the factors, we find that μ, and hence λ, can be pulled back to a polarization of degree 4 on E.

5.2. Proof of Proposition 5.1 when $\Delta = -16$. Schiemann's tables show two principal polarizations on E^4, the product polarization and the polarization given by the Hermitian matrix

$$P = \begin{bmatrix} 2 & 1 & 0 & 0 \\ 1 & 2 & 1-i & 1-i \\ 0 & 1+i & 2 & 1 \\ 0 & 1+i & 1 & 2 \end{bmatrix}.$$

Twice the product polarization can be pulled back to give a polarization of degree 4 on E, so we need only consider the polarization $2P$.

A computer calculation shows that there are 1024 subgroups of $E^4[2]$ that are maximal isotropic with respect to the Weil pairing determined by P and that generate all of $E^4[2]$ as an $\mathcal{O}$-module. For each such subgroup G, we compute the polarization on F^4 obtained by pushing the polarization $2P$ down through the isogeny with kernel G; this polarization can be represented by a Hermitian matrix in $M_4(R)$, which gives a Hermitian form on R^4. We can then compute the short vectors in this Hermitian lattice. We find that for each group G, the Hermitian lattice has a short vector of length 2. This means that the polarization on F^4 can be pulled back to a polarization of degree 4 on F.

5.3. Proof of Proposition 5.1 when $\Delta = -28$. Schiemann's tables show that there are three principal polarizations on E^4: the product polarization and the polarizations given by the Hermitian matrices

$$P_1 = \begin{bmatrix} 1 & 0 & 0 & 0 \\ 0 & 2 & \frac{1+\sqrt{-7}}{2} & -1 \\ 0 & \frac{1-\sqrt{-7}}{2} & 2 & \frac{1+\sqrt{-7}}{2} \\ 0 & -1 & \frac{1-\sqrt{-7}}{2} & 2 \end{bmatrix}$$

and

$$P_2 = \begin{bmatrix} 2 & 1 & 0 & \frac{1+\sqrt{-7}}{2} \\ 1 & 2 & \frac{1+\sqrt{-7}}{2} & \frac{1+\sqrt{-7}}{2} \\ 0 & \frac{1-\sqrt{-7}}{2} & 2 & 1 \\ \frac{1-\sqrt{-7}}{2} & \frac{1-\sqrt{-7}}{2} & 1 & 2 \end{bmatrix}.$$

There is a 2 on the diagonal of $2P_1$, so this polarization can be pulled back to a degree-4 polarization on E. Likewise, twice the product polarization can be pulled back to a degree-4 polarization on E. That leaves us to consider the polarization $2P_2$.

A computer calculation shows that there are 448 subgroups of $E^4[2]$ that are maximal isotropic with respect to the Weil pairing determined by P_2 and that generate all of $E^4[2]$ as an $\mathcal{O}$-module. Of these subgroups, 256 give rise to principal polarizations on F^4 whose associated Hermitian forms have short vectors of length 2. The other 192 subgroups give principal polarizations on F^4 that are isomorphic to the polarizations defined by one of the following three Hermitian matrices:

$$Q_1 = \begin{bmatrix} 4 & 2 & 2 & -\sqrt{-7} \\ 2 & 4 & -\sqrt{-7} & 2 \\ 2 & \sqrt{-7} & 4 & -2 \\ \sqrt{-7} & 2 & -2 & 4 \end{bmatrix}$$

$$Q_2 = \begin{bmatrix} 3 & 1 & 1 & -1-\sqrt{-7} \\ 1 & 3 & -1-\sqrt{-7} & 1 \\ 1 & -1+\sqrt{-7} & 4 & -2 \\ -1+\sqrt{-7} & 1 & -2 & 4 \end{bmatrix}$$

$$Q_3 = \begin{bmatrix} 3 & 1 & 0 & -\sqrt{-7} \\ 1 & 3 & -\sqrt{-7} & 0 \\ 0 & \sqrt{-7} & 3 & -1 \\ \sqrt{-7} & 0 & -1 & 3 \end{bmatrix}.$$

We check that for each of these three polarizations Q there is an involution of the polarized variety (F^4, Q) that fixes a 2-dimensional subvariety of F^4; such an involution can be represented by a matrix A such that $A^2 = I$ and $A^*QA = Q$, and such that $A - I$ has rank 2. For each Q we note that the matrix

$$A = \begin{bmatrix} 0 & 1 & 0 & 0 \\ 1 & 0 & 0 & 0 \\ 0 & 0 & 0 & 1 \\ 0 & 0 & 1 & 0 \end{bmatrix}$$

has these properties. It follows from Torelli's theorem [**15**, Thm. 12.1, p. 202] that if (F^4, Q) is the Jacobian of a curve C, then C has an involution α such that the quotient of C by α is a genus-2 curve whose Jacobian is necessarily isogenous to F^2.

This completes the proof of Proposition 5.1. □

6. Hermitian forms over $\mathbf{Z}[\zeta_5]$

Let $\mathcal{O} = \mathbf{Z}[\zeta_5]$. As Table 2 indicates, the programs in `IsogenyClasses.magma` show that if C is a genus-4 curve over $\mathbf{F}_{11}$ with 34 points, the center of the endomorphism ring of the Jacobian of C is isomorphic to $\mathcal{O}$. (At the end of this section we will explain how to derive this statement from the output of the programs.) Likewise, if C is a genus-4 curve over $\mathbf{F}_{61}$ with 120 points that is not a double cover of an elliptic curve, then the center of the endomorphism ring of its Jacobian is isomorphic to $\mathcal{O}$. In this section, we show that no ordinary genus-4 curve over a finite field can have $\mathcal{O}$ as the center of the endomorphism ring of its Jacobian.

PROPOSITION 6.1. *Let k be a finite field. There is no genus-4 curve over k with ordinary Jacobian J such that the center of* $\operatorname{End} J$ *is isomorphic to* $\mathbf{Z}[\zeta_5]$.

PROOF. Suppose, to obtain a contradiction, that such a curve C exists. Let $K = \mathbf{Q}(\zeta_5)$, let π be the Frobenius endomorphism of J, let $\overline{\pi}$ be the Verschiebung endomorphism of J, and let R be the ring $\mathbf{Z}[\pi, \overline{\pi}]$. Since R lies in the center of $\operatorname{End} J$ we may view R as a subring of $\mathcal{O}$, and since π generates the center of $(\operatorname{End} J) \otimes \mathbf{Q}$, we see that R is an order in $\mathcal{O}$. Let M be the Deligne module associated with J (see [**7**]), so that M is a finitely-generated R-module that is isomorphic to a submodule of K^2. Since $\operatorname{End} J$ contains $\mathcal{O}$, we see that M is in fact an $\mathcal{O}$-module, and since $\mathcal{O}$ has class number 1 we have $M \cong \mathcal{O} \oplus \mathcal{O}$. Translating this statement back from Deligne modules to ordinary abelian varieties, we see that $J \cong A \times A$ for a 2-dimensional abelian variety with $\operatorname{End} A \cong \mathcal{O}$.

The field K is ramified over its real subfield at a finite prime, so by [**7**, Cor. 11.4, p. 2391] the variety A has a principal polarization λ. Let μ be the canonical polarization on J, viewed as a polarization on $A \times A$. Then μ is equal to the product polarization $\lambda \times \lambda$ preceded by an endomorphism P of $A \times A$; viewing P as an element of $M_2(\mathcal{O})$, we find that P is a unimodular Hermitian matrix that is totally positive (meaning that all of the roots of its minimal polynomial are totally positive algebraic numbers). But Lemma 6.2 below says that all totally positive rank-2 unimodular Hermitian lattices over $\mathcal{O}$ are decomposable, so μ is isomorphic to the product polarization $\lambda \times \lambda$. This contradicts the fact that the polarization on a Jacobian is never a product. □

LEMMA 6.2. *Let $\mathcal{O} = \mathbf{Z}[\zeta_5]$. Suppose P is a totally positive unimodular Hermitian matrix in $M_2(\mathcal{O})$. Then there is an invertible $C \in M_2(\mathcal{O})$ such that $P = C^*C$, where C^* is the conjugate transpose of C.*

PROOF. Our proof follows the lines set out in [**8**, §8].

Let $K = \mathbf{Q}(\zeta_5)$, let φ be a fundamental unit of the maximal real subfield $K^+ = \mathbf{Q}(\sqrt{5})$ of K with $\operatorname{Tr}_{K^+/\mathbf{Q}} \varphi = 1$, and let ϕ be the real number $(1+\sqrt{5})/2$. Let ψ_1 and ψ_2 be embeddings of K into the complex numbers $\mathbf{C}$ with $\psi_1(\varphi) = \phi$ and $\psi_2(\varphi) = 1/\phi$. If z is a complex number, we let $|z|$ be its magnitude and we let $||z||$ be its norm, so that $||z|| = |z|^2 = z\overline{z}$.

Let q be the quadratic form on $\mathcal{O}$ that sends x to the trace from K to $\mathbf{Q}$ of $x\overline{x}$, so that $q(x) = 2||\psi_1(x)|| + 2||\psi_2(x)||$. Let Λ be the lattice $(\mathcal{O}, q)$. Using Magma, we compute the Voronoi cell for this lattice, and we find that the covering radius of the lattice is 2; that is, every element of $\Lambda \otimes \mathbf{Q}$ differs from a lattice point by an element x with $q(x) \le 2$. This means that every element of K differs from an element of $\mathcal{O}$ by an element x satisfying

$$||\psi_1(x)|| + ||\psi_2(x)|| \le 1.$$

In particular, we see that this x also satisfies

$$N_{K/\mathbf{Q}}(x) \le 1/4 \quad \text{and} \quad ||\psi_i(x)|| \le 1 \quad \text{for } i = 1, 2.$$

In turn, this statement about the lattice Λ gives us a Euclidean algorithm on $\mathcal{O}$: Given elements n and d of $\mathcal{O}$, there are elements q and r of $\mathcal{O}$ such that $n = qd + r$ with

$$N(r) \le 1/4 \quad \text{and} \quad ||\psi_i(r)|| \le ||\psi_i(n)|| \quad \text{for } i = 1, 2.$$

Write our totally positive unimodular Hermitian matrix P as

$$P = \begin{bmatrix} \alpha & \overline{\beta} \\ \beta & \gamma \end{bmatrix}$$

where α and γ are totally real and where α and $\alpha\gamma - \beta\overline{\beta}$ are totally positive. We will repeatedly choose invertible matrices C and replace P with C^*PC in order to reduce the size of the norm of the upper left element of P.

The determinant of P is a totally positive unit in K^+, and so is an even power of φ. By modifying P by a matrix C of the form

$$\begin{bmatrix} \varphi^i & 0 \\ 0 & 1 \end{bmatrix}$$

we may assume that P has determinant 1. Then by modifying P by a power of the matrix

$$\begin{bmatrix} \varphi & 0 \\ 0 & \varphi^{-1} \end{bmatrix}$$

we can ensure that

$$\frac{1}{\phi^2} \leq \frac{\psi_1(\alpha)}{\psi_2(\alpha)} \leq \phi^2.$$

Another way of expressing this is to say that

$$\frac{1}{\phi^2} \leq \frac{\psi_i(\alpha)^2}{\operatorname{Norm}_{K^+/\mathbf{Q}}(\alpha)} \leq \phi^2 \quad \text{for } i = 1, 2. \tag{6.1}$$

Applying our Euclidean algorithm to β and α, we find that $\beta = q\alpha + r$ for a q and an r with $||\psi_i(r)|| \leq ||\psi_i(\alpha)||$ and with $\operatorname{Norm}_{K/\mathbf{Q}}(r) \leq (1/4) \operatorname{Norm}_{K/\mathbf{Q}}(\alpha)$. If we set

$$C = \begin{bmatrix} 1 & -\overline{q} \\ 0 & 1 \end{bmatrix}$$

then

$$C^* P C = \begin{bmatrix} \alpha & \overline{r} \\ r & \gamma' \end{bmatrix}$$

for some integer γ' in K^+. Replace β with r and γ with γ', so that now we have

$$||\psi_i(\beta)|| \leq ||\psi_i(\alpha)|| \quad \text{for } i = 1, 2 \tag{6.2}$$

and

$$\operatorname{Norm}_{K/\mathbf{Q}}(\beta) \leq (1/4) \operatorname{Norm}_{K/\mathbf{Q}}(\alpha). \tag{6.3}$$

Let $B = \beta\overline{\beta}$, so that B is an integer in K^+. Note that we have $\alpha\gamma - B = 1$, so

$$\psi_i(\alpha)\psi_i(\gamma) = 1 + \psi_i(B) \quad \text{for } i = 1, 2$$

and therefore

$$\psi_i(\gamma)/\psi_i(\alpha) = 1/\psi_i(\alpha)^2 + \psi_i(B)/\psi_i(\alpha)^2 \quad \text{for } i = 1, 2. \tag{6.4}$$

Now let

$$\begin{aligned} b_1 &= \psi_1(B)/\psi_1(\alpha^2) \\ b_2 &= \psi_2(B)/\psi_2(\alpha^2) \\ c_1 &= 1/\psi_1(\alpha^2) \\ c_2 &= 1/\psi_2(\alpha^2) \end{aligned}$$

so that equation (6.4) becomes

$$\psi_1(\gamma)/\psi_1(\alpha) = b_1 + c_1 \qquad \text{and} \qquad \psi_2(\gamma)/\psi_2(\alpha) = b_2 + c_2.$$

Multiplying these last two equalities gives

$$\operatorname{Norm}_{K^+/\mathbf{Q}}(\gamma/\alpha) = b_1 b_2 + b_1 c_2 + b_2 c_1 + c_1 c_2. \tag{6.5}$$

Note that

$$b_1 b_2 = \frac{\operatorname{Norm}_{K^+/\mathbf{Q}}(B)}{\operatorname{Norm}_{K^+/\mathbf{Q}}(\alpha)^2} = \frac{\operatorname{Norm}_{K/\mathbf{Q}}(\beta)}{\operatorname{Norm}_{K/\mathbf{Q}}(\alpha)} \leq \frac{1}{4} \tag{6.6}$$

(where the final inequality comes from (6.3)) and

$$c_1 c_2 = \frac{1}{\operatorname{Norm}_{K^+/\mathbf{Q}}(\alpha^2)}. \tag{6.7}$$

Furthermore, from inequality (6.1) we see that

$$c_1 \le \frac{\phi^2}{\operatorname{Norm}_{K^+/\mathbf{Q}}(\alpha)} \quad \text{and} \quad c_2 \le \frac{\phi^2}{\operatorname{Norm}_{K^+/\mathbf{Q}}(\alpha)}, \tag{6.8}$$

and from inequality (6.2) we see that

$$b_1 = \frac{||\psi_1(\beta)||}{||\psi_1(\alpha)||} \le 1 \quad \text{and} \quad b_2 = \frac{||\psi_2(\beta)||}{||\psi_2(\alpha)||} \le 1. \tag{6.9}$$

If we view b_1, b_2, c_1, and c_2 as non-negative real variables subject only to the conditions expressed in relations (6.6), (6.7), (6.8), and (6.9), and if we maximize $b_1c_1 + b_2c_2$ subject to these conditions, we find that the maximum value occurs when $b_1 = 1$ and $c_1 = \phi^2/\operatorname{Norm}_{K^+/\mathbf{Q}}(\alpha)$. Thus we have

$$b_1c_1 + b_2c_2 \le 1 \cdot \frac{\phi^2}{\operatorname{Norm}_{K^+/\mathbf{Q}}(\alpha)} + \frac{1}{4} \cdot \frac{(1/\phi^2)}{\operatorname{Norm}_{K^+/\mathbf{Q}}(\alpha)} \le \frac{2.72}{\operatorname{Norm}_{K^+/\mathbf{Q}}(\alpha)}. \tag{6.10}$$

Let $\epsilon = 1/\operatorname{Norm}_{K^+/\mathbf{Q}}(\alpha)$. Then by combining the relations (6.5), (6.6), (6.7) and (6.10) we find that

$$\operatorname{Norm}_{K^+/\mathbf{Q}}(\gamma/\alpha) \le \epsilon^2 + 2.72\,\epsilon + 1/4.$$

If $\operatorname{Norm}_{K^+/\mathbf{Q}}(\alpha) \ge 4$ then $\epsilon \le 1/4$ and $\operatorname{Norm}_{K^+/\mathbf{Q}}(\gamma/\alpha) < 1$. Then we can modify P by

$$\begin{bmatrix} 0 & 1 \\ 1 & 0 \end{bmatrix}$$

to exchange α and γ, and this decreases the norm of the upper left element of P.

We repeat this procedure until we reach the point where $\operatorname{Norm}_{K^+/\mathbf{Q}}(\alpha) \le 3$. The only totally positive integer of K^+ with norm less than 4 is 1, so $\alpha = 1$ and we can reduce β to be 0. Then we find $\gamma = 1$, so that we have reduced P to the identity matrix. □

Let us turn to the specific pairs (q, N) that we must consider. For $q = 11$ and $N = 34$, the function `isogeny_classes` in the package `IsogenyClasses.magma` shows that a genus-4 curve C over $\mathbf{F}_{11}$ with 34 points must have real Weil polynomial equal to $(x^2 + 11x + 29)^2$; that is, the characteristic polynomial of Frobenius plus Verschiebung is the square of this polynomial. It follows that the characteristic polynomial of Frobenius is

$$(x^4 + 11x^3 + 51x^2 + 121x + 121)^2,$$

which means that the Jacobian of C is isogenous to the square of an abelian surface A with characteristic polynomial $x^4 + 11x^3 + 51x^2 + 121x + 121$. Since the middle coefficient of this characteristic polynomial is coprime to $q = 11$, we see that A is ordinary. Furthermore, the polynomial has a root π in $\mathbf{Q}(\zeta_5)$, namely $\pi = \zeta_5^2 + 2\zeta_5 - 2$. One checks that π and its complex conjugate generate the ring $\mathbf{Z}[\zeta_5]$, so the center of $\operatorname{End} J$, which contains Frobenius and Verschiebung, must equal $\mathbf{Z}[\zeta_5]$. As we have seen, this is impossible.

For $q = 61$ and $N = 120$, the function `isogeny_classes` tells us that if a genus-4 curve C over $\mathbf{F}_{61}$ with 120 points is not a double cover of an elliptic curve of trace -13, then it must have real Weil polynomial equal to $(x^2 + 29x + 209)^2$. As above, we see that this implies that C is ordinary and the center of the endomorphism ring of its Jacobian is $\mathbf{Z}[\zeta_5]$, which is impossible.

q	N	Equations for a genus-4 curve over $\mathbf{F}_q$ with N points	
13	38	$y^2 = x^3 + 4$	$z^2 = x^3 + x^2 - 4x - 3$
17	46	$y^2 = x^3 + x + 8$	$z^2 = x^3 - 5x^2 - 2x - 8$
19	48	$y^2 = x^3 + 8$	$z^2 = x^3 - 9x^2 - 5$
23	57	$y^2 = x^3 - 6x^2 - 3x - 7$	$z^2 = xy + 2x^3 - 11x^2 + 7x + 1$
29	67	$y^2 = x^3 - 3x + 18$	$z^2 = y + 7x^3 + 6x^2 + 2x - 10$
31	72	$y^2 = x^3 + x + 10$	$z^2 = x^3 + 7x^2 - 13x - 13$
37	82	$y^2 = x^3 + 2x$	$z^2 = x^3 + x^2 - 10x - 13$
41	88	$y^2 = x^3 + 6x + 5$	$z^2 = 3x^3 - 17x^2 - 11x$
43	92	$y^2 = x^3 + 2x + 1$	$z^2 = x^3 - 6x^2 + 11x$
47	98	$y^2 = x^5 - 6x^3 + 8x^2 - 5x + 12$	$z^2 = y + 6x^3 + 6x^2 - x - 3$
49	102	$y^2 = x^3 + x$	$z^2 = x^3 + x^2 + x + 4$
53	108	$y^2 = x^3 + 4x + 10$	$z^2 = x^3 + 12x^2 + 17x + 9$
59	116	$y^2 = x^3 + 2x + 22$	$z^2 = 2x^3 + x^2 - x + 9$
61	118	$y^2 = x^3 + 4$	$z^2 = x^3 + 23x^2 + 25x + 36$
67	129	$y^2 = x^3 + 25$	$z^2 = xy - 8x^3 - 27x + 4$
71	134	$y^2 = x^3 + x + 9$	$z^2 = x^3 + 9x^2 + 24x - 9$
73	138	$y^2 = x^3 + 3x + 11$	$z^2 = x^3 + 34x^2 + 18x + 40$
79	148	$y^2 = x^3 + x + 6$	$z^3 = y + 33x + 2$
83	152	$y^2 = x^3 + 2x + 19$	$z^2 = x^3 + 38x^2 - 6x + 39$
89	160	$y^2 = x^3 + 3x$	$z^2 = x^3 + 13x^2 - 22x + 28$
97	174	$y^2 = x^3 + 5x + 26$	$z^3 = y + 37x + 16$

TABLE 3. Genus-4 curves over small finite fields with many points. Note that for $q = 79$ and $q = 97$ the exponent on z is 3, not 2.

REMARK 6.3. For our particular cases $(q, N) = (11, 34)$ and $(q, N) = (61, 120)$, there is also a more computational approach to showing that there is no genus-4 curve C over $\mathbf{F}_q$ with N points. As we have seen, we may assume that there is a fifth root of unity in the center of the endomorphism ring of the Jacobian of C, and it follows that C has an automorphism of order 5. Since the finite fields we are concerned with contain the fifth roots of unity, this shows that C is a degree-5 Kummer extension of another curve, and by using the Riemann-Hurwitz formula we see that this second curve must be the projective line. It is not hard to enumerate such Kummer covers; doing so, we find no curves with N points.

7. Lower bounds from explicit examples

In this section we prove the new lower bounds for $N_q(4)$ given in Table 1 by providing examples of genus-4 curves with many points. Each line of Table 3 gives a prime power q, an integer N, and the equations for a genus-4 curve over $\mathbf{F}_q$ having N points.

Almost all of the examples in Table 3 were obtained by running our program `double_cover_given_trace`. The exceptions are the example for $q = 47$, which was found by running `double_cover_genus_4` on a carefully chosen genus-2 curve, and the examples for $q = 79$ and $q = 97$, which were found during a search of degree-3 Kummer covers of elliptic curves. While searching for such Kummer covers, we also found a particularly nice example for $q = 67$: the curve $y^6 = x^3 + x - 6$ attains $N_{67}(4)$.

The function `check_examples` in the file `Genus4.magma` verifies that all of these examples do have the number of points claimed.

References

[1] Z. I. Borevič and D. K. Faddeev, *Representations of orders with cyclic index*, Trudy Mat. Inst. Steklov **80** (1965), 51–65; English transl. in *Proceedings of the Steklov Institute of Mathematics. No. 80 (1965): Algebraic number theory and representations* (1968). MR0205980 (34:5805)

[2] Wieb Bosma, John Cannon, and Catherine Playoust, *The Magma algebra system. I. The user language*, J. Symbolic Comput. **24** (1997), no. 3-4, 235–265. Computational algebra and number theory (London, 1993). DOI: 10.1006/jsco.1996.0125. MR1484478

[3] Pierre Deligne, *Variétés abéliennes ordinaires sur un corps fini*, Invent. Math. **8** (1969), 238–243. DOI: 10.1007/BF01406076. MR0254059 (40:7270)

[4] Max Deuring, *Die Typen der Multiplikatorenringe elliptischer Funktionenkörper*, Abh. Math. Sem. Hansischen Univ. **14** (1941), 197–272. MR0005125 (3:104f)

[5] Robert M. Guralnick and Everett W. Howe, *Characteristic polynomials of automorphisms of hyperelliptic curves*, Arithmetic, geometry, cryptography and coding theory, Contemp. Math., vol. 487, Amer. Math. Soc., Providence, RI, 2009, pp. 101–111. MR2555989 (2010j:14060)

[6] H. Hasse, *Zur Theorie der abstrakten elliptischen Funktionkörper.* I: *Die Struktur der Gruppe der Divisorenklassen endlicher Ordnung*, J. Reine Angew. Math. **175** (1936), 55–62; II: *Automorphismen und Meromorphismen. Das Additionstheorem*, J. Reine Angew. Math. **175** (1936), 69–88; III: *Die Struktur des Meromorphismenringes. Die Riemannsche Vermutung*, J. Reine Angew. Math. **175** (1936), 193–208. DOI: 10.1515/crll.1936.175.55, 10.1515/crll.1936.175.69, 10.1515/crll.1936.175.193.

[7] Everett W. Howe, *Principally polarized ordinary abelian varieties over finite fields*, Trans. Amer. Math. Soc. **347** (1995), 2361–2401. DOI: 10.2307/2154828. MR1297531 (96i:11065)

[8] E. W. Howe and K. E. Lauter, *Improved upper bounds for the number of points on curves over finite fields*, Ann. Inst. Fourier (Grenoble) **53** (2003), no. 6, 1677–1737. DOI: 10.5802/aif.1990; Corrigendum, **57** (2007), no. 3, 1019–1021. DOI: 10.5802/aif.2284. MR2038778 (2005c:11079)

[9] Everett W. Howe and Kristin E. Lauter, *New methods for bounding the number of points on curves over finite fields*, Geometry and Arithmetic (Schiermonnikoog, 2010) (C. Faber, G. Farkas, and R. de Jong, eds.), European Mathematical Society. To appear. Preprint version: arXiv:1202.6308v2 [math.NT].

[10] Everett W. Howe, Enric Nart, and Christophe Ritzenthaler, *Jacobians in isogeny classes of abelian surfaces over finite fields*, Ann. Inst. Fourier (Grenoble) **59** (2009), no. 1, 239–289. DOI: 10.5802/aif.2430. MR2514865 (2010b:11064)

[11] Yasutaka Ihara, *Some remarks on the number of rational points of algebraic curves over finite fields*, J. Fac. Sci. Univ. Tokyo Sect. IA Math. **28** (1981), no. 3, 721–724 (1982). `http://hdl.handle.net/2261/6319`. MR656048 (84c:14016)

[12] Gilles Lachaud, Christophe Ritzenthaler, and Alexey Zykin, *Jacobians among abelian threefolds: a formula of Klein and a question of Serre*, Math. Res. Lett. **17** (2010), no. 2, 323–333. `http://www.mrlonline.org/mrl/2010-017-002/2010-017-002-011.html`. MR2644379 (2011f:14074)

[13] Kristin Lauter, *The maximum or minimum number of rational points on genus three curves over finite fields*, Compositio Math. **134** (2002), no. 1, 87–111. With an appendix by Jean-Pierre Serre. DOI:10.1023/A:1020246226326. MR1931964 (2003i:14025)

[14] Yu. I. Manin, *What is the maximum number of points on a curve over* $\mathbf{F}_2$*?*, J. Fac. Sci. Univ. Tokyo Sect. IA Math. **28** (1981), no. 3, 715–720 (1982). `http://hdl.handle.net/2261/6318`. MR656047 (84c:14015)

[15] J. S. Milne, *Jacobian varieties*, Arithmetic geometry (Storrs, Conn., 1984), Springer, New York, 1986, pp. 167–212. http://jmilne.org/math/articles/index.html#1986c. MR861976

[16] Alexander Schiemann, *Classification of Hermitian forms with the neighbour method*, J. Symbolic Comput. **26** (1998), no. 4, 487–508. DOI: 10.1006/jsco.1998.0225, online tables available at http://www.math.uni-sb.de/ag/schulze/Hermitian-lattices/. MR1646667 (99g:11053)

[17] Jean-Pierre Serre, *Sur le nombre des points rationnels d'une courbe algébrique sur un corps fini*, C. R. Acad. Sci. Paris Sér. I Math. **296** (1983), no. 9, 397–402. = Œuvres [128]. http://gallica.bnf.fr/ark:/12148/bpt6k55351747/f35. MR703906 (85b:14027)

[18] ———, *Nombres de points des courbes algébriques sur* $\mathbf{F}_q$, Seminar on number theory, 1982–1983 (Talence, 1982/1983), Univ. Bordeaux I, Talence, 1983, pp. Exp. No. 22, 8. = Œuvres [129]. MR750323 (86d:11051)

[19] ———, *Résumé des cours de 1983–1984*, Ann. Collège France (1984), 79–83. =Œuvres [132].

[20] S. G. Vlèduts and V. G. Drinfel'd, *The number of points of an algebraic curve*, Funktsional. Anal. i Prilozhen. **17** (1983), no. 1, 68–69; English transl., Functional Anal. Appl. **17** (1983), no. 1, 53–54. DOI: 10.1007/BF01083182.

[21] William C. Waterhouse, *Abelian varieties over finite fields*, Ann. Sci. École Norm. Sup. (4) **2** (1969), 521–560. http://www.numdam.org/item?id=ASENS_1969_4_2_4_521_0. MR0265369 (42:279)

[22] André Weil, *Sur les fonctions algébriques à corps de constantes fini*, C. R. Acad. Sci. Paris **210** (1940), 592–594. http://gallica.bnf.fr/ark:/12148/bpt6k31623/f592. MR0002863 (2:123d)

[23] ———, *On the Riemann hypothesis in function-fields*, Proc. Nat. Acad. Sci. U. S. A. **27** (1941), 345–347. http://www.pnas.org/content/27/7/345.short.

[24] ———, *Sur les courbes algébriques et les variétés qui s'en déduisent*, Actualités Sci. Ind., no. 1041 = Publ. Inst. Math. Univ. Strasbourg **7** (1945), Hermann et Cie., Paris, 1948. MR0027151 (10:262c)

[25] ———, *Variétés abéliennes et courbes algébriques*, Actualités Sci. Ind., no. 1064 = Publ. Inst. Math. Univ. Strasbourg 8 (1946), Hermann & Cie., Paris, 1948. MR0029522 (10:621d)

Center for Communications Research, 4320 Westerra Court, San Diego, California 92121-1967

E-mail address: however@alumni.caltech.edu

URL: http://www.alumni.caltech.edu/~however/

Contemporary Mathematics
Volume **574**, 2012
http://dx.doi.org/10.1090/conm/574/11425

Some planar maps and related function fields

Gohar M. Kyureghyan, Ferruh Özbudak, and Alexander Pott

ABSTRACT. Let $\mathbb{F}_q$ be a finite field with q elements, where q is odd. In this paper we classify *q-quadratic planar binomials* over $\mathbb{F}_{q^3}$ completely. We also obtain some results on the related algebraic function fields.

1. Introduction

Let q be a power of an odd prime and let $\mathbb{F}$ be a finite field containing $\mathbb{F}_q$. A map $f : \mathbb{F} \to \mathbb{F}$ is called *planar* on $\mathbb{F}$ if for each nonzero $a \in \mathbb{F}$, the corresponding difference map

$$\begin{aligned} D_{f,a} : \mathbb{F} &\to \mathbb{F} \\ x &\mapsto f(x+a) - f(x) - f(a) \end{aligned}$$

is bijective. There are interesting connections between planar maps, finite commutative semifields and finite geometry (see, for example, [**5**], [**6**], [**7**], [**10**]). An analogous notion in characteristic 2 is *almost perfect nonlinear maps* [**14**]. They have been investigated intensively and they have applications in cryptography (see, for example, [**2**], [**4**], [**8**], [**11**]).

There are some simple facts and notions which seem useful to recall here. Let $F(x) \in \mathbb{F}_q[x]$ be a polynomial. The map f on $\mathbb{F}_q$ defined by $F(x)$ is the map $f : x \mapsto F(x)$. We say that the polynomial $F(x)$ is planar (on $\mathbb{F}_q$) if the map defined by $F(x)$ is planar. Two polynomials $H_1(x), H_2(x) \in \mathbb{F}_q[x]$ define the same map if and only if $H_1(x) \equiv H_2(x) \mod (x^q - x)$. Therefore given any map f on $\mathbb{F}_q$, there exists a uniquely determined polynomial $F(x) \in \mathbb{F}_q[x]$ of degree at most $q-1$ such that $f(x) = F(x)$ for all $x \in \mathbb{F}_q$.

A map f is planar on $\mathbb{F}_q$ if and only if the composite map $g = L_1 \circ f \circ L_2$ on $\mathbb{F}_q$ is a planar map for any bijective additive maps L_1, L_2 on $\mathbb{F}_q$ (see Lemma 2.1 below). Here we call such f and g *equivalent*. Note that L_1 and L_2 can be considered as linearized permutation polynomials of degree strictly less than q without loss of generality (see, for example, [**12**, Section 3.4]). In particular a

---

2010 *Mathematics Subject Classification.* Primary: 12E20, 12K10, 11R58; Secondary: 94A60.

Key words and phrases. Planar maps, q-binomials, algebraic function fields.

A part of this paper was written while the second named author was visiting Faculty of Mathematics, Otto-von-Guericke University of Magdeburg, Germany with a research fellowship from the Alexander von Humboldt Foundation.

The second named author is partially supported by TÜBİTAK under Grant No. TBAG-109T672.

polynomial $F(x) \in \mathbb{F}_q[x]$ is planar if and only if the polynomial $\frac{1}{\alpha}F(x)$ is planar, where α is an arbitrary nonzero element of $\mathbb{F}_q$.

A monomial x^d is called q-quadratic if the q-ary weight of its exponent d is 2. The monic q-quadratic monomials yielding different maps on $\mathbb{F}_{q^3}$ are

$$x^2, \quad x^{q+1}, \quad x^{2q}, \quad x^{q^2+1}, \quad x^{q^2+q}, \quad x^{2q^2}. \tag{1.1}$$

It is known and easy to prove that all these q-quadratic monomials define planar maps on $\mathbb{F}_{q^3}$. In this paper we completely classify the *q-quadratic planar binomials* over $\mathbb{F}_{q^3}$. The following is the list of all q-quadratic binomials up to multiplication with a nonzero constant in $\mathbb{F}_{q^3}$ and up to modulo $x^{q^3} - x$:

(A.1) $x^2 + ux^{q^2+q}$.
(A.2) $x^{q+1} + ux^{2q^2} \equiv (x^{q^2+q} + ux^2) \circ x^{q^2} \mod (x^{q^3} - x)$.
(A.3) $x^{2q} + ux^{q^2+1} \equiv (x^2 + ux^{q^2+q}) \circ x^q \mod (x^{q^3} - x)$.

(B.1) $x^2 + ux^{q+1}$.
(B.2) $x^2 + ux^{2q} = (x + ux^q) \circ x^2$.
(B.3) $x^2 + ux^{q^2+1}$.
(B.4) $x^2 + ux^{2q^2} = (x + ux^{q^2}) \circ x^2$.
(B.5) $x^{q+1} + ux^{2q} \equiv (x^{q^2+1} + ux^2) \circ x^q \mod (x^{q^3} - x)$.
(B.6) $x^{q+1} + ux^{q^2+1} \equiv (x + ux^{q^2}) \circ x^{q+1} \mod (x^{q^3} - x)$.
(B.7) $x^{q+1} + ux^{q^2+q} = (x + ux^q) \circ x^{q+1}$.
(B.8) $x^{2q} + ux^{q^2+q} = (x^2 + ux^{q+1}) \circ x^q$.
(B.9) $x^{2q} + ux^{2q^2} = (x^q + ux^{q^2}) \circ x^2$.
(B.10) $x^{q^2+1} + ux^{q^2+q} \equiv (x^{q^2} + ux^q) \circ x^{q+1} \mod (x^{q^3} - x)$.
(B.11) $x^{q^2+1} + ux^{2q^2} \equiv (x^{q+1} + ux^2) \circ x^{q^2} \mod (x^{q^3} - x)$.
(B.12) $x^{q^2+q} + ux^{2q^2} \equiv (x^{q^2+1} + ux^2) \circ x^{q^2} \mod (x^{q^3} - x)$.

Here u is an arbitrary nonzero element in $\mathbb{F}_{q^3}$. It is easy to see whether the binomials in the part B of the list are planar. Indeed, first consider the types in B.1, B.3, B.5, B.8, B.11 and B.12. Up to composition with a linearized permutation polynomial and up to equivalence $\mod (x^{q^3} - x)$, they all are equivalent to a polynomial of the form

$$x^{q+1} + ux^2 \in \mathbb{F}_{q^3}[x] \quad \text{or} \quad x^{q^2+1} + ux^2 \in \mathbb{F}_{q^3}[x].$$

The polynomials in these forms are known to be not planar [**9**, Lemma 5.1 and page 643]. Also the remaining B types, B.2, B.4, B.6, B.7, B.9 and B.10, can be written as a composition of a simple linearized polynomial and the planar polynomials x^2 or x^{q+1} modulo $x^{q^3} - x$. Hence they are planar if and only if the corresponding linearized polynomials are permutation polynomials (see Lemma 2.1 below). It is easy to determine whether the corresponding linearized polynomials are permutation polynomials in these specific cases (see [**12**, Lemma 3.51 and page 362]).

However determining whether the A types are planar or not appears to be more challenging. Observe that the binomials in the A part of the list are all equivalent to the binomials of shape

$$F_u(x) = x^{q^2+q} + ux^2 \in \mathbb{F}_{q^3}[x] \quad \text{with } u \in \mathbb{F}_{q^3} \setminus \{0\}. \tag{1.2}$$

We show that the class in (1.2) is rather different from the binomials in the B part of the list above. Each type in the B part of the list may be equivalent only to one of the planar monomials x^2 and x^{q+1}, not both of them. We will show that this is not anymore true for the binomials in the A part of the list: Let $\mathbb{F}_{q^3}^*$ be the set of nonzero elements of $\mathbb{F}_{q^3}$. There are $u \in \mathbb{F}_{q^3}^*$ such that $F_u(x)$ is equivalent to x^2 and there are $u \in \mathbb{F}_{q^3}^*$ such that $F_u(x)$ is equivalent to x^{q+1} (and there are $u \in \mathbb{F}_{q^3}^*$ such that $F_u(x)$ is not planar).

In this paper we determine exactly when $F_u(x)$ is planar, when $F_u(x)$ is equivalent to x^2 and when $F_u(x)$ is equivalent to x^{q+1} as u runs through $\mathbb{F}_{q^3}^*$. Our main result is Theorem 4.5.

In Section 2 we prove that $F_u(x)$ is planar for certain $u \in \mathbb{F}_{q^3}^*$ when $q \equiv 1 \mod 3$. The conditions whether $F_u(x)$ is planar or not is closely related to the number of rational places of some algebraic function fields. Using this connection we show in Section 3 that $F_u(x)$ is never planar if $q \not\equiv 1 \mod 3$. Moreover we also prove that $F_u(x)$ is not planar for certain $u \in \mathbb{F}_{q^3}^*$ when $q \equiv 1 \mod 3$. In Section 4 we give explicit isotopisms for the planar maps $F_u(x)$ determined in Section 2 to x^2 and x^{q+1}. We also prove that the remaining values of $u \in \mathbb{F}_{q^3}^*$ do not give any planar map. Conversely, using such results on planar maps, we also obtain in Section 3 some results on the number of rational places for the related algebraic function fields.

Throughout the paper if $\mathbb{F}$ is a finite field, then $\mathbb{F}^*$ denotes the multiplicative subgroup (or subset) $\mathbb{F} \setminus \{0\}$. If also $\mathbb{E}$ is a subfield of $\mathbb{F}$, then the relative trace and the relative norm maps from $\mathbb{F}$ to $\mathbb{E}$ (and $\mathbb{F}^*$ to $\mathbb{E}^*$) are denoted as $\mathrm{Tr}_{\mathbb{F}/\mathbb{E}}$ and $\mathrm{Norm}_{\mathbb{F}/\mathbb{E}}$. Moreover $\mathbb{F}_q$ denotes a finite field with q elements. For a subset E of a finite field $\mathbb{F}^*$ and $c \in \mathbb{F}^*$, we use cE to denote the subset $\{ce : e \in E\}$ of $\mathbb{F}^*$.

2. Proving that $F_u(x)$ is planar for some $u \in \mathbb{F}_{q^3}^*$

In this section we give some preliminary results and we also prove that $F_u(x)$ is planar for certain $u \in \mathbb{F}_{q^3}^*$ and $q \equiv 1 \mod 3$.

First we prove a lemma that we have already referred in Section 1 above. The lemma is well known, nevertheless for the completeness we also provide a proof.

LEMMA 2.1. *Let f, g, L_1 and L_2 be maps on $\mathbb{F}_q$. Assume that L_1 and L_2 are additive, that is $L_i(a+b) = L_i(a) + L_i(b)$ for $a, b \in \mathbb{F}_q$ and $1 \le i \le 2$. Assume further that*

$$f(x) = (L_1 \circ g \circ L_2)(x) \text{ for } x \in \mathbb{F}_q. \tag{2.1}$$

Then

f is planar if and only if g is planar and both L_1, L_2 are bijective.

PROOF. For a map h on $\mathbb{F}_q$ and $a \in \mathbb{F}_q^*$, let $D_{h,a}$ denote the map $D_{h,a}(x) = h(x+a) - h(x) - h(a)$. Using (2.1) we get

$$\begin{aligned} D_{f,a}(x) &= f(x+a) - f(x) - f(a) \\ &= L_1 \circ g \circ L_2(x+a) - L_1 \circ g \circ L_2(x) - L_1 \circ g \circ L_2(a) \\ &= L_1 \circ (g(L_2(x) + L_2(a)) - g(L_2(x)) - g(L_2(a))) \\ &= L_1 \circ \left(D_{g,L_2(a)}(L_2(x))\right) \\ &= L_1 \circ D_{g,L_2(a)} \circ L_2(x). \end{aligned}$$

Then $D_{f,a}$ is bijective if and only if the maps $L_1, D_{g,L_2(a)}$ and L_2 are bijective. Moreover L_2 is bijective if and only if $\{L_2(a) : a \in \mathbb{F}_q^*\} = \mathbb{F}_q^*$. Using the definition of the planarity of f and g we complete the proof. □

Next we give the conditions to determine whether $F_u(x)$ is planar. For $y \in \mathbb{F}_{q^3}^*$, let $D_{F_u,y} : \mathbb{F}_{q^3} \to \mathbb{F}_{q^3}$ be the corresponding difference map at y, which is given by

$$D_{F_u,y}(x) = F_u(x+y) - F_u(x) - F_u(y) = x^{q^2}y^q + x^q y^{q^2} + 2uxy. \tag{2.2}$$

For each fixed $u, y \in \mathbb{F}_{q^3}^*$, considering the right hand side of (2.2), we observe that $D_{F_u,y}$ is an $\mathbb{F}_q$-linear map on $\mathbb{F}_{q^3}$. Therefore $D_{F_u,y}$ is bijective if and only if its kernel is trivial, or equivalently its kernel has no $x \in \mathbb{F}_{q^3}^*$. Using the definition of planar functions, we conclude that $F_u(x)$ is planar if and only if

$$F_u(x+y) - F_u(x) - F_u(y) = 0$$

has no solution $(x, y) \in \mathbb{F}_{q^3}^* \times \mathbb{F}_{q^3}^*$. Dividing the right hand side of (2.2) by xy we obtain the equation

$$x^{q^2-1}y^{q-1} + x^{q-1}y^{q^2-1} + 2u = 0.$$

Then dividing further by $y^{q^2-1}y^{q-1}$ and making the change of variable

$$(x, y) \mapsto (x/y, 1/y)$$

we get that

$$x^{q^2-1} + x^{q-1} + 2uy^{q^2+q-2} = 0.$$

Note that

$$\gcd(q^2+q-2, q^3-1) = \begin{cases} q-1 & \text{if } q \not\equiv 1 \mod 3, \\ 3(q-1) & \text{if } q \equiv 1 \mod 3. \end{cases}$$

These arguments imply the following lemma.

LEMMA 2.2. *Let q be a power of an odd prime, u be an arbitrary element of $\mathbb{F}_{q^3}^*$ and $F_u(x) \in \mathbb{F}_{q^3}[x]$ be the q-quadratic binomial given in (1.2). We have that $F_u(x)$ is planar if and only if*

when $q \not\equiv 1 \mod 3$: there is no $(x, y) \in \mathbb{F}_{q^3}^* \times \mathbb{F}_{q^3}^*$ satisfying

$$x^{q^2-1} + x^{q-1} + 2uy^{q-1} = 0,$$

(2.3) *when $q \equiv 1 \mod 3$: there is no solution $(x, y) \in \mathbb{F}_{q^3}^* \times \mathbb{F}_{q^3}^*$ satisfying*

$$x^{q^2-1} + x^{q-1} + 2uy^{3(q-1)} = 0.$$

In the rest of this section we will prove that $F_u(x)$ is planar for some $u \in \mathbb{F}_{q^3}^*$ when $q \equiv 1 \mod 3$. We continue with a simple lemma, which will be used later.

LEMMA 2.3. *Let q be a power of an odd prime and assume that $q \equiv 1 \mod 3$. Let G be the subgroup of $\mathbb{F}_{q^3}^*$ of order q^2+q+1. Let H be the subgroup of $\mathbb{F}_{q^3}^*$ of order $(q^2+q+1)/3$. Let ϵ be an element of $\mathbb{F}_{q^3}^*$, whose multiplicative order is 3. Then the following properties hold.*

1) *For $c \in \mathbb{F}_q^*$, we have $c^{(q^2+q+1)/3} = c$.*
2) *We have $\epsilon, \epsilon^2 \in G \setminus H$ and the disjoint union*

$$G = H \sqcup \epsilon H \sqcup \epsilon^2 H.$$

3) *We have $-2 \notin G$ and hence*

$$G \cap (-2G) = \emptyset.$$

PROOF. Note that $\frac{q^2+q+1}{3} = \frac{(q-1)q+2q+1}{3}$. For $c \in \mathbb{F}_q^*$ we obtain that

$$c^{(q^2+q+1)/3} = c^{q(q-1)/3}c^{(2q+1)/3} = c^{(q-1)/3}c^{(2q+1)/3} = c^q = c.$$

As 3 divides $(q-1)$, we have $\epsilon \in \mathbb{F}_q^*$. Then using item 1) we get that $\epsilon^{(q^2+q+1)/3} = \epsilon \neq 1$ and hence $\epsilon \notin H$. It is clear that $\epsilon \in G$ as $3 \mid (q^2+q+1)$. Therefore $\epsilon \in G \setminus H$ and similarly $\epsilon^2 \in G \setminus H$.

Finally we prove item 3). We have $(-2)^{q^2+q+1} = (-2)^3 = -8$. Assume that $-2 \in G$, or equivalently $-8 = 1$. Then the characteristic of $\mathbb{F}_q$ is 3, which is a contradiction as $3 \mid (q-1)$. Therefore $G \cap (-2G) = \emptyset$. □

The following lemma is used in the proofs of the theorems in this section.

LEMMA 2.4. *Let q be a power of an odd prime and assume that $q \equiv 1 \mod 3$. Let H be the subgroup of $\mathbb{F}_{q^3}^*$ of order $(q^2+q+1)/3$. Assume that $F_u(x)$ is planar, where $u \in \mathbb{F}_{q^3}^*$. Then $F_{uh}(x)$ is planar for each $h \in H$.*

PROOF. As $F_u(x)$ is planar, using (2.3) we obtain that

(2.4) there is no $(x, y) \in \mathbb{F}_{q^3}^* \times \mathbb{F}_{q^3}^*$ such that $x^{q^2-1} + x^{q-1} + 2uy^{3(q-1)} = 0$.

Let $h \in H$. We will prove by contradiction that $F_{uh}(x)$ is planar. Assume the contrary and let $(x_0, y_0) \in \mathbb{F}_{q^3}^* \times \mathbb{F}_{q^3}^*$ such that

$$x_0^{q^2-1} + x_0^{q-1} + 2uhy_0^{3(q-1)} = 0.$$

As $|H| = (q^2+q+1)/3$, there exists $v \in \mathbb{F}_{q^3}^*$ such that $h = v^{3(q-1)}$. Putting $y_1 = vy_0$ we obtain that

$$x_0^{q^2-1} + x_0^{q-1} + 2uy_1^{3(q-1)} = 0,$$

which is a contradiction to (2.4). This completes the proof. □

Now we give the main results of this section. Their proofs are based on a beautiful technique of Bierbrauer [**2**].

THEOREM 2.5. *Let q be a power of an odd prime and assume that $q \equiv 1 \mod 3$. Let G be the subgroup of $\mathbb{F}_{q^3}^*$ of order q^2+q+1. Let H be the subgroup of $\mathbb{F}_{q^3}^*$ of order $(q^2+q+1)/3$. Let ϵ be an element of $\mathbb{F}_{q^3}^*$, whose multiplicative order is 3. We have the following:*

If $-u \in G \setminus H$, then the polynomial $F_u(x) = x^{q^2+q} + ux^2 \in \mathbb{F}_{q^3}[x]$ is planar.

PROOF. We first prove that $F_u(x)$ is planar when $u = -\epsilon$. Using (2.3) it is enough to show that there is no $(x_0, y_0) \in \mathbb{F}_{q^3}^* \times \mathbb{F}_{q^3}^*$ such that

$$x_0^{q^2-1} + x_0^{q-1} - 2\epsilon y_0^{3(q-1)} = 0. \tag{2.5}$$

Assume the contrary and let $(x_0, y_0) \in \mathbb{F}_{q^3}^* \times \mathbb{F}_{q^3}^*$ satisfying (2.5). As $\epsilon y_0^{3(q-1)} \in G$, there exists $z_0 \in \mathbb{F}_{q^3}^*$ such that

$$\epsilon y_0^{3(q-1)} = z_0^{q-1}. \tag{2.6}$$

We consider the equation in one variable $x \in \mathbb{F}_{q^3}$ given by

$$x^{q^2} + x^q - 2z_0^{q-1}x = 0. \tag{2.7}$$

Let $\mathcal{S}$ be the set of solutions $x \in \mathbb{F}_{q^3}$ of (2.7). It is clear that $\mathcal{S}$ is an $\mathbb{F}_q$-linear subspace of $\mathbb{F}_{q^3}$ and $\dim_{\mathbb{F}_q} \mathcal{S} \geq 1$ as $x_0 \in \mathcal{S}$.

For $x \in \mathcal{S}$ and $c \in \mathbb{F}_{q^3}$ we have

$$\begin{aligned} 0 &= \mathrm{Tr}_{\mathbb{F}_{q^3}/\mathbb{F}_q}\left(c\left(x^{q^2}+x^q-2z_0^{q-1}x\right)\right) \\ &= \mathrm{Tr}_{\mathbb{F}_{q^3}/\mathbb{F}_q}\left(cx^{q^2}-cz_0^{q-1}x\right)+\mathrm{Tr}_{\mathbb{F}_{q^3}/\mathbb{F}_q}\left(cx^q-cz_0^{q-1}x\right) \\ &= \mathrm{Tr}_{\mathbb{F}_{q^3}/\mathbb{F}_q}\left((c^q-cz_0^{q-1})x\right)+\mathrm{Tr}_{\mathbb{F}_{q^3}/\mathbb{F}_q}\left((c^{q^2}-cz_0^{q-1})x\right), \end{aligned}$$

where we use the facts that $\mathrm{Tr}_{\mathbb{F}_{q^3}/\mathbb{F}_q}(cx^{q^2}) = \mathrm{Tr}_{\mathbb{F}_{q^3}/\mathbb{F}_q}(c^q x)$ and $\mathrm{Tr}_{\mathbb{F}_{q^3}/\mathbb{F}_q}(cx^q) = \mathrm{Tr}_{\mathbb{F}_{q^3}/\mathbb{F}_q}(c^{q^2}x)$. Considering the coefficients in the traces of the last equation we let

$$H_1(c) = c^q - cz_0^{q-1} \quad \text{and} \quad H_2(c) = c^{q^2} - cz_0^{q-1}.$$

We put $c_1 = z_0$, $c_2 = 1/z_0^q$ and

$$\mu_1 = z_0^{q^2} - z_0^q, \qquad \mu_2 = \frac{1}{z_0^{q^2}} - \frac{1}{z_0}. \tag{2.8}$$

Then we obtain $H_1(c_1) = H_2(c_2) = 0$ and

$$\mu_1 = H_2(c_1), \qquad \mu_2 = H_1(c_2).$$

Let $\mathcal{S}_1$ be the set of solutions $x \in \mathbb{F}_{q^3}$ of the system

$$\mathrm{Tr}_{\mathbb{F}_{q^3}/\mathbb{F}_q}(\mu_1 x) = \mathrm{Tr}_{\mathbb{F}_{q^3}/\mathbb{F}_q}(\mu_2 x) = 0. \tag{2.9}$$

It is clear that $\mathcal{S}_1$ is an $\mathbb{F}_q$-linear subspace of $\mathbb{F}_{q^3}$ with $\mathcal{S} \subseteq \mathcal{S}_1$. As $\dim_{\mathbb{F}_q} \mathcal{S} \geq 1$, in order to conclude that $\mathcal{S} = \mathcal{S}_1$ and $\dim_{\mathbb{F}_q} \mathcal{S} = 1$, it is enough to show that $\{\mu_1, \mu_2\}$ is linearly independent over $\mathbb{F}_q$.

For this purpose first note that $z_0 \notin \mathbb{F}_q$ as $z_0^{q-1} = \epsilon y_0^{3(q-1)} \notin H$ (see (2.6) and Lemma 2.3). Then $\mu_1 \neq 0$, $\mu_2 \neq 0$. Assume that the set $\{\mu_1, \mu_2\}$ is linearly dependent over $\mathbb{F}_q$. This means that $\mu_1^{q-1} = \mu_2^{q-1}$, or equivalently

$$\left(z_0^{q^2} - z_0^q\right)^{q-1} = \left(\frac{1}{z_0^{q^2}} - \frac{1}{z_0}\right)^{q-1}. \tag{2.10}$$

We have

$$\left(\frac{1}{z_0^{q^2}} - \frac{1}{z_0}\right)^{q-1} = \frac{\left(z_0 - z_0^{q^2}\right)^{q-1}}{z_0^{(q^2+1)(q-1)}} = \frac{(z_0^q - z_0)^{q^2(q-1)}}{z_0^{(q^2+1)(q-1)}}. \tag{2.11}$$

Combining (2.10) and (2.11) we conclude that

$$z_0^{(q^2+1)(q-1)} = (z_0^q - z_0)^{(q^2-q)(q-1)}.$$

Taking (q^2+1) powers of the both sides of (2.6) we obtain

$$z_0^{(q^2+1)(q-1)} = \epsilon^2 y_0^{3(q-1)(q^2+1)}.$$

Therefore we get

$$\epsilon^2 = \frac{(z_0^q - z_0)^{q(q-1)(q-1)}}{y_0^{3(q^2+1)(q-1)}} = \left(\frac{(z_0^q - z_0)^{q\frac{q-1}{3}}}{y_0^{q^2+1}}\right)^{3(q-1)},$$

which implies that $\epsilon \in H$. However this is a contradiction to Lemma 2.3. This shows that $\mathcal{S} = \mathcal{S}_1$ and $\dim_{\mathbb{F}_q} \mathcal{S} = 1$. Moreover it is easy to observe that $\mathrm{Tr}_{\mathbb{F}_{q^3}/\mathbb{F}_q}(\mu_1) =$

$\mathrm{Tr}_{\mathbb{F}_{q^3}/\mathbb{F}_q}(\mu_2) = 0$ and hence using (2.9) we conclude that $\mathbb{F}_q = \mathcal{S}$. In particular we obtain that $x_0 \in \mathbb{F}_q^*$. Then putting $x_0^{q-1} = 1$ in (2.5) we obtain that

$$1 + 1 - 2\epsilon y_0^{3(q-1)} = 0 \text{ and hence } y_0^{3(q-1)} = \epsilon^2.$$

This implies that $\epsilon^2 \in H$, which is a contradiction. This completes the proof of the statement that $F_u(x)$ is planar when $u = -\epsilon$.

Using Lemma 2.4 we immediately conclude that $F_u(x)$ is planar when $u \in -\epsilon H$.

Note that in the arguments above the only property of ϵ used is that its order is 3. These arguments also hold for ϵ^2 and hence $F_u(x)$ is planar also when $u \in -\epsilon^2 H$. This completes the proof. □

Using a similar method as in the proof of Theorem 2.5, we obtain the following theorem. Although we use a similar method, there are still important differences in its proof and we indicate them explicitly.

THEOREM 2.6. *Let q be a power of an odd prime and assume that $q \equiv 1 \mod 3$. Let G be the subgroup of $\mathbb{F}_{q^3}^*$ of order q^2+q+1. Let H be the subgroup of $\mathbb{F}_{q^3}^*$ of order $(q^2+q+1)/3$. Let ϵ be an element of $\mathbb{F}_{q^3}^*$, whose multiplicative order is 3. We have the following:*

If $2u \in G \setminus H$, then the polynomial $F_u(x) = x^{q^2+q} + ux^2 \in \mathbb{F}_{q^3}[x]$ is planar.

PROOF. We first prove that $F_u(x)$ is planar when $u = \epsilon/2$. Using (2.3) we need to show that there is no $(x_0, y_0) \in \mathbb{F}_{q^3}^* \times \mathbb{F}_{q^3}^*$ such that

$$x_0^{q^2-1} + x_0^{q-1} + \epsilon y_0^{3(q-1)} = 0. \tag{2.12}$$

We choose and fix $z_0 \in \mathbb{F}_{q^3}^*$ as in (2.6). We consider the equation in one variable $x \in \mathbb{F}_{q^3}$ given by

$$x^{q^2} + x^q + z_0^{q-1}x = 0. \tag{2.13}$$

Again let $\mathcal{S}$ be the set of solutions $x \in \mathbb{F}_{q^3}$ of (2.13). We have $x_0 \in \mathcal{S}$ and $\dim_{\mathbb{F}_q} \mathcal{S} \geq 1$ similarly.

Now we have the first difference in the proof. Note that

$$1 = -\epsilon - \epsilon^2. \tag{2.14}$$

For $x \in \mathcal{S}$ and $c \in \mathbb{F}_{q^3}$, using (2.14) we obtain that

$$\begin{aligned} 0 &= \mathrm{Tr}_{\mathbb{F}_{q^3}/\mathbb{F}_q}\left(c\left(x^{q^2} + x^q + z_0^{q-1}x\right)\right) \\ &= \mathrm{Tr}_{\mathbb{F}_{q^3}/\mathbb{F}_q}\left((c^q - \epsilon c z_0^{q-1})x\right) + \mathrm{Tr}_{\mathbb{F}_{q^3}/\mathbb{F}_q}\left((c^{q^2} - \epsilon^2 c z_0^{q-1})x\right). \end{aligned}$$

Let

$$H_1(c) = c^q - \epsilon c z_0^{q-1} \text{ and } H_2(c) = c^{q^2} - \epsilon^2 c z_0^{q-1}.$$

As $\epsilon \in G$, there exists $a \in \mathbb{F}_{q^3}^*$ such that

$$a^{q-1} = \epsilon. \tag{2.15}$$

We put

$$c_1 = az_0, \quad c_2 = \frac{\epsilon^2}{a^2}\frac{1}{z_0^q} \quad \text{and} \quad \mu_1 = a\epsilon^2\left(z_0^{q^2} - z_0^q\right), \quad \mu_2 = \frac{\epsilon^2}{a^2}\left(\frac{1}{z_0^{q^2}} - \frac{1}{z_0}\right).$$

We have that $H_1(c_1) = H_2(c_2) = 0$ and $\mu_1 = H_2(c_1)$, $\mu_2 = H_1(c_2)$. Again let $\mathcal{S}_1$ be the set of solutions $x \in \mathbb{F}_{q^3}$ of the system

$$\mathrm{Tr}_{\mathbb{F}_{q^3}/\mathbb{F}_q}(\mu_1 x) = \mathrm{Tr}_{\mathbb{F}_{q^3}/\mathbb{F}_q}(\mu_2 x) = 0. \tag{2.16}$$

Similarly it is enough to show that $\{\mu_1, \mu_2\}$ is linearly independent over $\mathbb{F}_q$ in order to conclude that $\mathcal{S} = \mathcal{S}_1$ and $\dim_{\mathbb{F}_q} \mathcal{S} = 1$. Note that $a^3 \in \mathbb{F}_q^*$ as $a^{3(q-1)} = \epsilon^3 = 1$ (see (2.15)). Hence $\{\mu_1, \mu_2\}$ is linearly independent over $\mathbb{F}_q$ if and only if the set

$$\left\{\frac{1}{\epsilon^2}\mu_1, \frac{a^3}{\epsilon^2}\mu_2\right\} = \left\{a\left(z_0^{q^2} - z_0^q\right), a\left(\frac{1}{z_0^{q^2}} - \frac{1}{z_0}\right)\right\}$$

is linearly independent over $\mathbb{F}_q$. In the proof of Theorem 2.5 we have already shown that the set $\left\{z_0^{q^2} - z_0^q, \frac{1}{z_0^{q^2}} - \frac{1}{z_0}\right\}$ is linearly independent over $\mathbb{F}_q$. Combining this with the arguments above we conclude that $\{\mu_1, \mu_2\}$ is linearly independent over $\mathbb{F}_q$. Hence we obtain that $\mathcal{S} = \mathcal{S}_1$ and $\dim_{\mathbb{F}_q} \mathcal{S} = 1$.

Next we note that $\left\{\frac{1}{a}x : x \in \mathbb{F}_q\right\} \subseteq \mathcal{S}_1$. Indeed, if $x \in \mathbb{F}_q$, then we have

$$\mathrm{Tr}_{\mathbb{F}_{q^3}/\mathbb{F}_q}\left(\mu_1 \frac{x}{a}\right) = \mathrm{Tr}_{\mathbb{F}_{q^3}/\mathbb{F}_q}\left(a\epsilon^2 \frac{x}{a}\left(z_0^{q^2} - z_0^q\right)\right) = \epsilon^2 x \mathrm{Tr}_{\mathbb{F}_{q^3}/\mathbb{F}_q}\left(z_0^{q^2} - z_0^q\right) = 0,$$

and

$$\mathrm{Tr}_{\mathbb{F}_{q^3}/\mathbb{F}_q}\left(\mu_2 \frac{x}{a}\right) = \mathrm{Tr}_{\mathbb{F}_{q^3}/\mathbb{F}_q}\left(\frac{\epsilon^2}{a^2}\frac{x}{a}\left(\frac{1}{z_0^{q^2}} - \frac{1}{z_0}\right)\right) = \frac{\epsilon^2 x}{a^3}\mathrm{Tr}_{\mathbb{F}_{q^3}/\mathbb{F}_q}\left(\frac{1}{z_0^{q^2}} - \frac{1}{z_0}\right) = 0.$$

This implies that for $(x_0, y_0) \in \mathbb{F}_{q^3}^* \times \mathbb{F}_{q^3}^*$ satisfying (2.12), there exists $x_1 \in \mathbb{F}_q^*$ such that $x_0 = x_1/a$. Putting $x_0 = x_1/a$ and noting $x_1^{q-1} = 1$ in (2.12) we obtain that

$$\frac{1}{a^{q^2-1}} + \frac{1}{a^{q-1}} + \epsilon y_0^{3(q-1)} = \frac{1}{\epsilon^{q+1}} + \frac{1}{\epsilon} + \epsilon y_0^{3(q-1)} = -1 + \epsilon y_0^{3(q-1)} = 0,$$

and hence $y_0^{3(q-1)} = \epsilon^2$. This gives the contradiction that $\epsilon^2 \in H$ and hence completes the proof of the statement that $F_{\epsilon/2}(x)$ is planar.

As in the proof of Theorem 2.5, using Lemma 2.4 we conclude that $F_u(x)$ is planar when $2u \in \epsilon H$. Again similarly we obtain the proof of the statement that $F_u(x)$ is planar when $2u \in \epsilon^2 H$ by changing ϵ to ϵ^2. □

REMARK 2.7. Using Lemma 2.3 item 3) we observe that the set of polynomials shown to be planar by Theorem 2.5 is disjoint from the set of polynomials shown to be planar by Theorem 2.6.

3. Related function fields and proving that $F_u(x)$ is not planar for some $u \in \mathbb{F}_{q^3}^*$

In this section we consider some algebraic function fields related to the planarity of the polynomial $F_u(x) = x^{q^2+q} + ux^2 \in \mathbb{F}_{q^3}[x]$, where $u \in \mathbb{F}_{q^3}^*$. Using algebraic function fields we show that $F_u(x)$ is not planar if $q \not\equiv 1 \mod 3$. We also show that $F_u(x)$ is not planar if $q \equiv 1 \mod 3$ and $u \in (1/2H \cup (-H))$, where H is the (multiplicative) subgroup of $\mathbb{F}_{q^3}^*$ with $|H| = (q^2 + q + 1)/3$. Conversely using planarity of some $F_u(x)$ we obtain some results on the number of rational places of the related algebraic function fields. For notation and definitions of algebraic function fields we refer to [**16**].

For $u \in \mathbb{F}_{q^3}^*$ let $\mathbb{K}_u$, $\mathbb{E}_u$ and $\mathbb{L}_u$ be the algebraic function fields given as follows:

- $\mathbb{K}_u = \mathbb{F}_{q^3}(X, Y)$ with $X^{q^2-1} + X^{q-1} + 2uY^{q-1} = 0$.
- $\mathbb{E}_u = \mathbb{F}_{q^3}(X, Y_1)$ with $X^{q-1} + 2uY_1^{q-1} = -1$.
- If $q \equiv 1 \mod 3$, then $\mathbb{L}_u = \mathbb{F}_{q^3}(X, Z)$ with $X^{q^2-1}+X^{q-1}+2uZ^{3(q-1)} = 0$.

We consider $\mathbb{K}_u$, $\mathbb{E}_u$ and $\mathbb{L}_u$ as finite extension of the rational function field $\mathbb{F}_{q^3}(X)$. If $q \equiv 1 \mod 3$, then it is easy to observe that $\mathbb{K}_u$ is the subfield of $\mathbb{L}_u$ given by $\mathbb{K}_u = \mathbb{F}_{q^3}(X, Z^3)$.

We will show that $\mathbb{F}_{q^3}$ is the full constant of field $\mathbb{K}_u$, $\mathbb{E}_u$, $\mathbb{L}_u$ and we will determine their genera in Lemma 3.2 below.

Let $N(\mathbb{K}_u)$ (respectively $N(\mathbb{E}_u)$ and $N(\mathbb{L}_u)$) denote the number of rational places of $\mathbb{K}_u$ (resp. $\mathbb{E}_u$ and $\mathbb{L}_u$). We further denote the cardinality of a specified subset of the set of rational places of $\mathbb{K}_u$ (resp. $\mathbb{E}_u$ and $\mathbb{L}_u$) as follows:

Let P_0 and P_∞ be the rational places of $\mathbb{F}_{q^3}(X)$ corresponding to the zero and the pole of the function X. Recall that $\mathbb{K}_u$, $\mathbb{E}_u$, $\mathbb{L}_u$ are finite extensions of $\mathbb{F}_{q^3}(X)$ and hence any rational place of them is over a rational place of $\mathbb{F}_{q^3}(X)$. Let $N^*(\mathbb{K}_u)$ (resp. $N^*(\mathbb{E}_u)$ and $N^*(\mathbb{L}_u)$) denote the number of rational places of $\mathbb{K}_u$ (resp. $\mathbb{E}_u$ and $\mathbb{L}_u$) which are neither over P_0 nor over P_∞.

Lemma 2.2 giving the planarity conditions of $F_u(x)$ is equivalent to the following.

LEMMA 3.1. *$F_u(x)$ is planar if and only if*

- *when $q \not\equiv 1 \mod 3$: $N^*(\mathbb{K}_u) = 0$.*
- *when $q \equiv 1 \mod 3$: $N^*(\mathbb{L}_u) = 0$.*

PROOF. Assume that $q \not\equiv 1 \mod 3$. It is enough to prove that there is no place P in $\mathbb{K}_u$ with the valuations $v_P(X) = 0$ and $v_P(Y) \geq 1$. Assume the contrary and hence we have a solution $(x, y) \in \mathbb{F}_{q^3} \times \mathbb{F}_{q^3}$ of the equation

$$x^{q^2-1} + x^{q-1} + 2uy^{q-1} = 0$$

with $x \neq 0$ and $y = 0$. Then we have $x^{q-1}\left(x^{q^2-q} + 1\right) = 0$ and $x^{q(q-1)} = -1$. Taking the norm of the both of the sides we get $1 = \mathrm{Norm}_{\mathbb{F}_{q^3}/\mathbb{F}_q}(x^{q(q-1)}) = \mathrm{Norm}_{\mathbb{F}_{q^3}/\mathbb{F}_q}(-1) = -1$, a contradiction. This completes the proof of the statement regarding $\mathbb{K}_u$ when $q \not\equiv 1 \mod 3$. The same argument also proves the statement regarding $\mathbb{L}_u$ when $q \equiv 1 \mod 3$. $\square$

In fact $N^*(\mathbb{K}_u)$, $N^*(\mathbb{E}_u)$, $N^*(\mathbb{L}_u)$ and $N(\mathbb{K}_u)$, $N(\mathbb{E}_u)$, $N(\mathbb{L}_u)$ are related to each other. In the following we give this simple relation and we prove some results that we referred earlier in this section before Lemma 3.1.

LEMMA 3.2. *$\mathbb{F}_{q^3}$ is the full constant field of $\mathbb{K}_u$, $\mathbb{E}_u$ and $\mathbb{L}_u$. For their genera we have*

$$g(\mathbb{K}_u) = g(\mathbb{E}_u) = 1 + \frac{(q-1)(q-4)}{2},$$

and

$$g(\mathbb{L}_u) = 1 + \frac{3(q-1)(q-2)}{2}.$$

We also have

$$N(\mathbb{K}_u) - N^*(\mathbb{K}_u) = N(\mathbb{E}_u) - N^*(\mathbb{E}_u) = \begin{cases} 2(q-1) & \text{if } \mathrm{Norm}_{\mathbb{F}_{q^3}/\mathbb{F}_q}(2u) = -1, \\ 0 & \text{otherwise.} \end{cases}$$

Moreover if $q \equiv 1 \mod 3$, *again it holds that*

$$N(\mathbb{L}_u) - N^*(\mathbb{L}_u) = \begin{cases} 2(q-1) & \text{if } \mathrm{Norm}_{\mathbb{F}_{q^3}/\mathbb{F}_q}(2u) = -1, \\ 0 & \text{otherwise.} \end{cases} \tag{3.1}$$

PROOF. The polynomial $X^q + X \in \mathbb{F}_q[X]$ splits into linear factors over $\mathbb{F}_{q^2}$, it is separable and it has no nonzero root in $\mathbb{F}_q$. Therefore the canonical factorization of $X^{q-1}+1$ over $\mathbb{F}_q$ is

$$X^{q-1} + 1 = r_1(X) r_2(X) \ldots r_{\frac{q-1}{2}}(X), \tag{3.2}$$

where $r_1(X), r_2(X), \ldots, r_{\frac{q-1}{2}}(X)$ are distinct and monic irreducible polynomials of degree 2 over $\mathbb{F}_q$. These factors are still irreducible over $\mathbb{F}_{q^3}$.

First we consider the algebraic function field $\mathbb{K}_u = \mathbb{F}_{q^3}(X)(Y)$. We have

$$Y^{q-1} = \frac{-1}{2u} X^{q-1}\left(X^{q-1}+1\right)^q = \frac{-1}{2u} X^{q-1} r_1(X)^q r_2(X)^q \ldots r_{\frac{q-1}{2}}(X)^q, \tag{3.3}$$

where we use (3.2).

For a place P of $\mathbb{F}_{q^3}(X)$ and a function $f(X) \in \mathbb{F}_{q^3}(X)$, recall that $v_P(f(X))$ denotes the valuation of $f(X)$ at P. Let $h(X) = \frac{-1}{2u} X^{q-1}\left(X^{q-1}+1\right)^q$ and let Q_i denote the degree 2 place of $\mathbb{F}_{q^3}(X)$ corresponding to the zero of $r_i(X)$ for $1 \le i \le (q-1)/2$. Using (3.3) and following the notation of [**16**, Proposition 3.7.3] we obtain that

- $r_{P_0} = \gcd(q-1, v_{P_0}(h(x))) = \gcd(q-1, q-1) = q-1$,
- $r_{Q_i} = \gcd(q-1, v_{Q_i}(h(x))) = \gcd(q-1, q) = 1$ for $1 \le i \le (q-1)/2$,
- $r_{P_\infty} = \gcd(q-1, v_{P_\infty}(h(x))) = \gcd(q-1, q^2-1) = q-1$.

As $r_{Q_i} = 1$, using [**16**, Corollary 3.7.4] we conclude that $\mathbb{F}_{q^3}$ is the full constant field of $\mathbb{K}_u$. For the genus of $\mathbb{K}_u$ we have (see [**16**, Corollary 3.7.4])

$$g(\mathbb{K}_u) = 1 + (q-1)(-1) + \frac{1}{2}\left(0 + \frac{q-1}{2}(q-1-1)2 + 0\right) = 1 + \frac{(q-1)(q-4)}{2}.$$

Let $\hat{h}(X) = \frac{h(X)}{X^{q-1}}$. The valuation of $\hat{h}(X)$ at P_0 is $\frac{-1}{2u}$. Therefore the number of rational places of $\mathbb{K}_u$ over P_0 is

$$\begin{cases} q-1 & \text{if } \frac{-1}{2u} \text{ is a } (q-1)\text{-power in } \mathbb{F}_{q^3}^*, \\ 0 & \text{otherwise.} \end{cases} \tag{3.4}$$

The leading coefficient of $h(X)$ is also $-1/2u$. Therefore the number of rational places of $\mathbb{K}_u$ over P_∞ is again given by the number in (3.4). These imply that

$$N(\mathbb{K}_u) - N^*(\mathbb{K}_u) = \begin{cases} 2(q-1) & \text{if } \mathrm{Norm}_{\mathbb{F}_{q^3}/\mathbb{F}_q}(2u) = -1, \\ 0 & \text{otherwise.} \end{cases}$$

Next we consider the algebraic function field $\mathbb{E}_u = \mathbb{F}_{q^3}(X)(Y_1)$. We have

$$Y_1^{q-1} = \frac{-1}{2u}\left(X^{q-1}+1\right).$$

Following the same arguments we obtain that $\mathbb{F}_{q^3}$ is the full constant field of $\mathbb{E}_u$, $g(\mathbb{E}_u) = g(\mathbb{K}_u)$ and

$$N(\mathbb{E}_u) - N^*(\mathbb{E}_u) = N(\mathbb{K}_u) - N^*(\mathbb{K}_u).$$

Finally we assume that $q \equiv 1 \mod 3$ and we consider the algebraic function field $\mathbb{L}_u = \mathbb{F}_{q^3}(X)(Z)$. We have

$$Z^{3(q-1)} = \frac{-1}{2u}X^{q-1}\left(X^{q-1}+1\right)^q = \frac{-1}{2u}X^{q-1}r_1(X)^q r_2(X)^q \dots r_{\frac{q-1}{2}}(X)^q.$$

Using the same notation and the same arguments as in the case of $\mathbb{K}_u$ above, we conclude that

- $r_{P_0} = \gcd(3(q-1), v_{P_0}(h(x))) = \gcd(3(q-1),(q-1)) = q-1$,
- $r_{Q_i} = \gcd(3(q-1), v_{Q_i}(h(x))) = \gcd(3(q-1), q) = 1$ for $1 \le i \le (q-1)/2$,
- $r_{P_\infty} = \gcd(3(q-1), v_{P_\infty}(h(x))) = \gcd(3(q-1), q^2-1) = q-1$.

Hence $\mathbb{F}_{q^3}$ is the full constant field of $\mathbb{L}_u$. For the genus we have

$$\begin{aligned} g(\mathbb{L}_u) &= 1 + 3(q-1)(-1) \\ &\quad + \frac{1}{2}\left((3(q-1)-(q-1)) + \frac{q-1}{2}(3(q-1)-1)\,2 + (3(q-1)-(q-1))\right) \\ &= 1 + \frac{3(q-1)(q-2)}{2}. \end{aligned}$$

Note that $\mathbb{L}_u$ is an extension of $\mathbb{K}_u$ with $[\mathbb{L}_u : \mathbb{K}_u] = 3$. Moreover if P' is a rational place of $\mathbb{K}_u$ over P_0 or P_∞, then P' is totaly ramified in the extension $\mathbb{L}_u/\mathbb{K}_u$ and hence there is a unique rational place of $\mathbb{L}_u$ over P'. This implies that the relation in (3.1) holds. □

We use some results from [**9**] in the following lemma.

LEMMA 3.3. *We have*

$$N^*(\mathbb{K}_u) = N^*(\mathbb{E}_u) \ge 6(q-1)^2. \tag{3.5}$$

Moreover we have the following:

- *If* $\mathrm{Norm}_{\mathbb{F}_{q^3}/\mathbb{F}_q}(2u) = 1$, *then* $N^*(\mathbb{K}_u) = (q-1)^2\left(3\left\lfloor\frac{q+1}{3}\right\rfloor + q + 1\right)$.
- *If* $\mathrm{Norm}_{\mathbb{F}_{q^3}/\mathbb{F}_q}(u) = -1$ *and* $\mathrm{char}\ \mathbb{F}_q \neq 3$, *then* $N^*(\mathbb{K}_u) = (q-1)^2\left(3\left\lfloor\frac{q+1}{3}\right\rfloor + 1\right)$.

If $\mathrm{char}\ \mathbb{F}_q = 3$ *and* $\mathrm{Norm}_{\mathbb{F}_{q^3}/\mathbb{F}_q}(u) = -1$, *then it is clear that* $\mathrm{Norm}_{\mathbb{F}_{q^3}/\mathbb{F}_q}(2u) = 1$ *and the first formula holds.*

PROOF. Note that $N^*(\mathbb{K}_u)$ is equal to the number of $(x, y) \in \mathbb{F}_{q^3}^* \times \mathbb{F}_{q^3}^*$ such that

$$x^{q^2-1} + x^{q-1} + 2uy^{q-1} = 0.$$

Dividing by x^{q-1} we obtain

$$x^{q^2-q} + 2u\left(\frac{y}{x}\right)^{q-1} = -1.$$

Then making the change of variable

$$(x,y) \mapsto \left(x^q, \frac{y}{x}\right)$$

we get

$$x^{q-1} + 2uy^{q-1} = -1. \tag{3.6}$$

Recall that $N^*(\mathbb{E}_u)$ is the number of $(x,y) \in \mathbb{F}_{q^3}^* \times \mathbb{F}_{q^3}^*$ satisfying (3.6). Hence we obtain that $N^*(\mathbb{K}_u) = N^*(\mathbb{E}_u)$.

Next we introduce a notation which also complies with the one in [**9**]. For $a \in \mathbb{F}_q^*$, let $I_3(a,-1)$ denote the cardinality

$$I_3(a,-1) = \left|\{f \in \mathbb{F}_q[x] : f \text{ is monic, irreducible, } \deg f = 3,\ f(0) = a,\ f(1) = -1\}\right|.$$

Then using [**9**, Equation 3.3] we obtain that

$$N^*(\mathbb{E}_u) = (q-1)^2\left(3I_3(a,-1) + \left|\left\{a_1 \in \mathbb{F}_q^* : a_1^3 = a,\ (a_1+1)^3 = -1\right\}\right|\right), \tag{3.7}$$

where $a = \mathrm{Norm}_{\mathbb{F}_{q^3}/\mathbb{F}_q}(2u)$. Moreover we have

$$I_3(a,-1) \geq 2, \tag{3.8}$$

which follows immediately from [**9**, Equation 3.5] and [**9**, Equation 3.6]. Combining (3.7) and (3.8) we complete the proof of (3.5).

For the rest of the proof we use [**9**, Equation 3.13] and [**9**, Equation 3.17]. Note that

$$(-1-a-1)^3 = -27a \iff (a-1)^2(a+8) = 0 \iff a \in \{1,-8\}. \tag{3.9}$$

Then combining [**9**, Equation 3.13] and [**9**, Equation 3.17] we determine $N^*(\mathbb{K}_u)$ when $\mathrm{Norm}_{\mathbb{F}_{q^3}/\mathbb{F}_q}(2u) = 1$ or $\mathrm{Norm}_{\mathbb{F}_{q^3}/\mathbb{F}_q}(u) = -1$. □

From Lemma 3.1 and Lemma 3.3 we get the following.

COROLLARY 3.4. *Let q be a power of an odd prime and $u \in \mathbb{F}_{q^3}^*$. Then the polynomial $F_u(x) = x^{q^2+q} + ux^2 \in \mathbb{F}_{q^3}[x]$ is not planar if $q \not\equiv 1 \mod 3$.*

It seems difficult to determine $N^*(\mathbb{K}_u)$ exactly when $\mathrm{Norm}_{\mathbb{F}_{q^3}/\mathbb{F}_q} \not\in \{1/2^3, -1\}$. Here we give a related result for such u using [**9**]. First we introduce a new notation. For $a \in \mathbb{F}_q^*$, let (see [**9**, page 634])

$$J_3(-a-2,-a) = \left|\left\{\nu \in \mathbb{F}_q : x^3 + (a+2)x^2 + \nu x + a \in \mathbb{F}_q[x] \text{ is irreducible}\right\}\right|.$$

LEMMA 3.5. *If $\mathrm{Norm}_{\mathbb{F}_{q^3}/\mathbb{F}_q}(2u) \neq 1$ and $\mathrm{Norm}_{\mathbb{F}_{q^3}/\mathbb{F}_q}(u) \neq -1$, then*

$$N^*(\mathbb{K}_u) = 3(q-1)^2 J_3(-a-2,-a), \tag{3.10}$$

where $a = \mathrm{Norm}_{\mathbb{F}_{q^3}/\mathbb{F}_q}(2u)$. Moreover if $a \neq 1$ and $a \neq -8$, then

$$J_3(-a-2,-a) \geq 2. \tag{3.11}$$

PROOF. The proof of (3.10) follows from Lemma 3.3, (3.9), [**9**, Equation 3.13] and [**9**, Equation 3.17]. Combining the lower bound in (3.5) with (3.10) we immediately obtain (3.11). □

The following lemma is useful in connecting $N^*(\mathbb{L}_u)$ and $N^*(\mathbb{K}_u)$.

LEMMA 3.6. *Assume that $q \equiv 1 \mod 3$. Let $\epsilon \in \mathbb{F}_q^*$ with $\epsilon^2 + \epsilon + 1 = 0$. We have*

$$3N^*(\mathbb{K}_u) = N^*(\mathbb{L}_u) + N^*(\mathbb{L}_{\epsilon u}) + N^*(\mathbb{L}_{\epsilon^2 u}).$$

PROOF. Recall that G and H are subgroups with $|G| = q^2 + q + 1$, $|H| = (q^2 + q + 1)/3$ and $G = H \sqcup \epsilon H \sqcup \epsilon^2 H$ (see Lemma 2.3). Let $A \in G$ and consider the three equations

$$z^{3(q-1)} = A, \tag{3.12}$$

$$z^{3(q-1)} = \epsilon A, \text{ and} \tag{3.13}$$

$$z^{3(q-1)} = \epsilon^2 A \tag{3.14}$$

in one variable $z \in \mathbb{F}_{q^3}^*$. As $3(q-1) \mid (q^3 - 1)$, it is clear that the number of solutions $z \in \mathbb{F}_{q^3}^*$ for each of these equations is either 0 or exactly $3(q-1)$. Moreover as $G = H \sqcup \epsilon H \sqcup \epsilon^2 H$ we have

i) (3.12) is solvable if and only if $A \in H$,

ii) (3.13) is solvable if and only if $A \in \epsilon^2 H$,

iii) (3.14) is solvable if and only if $A \in \epsilon H$.

Let $M^*(u)$ denote the number of elements $(x, A) \in \mathbb{F}_{q^3}^* \times G$ satisfying

$$x^{q^2-1} + x^{q-1} + 2uA = 0. \tag{3.15}$$

As $(q-1) \mid (q^3 - 1)$, it is easy to observe that

$$N^*(\mathbb{K}_u) = (q-1)M^*(u). \tag{3.16}$$

Note that $N^*(\mathbb{L}_u)$, $N^*(\mathbb{L}_{\epsilon u})$ and $N^*(\mathbb{L}_{\epsilon^2 u})$ denote the number of solutions $(x, z) \in \mathbb{F}_{q^3}^* \times \mathbb{F}_{q^3}^*$ of the following equations, respectively:

$$x^{q^2-1} + x^{q-1} + 2uz^{3(q-1)} = 0, \tag{3.17}$$

$$x^{q^2-1} + x^{q-1} + 2\epsilon u z^{3(q-1)} = 0, \quad \text{and} \tag{3.18}$$

$$x^{q^2-1} + x^{q-1} + 2\epsilon^2 u z^{3(q-1)} = 0. \tag{3.19}$$

Combining (3.12), (3.13), (3.14), (3.15), (3.17), (3.18), (3.19) and using the arguments following (3.14) we conclude that

$$3(q-1)M^*(u) = N^*(\mathbb{L}_u) + N^*(\mathbb{L}_{\epsilon u}) + N^*(\mathbb{L}_{\epsilon^2 u}). \tag{3.20}$$

The proof follows from (3.16) and (3.20). □

From the results above we conclude the following corollary, which in particular shows that the polynomial $F_u(x)$ is not planar for some values of $u \in \mathbb{F}_{q^3}^*$ when $q \equiv 1 \mod 3$. It also determines the number of rational places of the algebraic function field $\mathbb{L}_u$ for some values of $u \in \mathbb{F}_{q^3}$.

COROLLARY 3.7. *Let q be a power of an odd prime and assume that $q \equiv 1 \mod 3$. Let H be the subgroup of $\mathbb{F}_{q^3}^*$ of order $(q^2 + q + 1)/3$. Let ϵ be an element of $\mathbb{F}_{q^3}^*$, whose multiplicative order is 3.*

- *If* $-u \in H$*, then* $N^*(\mathbb{L}_u) = 3q(q-1)^2$*, the polynomial* $F_u(x) = x^{q^2+q} + ux^2$ *is not planar and*

$$N(\mathbb{L}_u) = \begin{cases} 3q(q-1)^2 + 2(q-1) & \text{if char } \mathbb{F}_q = 7, \\ 3q(q-1)^2 & \text{otherwise.} \end{cases} \tag{3.21}$$

- *If* $2u \in H$*, then* $N^*(\mathbb{L}_u) = 6q(q-1)^2$*, the polynomial* $F_u(x)$ *is not planar and*

$$N(\mathbb{L}_u) = 6q(q-1)^2.$$

- *If* $-u \in (\epsilon H \sqcup \epsilon^2 H)$*, then*

$$N(\mathbb{L}_u) = \begin{cases} 2(q-1) & \text{if char } \mathbb{F}_q = 7, \\ 0 & \text{otherwise.} \end{cases}$$

- *If* $2u \in (\epsilon H \sqcup \epsilon^2 H)$*, then*

$$N(\mathbb{L}_u) = 0.$$

PROOF. Assume that $-u \in H$. Note that $\left\lfloor \frac{q+1}{3} \right\rfloor = \frac{q-1}{3}$ and hence using Lemma 3.3 we obtain that $N^*(\mathbb{K}_u) = q(q-1)^2$. By Theorem 2.5 and Lemma 3.1 we have $N^*(\mathbb{L}_{\epsilon u}) = N^*(\mathbb{L}_{\epsilon^2 u}) = 0$. Hence using Lemma 3.6 we conclude that

$$N^*(\mathbb{L}_u) = 3N^*(\mathbb{K}_u) = 3q(q-1)^2. \tag{3.22}$$

This implies that $F_u(x)$ is not planar by Lemma 3.1.

Note that $\mathrm{Norm}_{\mathbb{F}_{q^3}/\mathbb{F}_q}(2u) = -8$. Then $\mathrm{Norm}_{\mathbb{F}_{q^3}/\mathbb{F}_q}(2u) = -1$ if and only if char $\mathbb{F}_q = 7$. Using (3.22) and Lemma 3.2 we obtain (3.21).

The proof of the statement for the case $2u \in H$ is similar.

If $-u \in (\epsilon H \sqcup \epsilon^2 H)$ or if $2u \in (\epsilon H \sqcup \epsilon^2 H)$, then by Theorem 2.5 or by Theorem 2.6 we get that $N^*(\mathbb{L}_u) = 0$. Then we determine $N(\mathbb{L}_u)$ using Lemma 3.2. □

It remains to consider the planarity of the polynomial $F_u(x)$ when $q \equiv 1 \mod 3$, $\mathrm{Norm}_{\mathbb{F}_{q^3}/\mathbb{F}_q}(2u) \neq 1$ and $\mathrm{Norm}_{\mathbb{F}_{q^3}/\mathbb{F}_q}(u) \neq -1$. We will prove that the polynomial $F_u(x)$ is not planar in the remaining cases in Theorem 4.5 below, using different methods.

Note that Corollary 3.7 determines the exact values of $N^*(\mathbb{L}_u)$ when $q \equiv 1 \mod 3$ and $\mathrm{Norm}_{\mathbb{F}_{q^3}/\mathbb{F}_q}(u) \in \{-1, 1/8\}$. Determining $N^*(\mathbb{L}_u)$ is the same as determining the number $N(\mathbb{L}_u)$ of rational places of $\mathbb{L}_u$ (see (3.1) in Lemma 3.2). Using Lemma 3.5, Lemma 3.6 and Theorem 4.5 below, we will obtain certain bounds on $N^*(\mathbb{L}_u)$ when $q \equiv 1 \mod 3$ and $\mathrm{Norm}_{\mathbb{F}_{q^3}/\mathbb{F}_q}(u) \notin \{-1, 1/8\}$ in Corollary 4.6. As the proof of Corollary 4.6 will use some results from Theorem 4.5, we will present the corollary in Section 4 after the proof of Theorem 4.5.

REMARK 3.8. Apart from the method of Section 4 that we use in order to prove the planarity of the polynomial $F_u(x)$ in the remaining cases when $q \equiv 1 \mod 3$, we have also considered an approach using the Hasse-Weil-Serre bounds. The following approach of using Hasse-Weil-Serre bounds does not work. The lower and the upper Hasse-Weil-Serre bounds for the algebraic function field $\mathbb{L}_u$ are (see [**16**, Theorem 5.3.1])

$$\mathrm{HWS1} = q^3 + 1 - \left(1 + \frac{3(q-1)(q-2)}{2}\right) \lfloor 2q^{3/2} \rfloor,$$

and

$$\text{HWS2} = q^3 + 1 + \left(1 + \frac{3(q-1)(q-2)}{2}\right) \lfloor 2q^{3/2} \rfloor,$$

where $\lfloor 2q^{3/2} \rfloor$ is the integer part of $2q^{3/2}$. Here we use Lemma 3.2 for the genus formula. Unfortunately HWS1 < 0 and HWS2 is much bigger than $2(q-1)$. By Lemmas 3.1 and 3.2, the polynomial $F_u(x)$ is planar if and only if $N(\mathbb{L}_u) = 0$ or $N(\mathbb{L}_u) = 2(q-1)$ depending on whether $\text{Norm}_{\mathbb{F}_{q^3}/\mathbb{F}_q}(2u)$ is -1 or not. Both of these values are in the range in between HWS1 and HWS2. Therefore Hasse-Weil-Serre bounds do not help in deciding the planarity in this question using this approach.

REMARK 3.9. Note that the algebraic function field $\mathbb{E}_u$ is defined by a diagonal equation over finite fields. There are some results in the literature to determine the number of solutions of diagonal equations (see [**15**]). Assume that $q \equiv 1 \mod 3$ and q is a power of an odd prime. The planarity of the polynomial $F_u(x)$ depends on the number of rational places of the function field $\mathbb{L}_u$, and the function fields $\mathbb{E}_u$ and $\mathbb{L}_u$ are related. Therefore it would be interesting to consider the following problem. Give an alternative proof (for example using exponential sums or diagonal equations) of the fact: Assume that q is a power of an odd prime and $q \equiv 1 \mod 3$. Assume also that $u \in \mathbb{F}_{q^3}^*$, $\text{Norm}_{\mathbb{F}_{q^3}/\mathbb{F}_q}(2u) \neq 1$ and $\text{Norm}_{\mathbb{F}_{q^3}/\mathbb{F}_q}(u) = -1$. Then the polynomial $F_u(x) = x^{q^2+q} + ux^2 \in \mathbb{F}_{q^3}[x]$ is not planar. We refer to Theorem 4.5 below, in which we give a proof of this fact using some results from the theory of finite semifields.

4. Completing the classification of planar maps $F_u(x)$

In this section we complete the classification of planar polynomials $F_u(x)$ given in (1.2). By Corollary 3.4, we can assume that $q \equiv 1 \mod 3$ (and q is odd) without loss of generality. Here we use some known results from the theory of finite semifields. Therefore we start by giving some definitions and the results from the theory of finite semifields that we will use below.

A finite *presemifield* is a finite set S with two binary operations $+$ and $*$ satisfying the following axioms:

- $(S, +)$ is an Abelian group with identity 0.
- $a * (b + c) = a * b + a * c$ and $(a + b) * c = a * c + b * c$ for all $a, b, c \in S$.
- If $a * b = 0$, then a or b is 0.

If, in addition to these, we also have

- there exists an element $e \neq 0$ such that $e * a = a = a * e$ for all $a \in S$,

then the presemifield is called a *semifield*.

Presemifields are commutative if $a * b = b * a$ for all $a, b \in S$. The additive group of a finite presemifield is necessarily elementary Abelian. Hence any finite presemifield can be represented by $(\mathbb{F}, +, *)$, where $\mathbb{F}$ is the underlying set of a finite field $\mathbb{F}_{p^n}$ and the addition of the semifield coincides with the one of the finite field. Two finite presemifields $(\mathbb{F}, +, *)$ and $(\mathbb{F}, +, \star)$ are called *isotopic* if there exist linearized permutation polynomials L, M, N over $\mathbb{F}_{p^n}$ such that

$$M(x) \star N(y) = L(x * y) \quad \text{for any } x, y \in \mathbb{F}.$$

Any presemifield $S = (\mathbb{F}, +, *)$ is isotopic to a semifield. Indeed, fix any nonzero $a \in \mathbb{F}$ and define $\star : \mathbb{F} \times \mathbb{F} \to \mathbb{F}$ as follows

$$x * y = (x * a) \star (a * y). \tag{4.1}$$

Then the element $a * a$ is the identity element of $(\mathbb{F}, +, \star)$. In (4.1), if $*$ is commutative then $\star$ is also commutative, although in general isotopism does not preserve commutativity.

Let $S = (\mathbb{F}, +, *)$ be a semifield. The subsets

$$\begin{aligned}\mathcal{N}_\ell(S) &= \{a \in S : (a * x) * y = a * (x * y) \text{ for all } x, y \in S\},\\ \mathcal{N}_m(S) &= \{a \in S : (x * a) * y = x * (a * y) \text{ for all } x, y \in S\},\\ \mathcal{N}_r(S) &= \{a \in S : (x * y) * a = x * (y * a) \text{ for all } x, y \in S\}\end{aligned}$$

are called the *left*, *middle*, and *right nucleus* of S, respectively. It is easy to verify that these are finite fields. The subset

$$\mathcal{N}(S) = \mathcal{N}_\ell(S) \cap \mathcal{N}_m(S) \cap \mathcal{N}_r(S)$$

is called the *nucleus* (or *associative center*) of S. The *center* $\mathcal{C}(S)$ of S is defined by

$$\mathcal{C}(S) = \mathcal{N}(S) \cap \{c \in \mathbb{F} \mid c * a = a * c \text{ for any } a \in \mathbb{F}\},$$

which is again a finite field. It is well known that S is a division algebra over its center. Moreover the nuclei are invariant under isotopisms (see, for example, [**5**]).

Given a planar p-quadratic polynomial f over $\mathbb{F}_{p^n}$, the multiplication $*$ defined by

$$x * y = \frac{1}{2}(f(x + y) - f(x) - f(y)) \tag{4.2}$$

yields a commutative presemifield, which we denote by $S_f = (\mathbb{F}_{p^n}, +, *)$. Conversely, any commutative presemifield $S = (\mathbb{F}_{p^n}, +, \star)$ of odd order p^n gives a planar p-quadratic polynomial f over $\mathbb{F}_{p^n}$ such that S is isotopic to the presemifield S_f obtained by the polynomial f [**5**]. Hence the classification of finite commutative presemifields of an odd order and the one of planar p-quadratic polynomials are equivalent.

For an odd prime p and positive integer e, let $q = p^e$ and recall that $F_u(x)$ is the polynomial $x^{q^2+q} + ux^2$ in $\mathbb{F}_{q^3}[x]$, where $u \in \mathbb{F}_{q^3}^*$. Assume that $F_u(x)$ is planar. Let S be a semifield isotopic to the presemifield obtained from $F_u(x)$ by the operation as in (4.2). Using [**3**, Theorems 4 and 5] we conclude that $\mathbb{F}_q$ is contained in the center of S. This implies that $\mathcal{C}(S)$ is either $\mathbb{F}_{q^3}$ or $\mathbb{F}_q$. In the first case we have that S is isotopic to $\mathbb{F}_{q^3}$. In the second case we have that S is a three dimensional division algebra over its center $\mathbb{F}_q$. It is known that a three dimensional division algebra over $\mathbb{F}_q$ is either a finite field or a generalized twisted field [**13**]. A commutative generalized twisted semifield over $\mathbb{F}_{q^3}$ whose left nucleus is $\mathbb{F}_q$ is isotopic to the presemifield obtained by the planar polynomial x^{q+1} (see [**1**, Theorems 1 and 2], see also [**5**]). Furthermore it is known that if S' is a generalized twisted semifield which is isotopic to a commutative semifield, then S' is also isotopic to a commutative generalized twisted semifield (see [**10**, Proposition 5.3]). It is also clear that S is isotopic to a finite field if and only if S is isotopic to the presemifield obtained by the planar polynomial x^2.

Therefore using the arguments above and [**5**, Theorems 2.6 and 3.5] we conclude that if $F_u(X)$ is planar, then there exists $\mathbb{F}_p$-linearized permutation polynomials L_1, L_2 in $\mathbb{F}_{q^3}[X]$ such that either $L_1 \circ F_u \circ L_2 = X^2$ or $L_1 \circ F_u \circ L_2 = X^{q+1}$. We use this conclusion to complete the classification of the planar maps F_u in the rest of the paper.

Next we give two lemmas which essentially mean that we can consider the $\mathbb{F}_p$-linearized polynomials L_1 and L_2 above as $\mathbb{F}_q$-linearized permutation polynomials, up to the action of an automorphism of $\mathbb{F}_{q^3}$ over $\mathbb{F}_p$ on u, without loss of generality.

LEMMA 4.1. *Let $q = p^e$, where p is an odd prime and $e \geq 1$. Let $A_0, A_1, \ldots, A_{3e-1}, B_0, B_1, \ldots, B_{3e-1}, u \in \mathbb{F}_{q^3}$ such that at least one of $A_0, A_1, \ldots, A_{3e-1}$ is nonzero and the equivalence*

$$\begin{aligned} &x^2 \circ \left(A_0x + A_1x^p + \cdots + A_{3e-1}x^{p^{3e-1}}\right) \\ (4.3) \qquad &\equiv \left(B_0x + B_1x^p + \cdots + B_{3e-1}x^{p^{3e-1}}\right) \circ \left(x^{q^2+q} + ux^2\right) \\ &\qquad \bmod \left(x^{q^3} - x\right) \end{aligned}$$

holds. Then there exists an integer $0 \leq i \leq 3e-1$ and elements $C_0, C_e, C_{2e}, D_0, D_e, D_{2e}, v \in \mathbb{F}_{q^3}$ such that $C_0 \neq 0$ and each of the following properties are satisfied:

- $\left(A_0x + A_1x^p + \cdots + A_{3e-1}x^{p^{3e-1}}\right) \circ x^{p^i} \equiv C_0x + C_ex^{p^e} + C_{2e}x^{p^{2e}} \bmod \left(x^{q^3} - x\right)$,
- $\left(B_0x + B_1x^p + \cdots + B_{3e-1}x^{p^{3e-1}}\right) \circ x^{p^i} \equiv D_0x + D_ex^{p^e} + D_{2e}x^{p^{2e}} \bmod \left(x^{q^3} - x\right)$,
- $u = v^{p^i}$, *and*
- $x^2 \circ \left(C_0x + C_ex^{p^e} + C_{2e}x^{2e}\right) \equiv \left(D_0x + D_ex^{p^e} + D_{2e}x^{2e}\right) \circ \left(x^{q^2+q} + vx^2\right)$.

PROOF. We have

$$\begin{aligned} &\left(A_0x + A_1x^p + \cdots + A_{3e-1}x^{p^{3e-1}}\right) \circ x^p \\ (4.4) \qquad &\equiv \left(A_{3e-1}x + A_0x^p + \cdots + A_{3e-2}x^{p^{3e-1}}\right) \quad \bmod \left(x^{q^3} - x\right). \end{aligned}$$

Let σ be the cyclic shift permutation acting on the set of $3e$-tuples of $\mathbb{F}_{q^3}$ via

$$\sigma(a_0, a_1, \ldots, a_{3e-1}) = (a_{3e-1}, a_0, \ldots, a_{3e-2}).$$

The ordered coefficients $(A_{3e-1}, A_0, \ldots, A_{3e-2})$ of the reduced polynomial in the right hand side of (4.4) is nothing but $\sigma(A_0, A_1, \ldots, A_{3e-1})$. Let σ^0 denote the identity operator on the set of $3e$-tuples of $\mathbb{F}_{q^3}$, and for $1 \leq j \leq 3e-1$ let σ^j denote the operator obtained by composition of σ with itself j times. As not all of $A_0, A_1, \ldots, A_{3e-1}$ are zero, there exists $0 \leq j \leq 3e-1$ such that the first entry of $\sigma^j(A_0, A_1, \ldots, A_{3e-1})$ is nonzero. Let i be the smallest integer with $0 \leq i \leq 3e-1$ such that the first entry of $\sigma^i(A_0, A_1, \ldots, A_{3e-1})$ is nonzero.

We define $C_0, C_1, \ldots, C_{3e-1}, D_0, D_1, \ldots, D_{3e-1} \in \mathbb{F}_{q^3}$ as

$$\begin{aligned} &C_0x + C_1x^p + \cdots + C_{3e-1}x^{p^{3e-1}} \\ (4.5) \qquad &\equiv \left(A_0x + A_1x^p + \cdots + A_{3e-1}x^{p^{3e-1}}\right) \circ x^{p^i} \quad \bmod \left(x^{q^3} - x\right), \end{aligned}$$

and

$$\begin{aligned} &D_0x + D_1x^p + \cdots + D_{3e-1}x^{p^{3e-1}} \\ (4.6) \qquad &\equiv \left(B_0x + B_1x^p + \cdots + D_{3e-1}x^{p^{3e-1}}\right) \circ x^{p^i} \quad \bmod \left(x^{q^3} - x\right). \end{aligned}$$

Note that $C_0 \neq 0$. Moreover we define $v \in \mathbb{F}_{q^3}$ by the identity

$$(4.7) \qquad u = v^{p^i}.$$

We have

$$(4.8) \qquad \left(x^{q^2+q} + ux^2\right) \circ x^{p^i} = x^{p^i} \circ \left(x^{q^2+q} + vx^2\right),$$

where we use (4.7). Applying x^{p^i} as a composition from the right to the both sides of the equivalence in (4.3) and using (4.5), (4.6) and (4.8) we obtain that

$$\begin{aligned} &x^2 \circ \left(C_0 x + C_1 x^p + \cdots + C_{3e-1} x^{p^{3e-1}}\right) \\ \equiv &\left(D_0 x + D_1 x^p + \cdots + D_{3e-1} x^{p^{3e-1}}\right) \circ \left(x^{q^2+q} + v x^2\right) \\ &\bmod \left(x^{q^3} - x\right). \end{aligned} \tag{4.9}$$

For the left hand side of (4.9) we have

$$x^2 \circ \left(C_0 x + C_1 x^p + \cdots + C_{3e-1} x^{p^{3e-1}}\right) = \sum_{j_1=0}^{3e-1} \sum_{j_2=0}^{3e-1} C_{j_1} C_{j_2} x^{p^{j_1} + p^{j_2}}. \tag{4.10}$$

For the right hand side of (4.9) we have

$$\begin{aligned} &\left(D_0 x + D_1 x^p + \cdots + D_{3e-1} x^{p^{3e-1}}\right) \circ \left(x^{q^2+q} + v x^2\right) \\ \equiv &\sum_{k=0}^{e-1} \left(D_k x^{p^{2e+k} + p^{e+k}} + D_k v^{p^k} x^{2p^k}\right) \\ &+ \sum_{k=0}^{e-1} \left(D_{e+k} x^{p^{2e+k} + p^k} + D_{e+k} v^{p^{e+k}} x^{2p^{e+k}}\right) \\ &+ \sum_{k=0}^{e-1} \left(D_{2e+k} x^{p^{e+k} + p^k} + D_{2e+k} v^{p^{2e+k}} x^{2p^{2e+k}}\right) \quad \bmod \left(x^{q^3} - x\right). \end{aligned} \tag{4.11}$$

Note that the degrees of the terms in the right hand side of (4.10) are mutually distinct and they are strictly smaller than p^{3e}. This property also holds for the right hand side of (4.11). Therefore using (4.9) we conclude that for each integer $0 \le \ell \le p^{3e} - 1$, the coefficients of the terms of degree ℓ in the right hand sides of (4.10) and (4.11) are equal to each other.

Let j be an arbitrary integer with $0 \le j \le e-1$. We consider the coefficients of the terms x^{1+p^j}, $x^{1+p^{j+e}}$ and $x^{1+p^{j+2e}}$ in the right hand sides of (4.10) and (4.11). The corresponding coefficients in the right hand side of (4.10) are $2C_0C_j$, $2C_0C_{j+e}$ and $2C_0C_{j+2e}$, respectively. The corresponding coefficients in the right hand side of (4.11) are all 0, as there are no such terms. Using the facts that $C_0 \neq 0$ and $\operatorname{char} \mathbb{F}_q \neq 2$ we conclude that $C_j = C_{j+e} = C_{j+2e} = 0$. Therefore we simplify (4.10) as

$$\begin{aligned} &x^2 \circ \left(C_0 x + C_1 x^p + \cdots + C_{3e-1} x^{p^{3e-1}}\right) \\ = &C_0^2 x^2 + C_e^2 x^{2p^e} + C_{2e}^2 x^{2p^{2e}} \\ &+ 2c_0 C_e x^{p^e+1} + 2C_0 C_{2e} x^{p^{2e}+1} + 2C_e C_{2e} x^{p^{2e}+p^e}. \end{aligned} \tag{4.12}$$

Let k be an arbitrary integer with $0 \le k \le e-1$. We consider the coefficients of the terms $x^{p^{2e+k}+p^{e+k}}$, $x^{p^{2e+k}+p^k}$ and $x^{p^{e+k}+p^k}$ in the right hand sides of (4.11) and (4.12). The corresponding coefficients in the right hand side of (4.11) are D_k, D_{e+k} and D_{2e+k}, respectively. The corresponding coefficients in the right hand side of (4.12) are all 0, as there are no such terms. Hence we conclude that $D_k = D_{k+e} = D_{k+2e} = 0$. This completes the proof. □

Using similar but more involved arguments than the ones in the proof of Lemma 4.1 we also prove the following lemma, in which x^2 in (4.3) is changed to x^{q+1}.

LEMMA 4.2. *Let $q = p^e$, where p is an odd prime and $e \geq 1$. Let $A_0, A_1, \ldots, A_{3e-1}, B_0, B_1, \ldots, B_{3e-1}, u \in \mathbb{F}_{q^3}$ such that at least one of $A_0, A_1, \ldots, A_{3e-1}$ is nonzero and the equivalence*

$$x^{q+1} \circ \left(A_0 x + A_1 x^p + \cdots + A_{3e-1} x^{p^{3e-1}}\right) \equiv \left(B_0 x + B_1 x^p + \cdots + B_{3e-1} x^{p^{3e-1}}\right) \circ \left(x^{q^2+q} + ux^2\right) \mod \left(x^{q^3} - x\right)$$

holds. Then there exists an integer $0 \leq i \leq 3e-1$ and elements $C_0, C_e, C_{2e}, D_0, D_e, D_{2e}, v \in \mathbb{F}_{q^3}$ such that $C_0 \neq 0$ and each of the following properties are satisfied:

- $\left(A_0 x + A_1 x^p + \cdots + A_{3e-1} x^{p^{3e-1}}\right) \circ x^{p^i} \equiv C_0 x + C_e x^{p^e} + C_{2e} x^{p^{2e}} \mod \left(x^{q^3} - x\right)$,
- $\left(B_0 x + B_1 x^p + \cdots + B_{3e-1} x^{p^{3e-1}}\right) \circ x^{p^i} \equiv D_0 x + D_e x^{p^e} + D_{2e} x^{p^{2e}} \mod \left(x^{q^3} - x\right)$,
- $u = v^{p^i}$, *and*
- $x^{q+1} \circ \left(C_0 x + C_e x^{p^e} + C_{2e} x^{2e}\right) \equiv \left(D_0 x + D_e x^{p^e} + D_{2e} x^{2e}\right) \circ \left(x^{q^2+q} + vx^2\right)$.

PROOF. We start as in the proof of Lemma 4.1 and define the integer i and the elements $C_0, C_1, \ldots, C_{3e-1}, D_0, D_1, \ldots, D_{3e-1}, v \in \mathbb{F}_{q^3}$ in the same way. Then we have (compare with (4.9))

$$\begin{aligned} & x^{q+1} \circ \left(C_0 x + C_1 x^p + \cdots + C_{3e-1} x^{p^{3e-1}}\right) \\ \equiv & \left(D_0 x + D_1 x^p + \cdots + D_{3e-1} x^{p^{3e-1}}\right) \circ \left(x^{q^2+q} + ux^2\right) \\ & \mod \left(x^{q^3} - x\right), \end{aligned} \tag{4.13}$$

and $C_0 \neq 0$. Let j be an arbitrary integer with $0 \leq j \leq e-1$. We will prove the following:

$$\text{Claim: } C_j = C_{j+e} = C_{j+2e} = 0. \tag{4.14}$$

The part of the proof in Lemma 4.1 that corresponds to (4.14) is simpler than the Claim in (4.14) here.

For the left hand side of (4.14) we have

$$\begin{aligned} & x^{p^e+1} \circ \left(C_0 x + C_1 x^p + \cdots + C_{3e-1} x^{p^{3e-1}}\right) \\ \equiv & \sum_{j_1=0}^{3e-1} \sum_{j_2=0}^{3e-1} C_{j_1} C_{j_2}^{p^e} x^{p^{j_1} + p^{j_2+e}} \mod \left(x^{q^3} - x\right). \end{aligned} \tag{4.15}$$

For the right hand side of (4.14) we again have (see (4.11))

$$\begin{aligned} & \left(D_0 x + D_1 x^p + \cdots + D_{3e-1} x^{p^{3e-1}}\right) \circ \left(x^{q^2+q} + vx^2\right) \\ \equiv & \sum_{k=0}^{e-1} \left(D_k x^{p^{2e+k} + p^{e+k}} + D_k v^{p^k} x^{2p^k}\right) \\ & + \sum_{k=0}^{e-1} \left(D_{e+k} x^{p^{2e+k} + p^k} + D_{e+k} v^{p^{e+k}} x^{2p^{e+k}}\right) \\ & + \sum_{k=0}^{e-1} \left(D_{2e+k} x^{p^{e+k} + p^k} + D_{2e+k} v^{p^{2e+k}} x^{2p^{2e+k}}\right) \mod \left(x^{q^3} - x\right). \end{aligned} \tag{4.16}$$

The equivalence in (4.13) implies that the coefficients of the same terms in the right hand sides of (4.15) and (4.16) are the same. The difficulty here is that the coefficients of the terms in the right hand side of (4.15) are more complicated than the ones in the right hand side of (4.10) in the proof of Lemma 4.1 and this makes the current proof more complicated.

Let j be an arbitrary integer with $0 \leq j \leq e-1$. Firstly we consider the coefficients of the terms x^{1+p^j}, $x^{1+p^{j+e}}$ and $x^{1+p^{j+2e}}$ in the right hand sides of (4.15) and (4.16). There are no such terms in the right hand side of (4.16). Considering these terms also in the right hand side of (4.15) we obtain the following identities:

$$\begin{aligned} \text{Coefficient of } x^{1+p^j}: \quad & C_0 C_{j+2e}^{p^e} + C_j C_{2e}^{p^e} &= 0, \\ \text{Coefficient of } x^{1+p^{j+e}}: \quad & C_0 C_j^{p^e} + C_{j+e} C_{2e}^{p^e} &= 0, \\ \text{Coefficient of } x^{1+p^{j+2e}}: \quad & C_0 C_{j+e}^{p^e} + C_{j+2e} C_{2e}^{p^e} &= 0. \end{aligned}$$

As $C_0 \neq 0$, then we get that

$$C_{j+2e}^q = -\frac{C_{2e}^q}{C_0} C_j, \quad C_j^q = -\frac{C_{2e}^q}{C_0} C_{j+e}, \quad C_{j+e}^q = -\frac{C_{2e}^q}{C_0} C_{j+2e}.$$

If $C_{2e} = 0$, or $C_j = 0$, or $C_{j+e} = 0$ or $C_{j+2e} = 0$, then by the identities above we conclude that $C_j = C_{j+e} + C_{j+2e}$, which proves the Claim in (4.14). Hence, for the remaining case in the proof of the Claim, we assume that $C_{2e} \neq 0$, $C_j \neq 0$, $C_{j+e} \neq 0$, and $C_{j+2e} \neq 0$. Let $c_1 = -\frac{C_{2e}^q}{C_0}$. Then we have

$$C_{j+2e}^q = c_1 C_j, \quad C_j^q = c_1 C_{j+e}, \quad C_{j+e}^q = c_1 C_{j+2e}.$$

These identities imply that

$$C_{j+2e}^{q^2} = c_1^q C_j^q = c_1^{q+1} C_{j+e}, \quad C_{j+2e} = c_1^{q^2+q} C_{j+e}^q = c_1^{q^2+q+1} C_{j+2e},$$

and hence $c_1^{q^2+q+1} = 1$. Then we have

$$\text{Norm}_{\mathbb{F}_{q^3}/\mathbb{F}_q}\left(\frac{C_{2e}^q}{C_0}\right) = \text{Norm}_{\mathbb{F}_{q^3}/\mathbb{F}_q}\left(\frac{C_{2e}}{C_0}\right) = -1. \tag{4.17}$$

Secondly we consider the coefficients of the terms $x^{p^e+p^j}$, $x^{p^e+p^{j+e}}$ and $x^{p^e+p^{j+2e}}$ in the right hand sides of (4.15) and (4.16). Again there are no such terms in the right hand side of (4.16). Considering these terms also in the right hand side of (4.15) we obtain the following identities:

$$\begin{aligned} \text{Coefficient of } x^{p^e+p^j}: \quad & C_e C_{j+2e}^{p^e} + C_j C_0^{p^e} &= 0, \\ \text{Coefficient of } x^{p^e+p^{j+e}}: \quad & C_e C_j^{p^e} + C_{j+e} C_0^{p^e} &= 0, \\ \text{Coefficient of } x^{p^e+p^{j+2e}}: \quad & C_e C_{j+e}^{p^e} + C_{j+2e} C_0^{p^e} &= 0. \end{aligned}$$

Then we have

$$C_j = -\frac{C_e}{C_0^q} C_{j+2e}^q, \quad C_{j+e} = -\frac{C_e}{C_0^q} C_j^q, \quad C_{j+2e} = -\frac{C_e}{C_0^q} C_{j+2e}^q.$$

If $C_e = 0$, then by the identities above we conclude that $C_j = C_{j+e} + C_{j+2e}$, which proves the Claim in (4.14). Hence, for the remaining case in the proof of the Claim, we further assume that $C_e \neq 0$. Let $c_2 = -\frac{C_e}{C_0^q}$. Then we have

$$C_j = c_2 C_{j+2e}^q, \quad C_{j+e} = c_2 C_0^q C_j^q, \quad C_{j+2e} = c_2 C_{j+2e}^q.$$

These identities imply that

$$C_j = c_2^{1+q} C_{j+e}^{q^2}, \quad C_{j+e} = c_2 C_j^q = c_2^{q^2+q+1} C_{j+e},$$

and hence $c_2^{q^2+q+1} = 1$. Then we have

$$\text{Norm}_{\mathbb{F}_{q^3}/\mathbb{F}_q}\left(\frac{C_e}{C_0^q}\right) = \text{Norm}_{\mathbb{F}_{q^3}/\mathbb{F}_q}\left(\frac{C_e}{C_0}\right) = -1. \tag{4.18}$$

Thirdly we consider the coefficients of the terms $x^{p^{2e}+p^j}$, $x^{p^{2e}+p^{j+e}}$ and $x^{p^{2e}+p^{j+2e}}$ in the right hand sides of (4.15) and (4.16). Again there are no such terms in the right hand side of (4.16). Considering these terms also in the right hand side of (4.15) we obtain the following identities:

$$\begin{array}{llcl} \text{Coefficient of } x^{p^{2e}+p^j}: & C_{2e}C_{j+2e}^{p^e} + C_j C_e^{p^e} & = & 0, \\ \text{Coefficient of } x^{p^{2e}+p^{j+e}}: & C_{2e}C_j^{p^e} + C_{j+e} C_e^{p^e} & = & 0, \\ \text{Coefficient of } x^{p^{2e}+p^{j+2e}}: & C_{2e}C_{j+e}^{p^e} + C_{j+2e} C_e^{p^e} & = & 0. \end{array}$$

Then, as $C_e \neq 0$, we have

$$C_j = -\frac{C_{2e}}{C_e^q} C_{j+2e}^q, \quad C_{j+e} = -\frac{C_{2e}}{C_e^q} C_j^q, \quad C_{j+2e} = -\frac{C_{2e}}{C_e^q} C_{j+2e}^q.$$

Let $c_3 = -\dfrac{C_{2e}}{C_e^q}$. Then we have

$$C_j = c_3 C_{j+2e}^q, \quad C_{j+e} = c_3 C_0^q C_j^q, \quad C_{j+2e} = c_3 C_{j+2e}^q.$$

These identities imply that

$$C_j = c_3^{1+q} C_{j+e}^{q^2}, \quad C_{j+e} = c_3^{q^2+q+1} C_{j+e},$$

and hence $c_3^{q^2+q+1} = 1$. Then we have

$$\text{Norm}_{\mathbb{F}_{q^3}/\mathbb{F}_q}\left(\frac{C_{2e}}{C_e^q}\right) = \text{Norm}_{\mathbb{F}_{q^3}/\mathbb{F}_q}\left(\frac{C_{2e}}{C_e}\right) = -1. \tag{4.19}$$

Combining (4.17), (4.18) and (4.19) we conclude that

$$\text{Norm}_{\mathbb{F}_{q^3}/\mathbb{F}_q}\left(\frac{C_{2e}}{C_0}\frac{C_0}{C_e}\frac{C_e}{C_{2e}}\right) = (-1)^3 = -1,$$

which is a contradiction as $\dfrac{C_{2e}}{C_0}\dfrac{C_0}{C_e}\dfrac{C_e}{C_{2e}} = 1$. This completes the proof of the Claim in (4.14).

Using the Claim in (4.14) we simplify (4.14) as

$$\begin{aligned} & x^{p^e+1} \circ \left(C_0 x + C_1 x^p + \cdots + C_{3e-1} x^{p^{3e-1}}\right) \\ & = x^{p^e+1} \circ \left(C_0 x + C_e x^{p^e} + C_{2e} x^{p^{2e}}\right) \\ & \equiv C_0 C_{2e}^{p^e} x^2 + C_e C_0^{p^e} x^{2p^e} + C_{2e} C_e^{p^e} x^{2p^{2e}} \\ & \quad + x^{p^e+1}\left(C_0^{1+p^e} + C_e C_{2e}^{p^e}\right) + x^{p^{2e}+1}\left(C_0 C_e^{p^e} + C_{2e}^{1+p^e}\right) \\ & \quad + x^{p^{2e}+p^e}\left(C_e^{1+p^e} + C_{2e} C_0^{p^e}\right) \mod \left(x^{3q} - x\right). \end{aligned} \tag{4.20}$$

We conclude as in the proof of Lemma 4.1. Let k be an arbitrary integer with $0 \le k \le e-1$. We consider the coefficients of the terms $x^{p^{2e+k}+p^{e+k}}$, $x^{p^{2e+k}+p^k}$ and $x^{p^{e+k}+p^k}$ in the right hand sides of (4.16) and (4.20). The corresponding coefficients in the right hand side of (4.16) are D_k, D_{e+k} and D_{2e+k}, respectively. The corresponding coefficients in the right hand side of (4.20) are all 0. Hence we conclude that $D_k = D_{k+e} = D_{k+2e} = 0$. This completes the proof. $\square$

The following two technical lemmas are important in obtaining the main result of this section.

LEMMA 4.3. *Let $q \equiv 1 \mod 3$ and $u \in \mathbb{F}_{q^3}^*$. Assume that there exist linearized polynomials $A_0x + A_1x^q + A_2x^{q^2}, B_0x + B_1x^q + B_2x^{q^2} \in \mathbb{F}_{q^3}[x]$, which are permutation polynomials on $\mathbb{F}_{q^3}$ and which also satisfy*

$$\begin{aligned} &x^2 \circ \left(A_0x + A_1x^q + A_2x^{q^2}\right) \\ &= \left(B_0x + B_1x^q + B_2x^{q^2}\right) \circ \left(x^{q^2+q} + ux^2\right) \mod \left(x^{q^3} - x\right). \end{aligned} \tag{4.21}$$

Then $\mathrm{Norm}_{\mathbb{F}_{q^3}/\mathbb{F}_q}(u) = \dfrac{1}{2^3}$.

PROOF. Using (4.21) and considering the coefficients of the polynomials in the both sides we obtain that

$$\begin{array}{lll} A_0^2 = B_0u, & A_1^2 = B_1u^q, & A_2^2 = B_2u^{q^2}, \\ 2A_0A_1 = B_2, & 2A_0A_2 = B_1, & 2A_1A_2 = B_0. \end{array} \tag{4.22}$$

Then we also have

$$A_0^2 = 2A_1A_2u, \qquad A_1^2 = 2A_0A_2u^q, \qquad A_2^2 = 2A_0A_1u^{q^2}. \tag{4.23}$$

As $u \neq 0$ and the polynomials $A_0x + A_1x^q + A_2x^{q^2}, B_0x + B_1x^q + B_2x^{q^2} \in \mathbb{F}_{q^3}[x]$ are permutation polynomials on $\mathbb{F}_{q^3}$, we obtain that $A_0 \neq 0$, $A_1 \neq 0$ and $A_2 \neq 0$. For example if $A_0 = 0$, then using (4.22) and $u \neq 0$ we get that $B_0 = 0$, $B_2 = 0$ and $B_1 = 0$. This means that the polynomial $B_0x + B_1x^q + B_2x^{q^2}$ is the zero polynomial, which is a contradiction to the assumption that $B_0x + B_1x^q + B_2x^{q^2}$ is a permutation polynomial on $\mathbb{F}_{q^3}$. Hence $A_0 \neq 0$ and similarly we conclude that $A_1 \neq 0$ and $A_2 \neq 0$.

Multiplying the equations in (4.23) we obtain that

$$A_0^2A_1^2A_2^2 = 2^3A_0^2A_1^2A_2^2u^{q^2+q+1}.$$

This implies that $\mathrm{Norm}_{\mathbb{F}_{q^3}/\mathbb{F}_q}(u) = \dfrac{1}{2^3}$. $\square$

LEMMA 4.4. *Let $q \equiv 1 \mod 3$ and $u \in \mathbb{F}_{q^3}^*$. Assume that there exist linearized polynomials $A_0x + A_1x^q + A_2x^{q^2}, B_0x + B_1x^q + B_2x^{q^2} \in \mathbb{F}_{q^3}[x]$, which are permutation polynomials on $\mathbb{F}_{q^3}$ and which also satisfy*

$$\begin{aligned} &x^{q+1} \circ \left(A_0x + A_1x^q + A_2x^{q^2}\right) \\ &= \left(B_0x + B_1x^q + b_2x^{q^2}\right) \circ \left(x^{q^2+q} + ux^2\right) \mod \left(x^{q^3} - x\right). \end{aligned} \tag{4.24}$$

Then $\mathrm{Norm}_{\mathbb{F}_{q^3}/\mathbb{F}_q}(u) = -1$.

PROOF. We start as in the proof of Lemma 4.3. Using (4.24) and considering the coefficients of the polynomials in the both sides we obtain that

$$\begin{aligned} &A_0A_2^q = B_0u, && A_1A_0^q = B_1u^q, && A_1^qA_2 = B_2u^{q^2}, \\ &A_0^{q+1} + A_1A_2^q = B_2, && A_0A_1^q + A_2^{q+1} = B_1, && A_1^{q+1} + A_2A_0^q = B_0. \end{aligned} \tag{4.25}$$

As $u \neq 0$ and the polynomials $A_0x + A_1x^q + A_2x^{q^2}, B_0x + B_1x^q + B_2x^{q^2} \in \mathbb{F}_{q^3}[x]$ are permutation polynomials on $\mathbb{F}_{q^3}$, we obtain that $A_0 \neq 0$, $A_1 \neq 0$ and $A_2 \neq 0$. For example if $A_0 = 0$, then using (4.25) and $u \neq 0$ we get that $B_0 = 0$, $B_1 = 0$, $A_1 = 0$ and $B_2 = 0$. This means that the polynomial $B_0x+B_1x^q+B_2x^{q^2}$ is the zero polynomial, which is a contradiction to the assumption that $B_0x + B_1x^q + B_2x^{q^2}$ is a permutation polynomial on $\mathbb{F}_{q^3}$. Hence $A_0 \neq 0$ and similarly we conclude that $A_1 \neq 0$ and $A_2 \neq 0$.

Putting $A_0' = 1$, $A_1' = \frac{A_1}{A_0}$, $A_2' = \frac{A_2}{A_0}$, $B_0' = \frac{B_0}{A_0^{q+1}}$, $B_1' = \frac{B_1}{A_0^{q+1}}$, $B_2' = \frac{B_2}{A_0^{q+1}}$ and dividing the equations in (4.25) by A_0^{q+1} we reach to the equations which are obtained from the equations in (4.25) just by changing $A_0, A_1, A_2, B_0, B_1, B_2$ to $A_0' = 1, A_1', A_2', B_0', B_1', B_2'$, respectively. This implies that we can assume $A_0 = 1$ in (4.25) without loss of generality.

Assume that $A_0 = 1$. Using (4.25) we obtain

$$A_2^q = \left(A_2 + A_1^{q+1}\right)u, \tag{4.26}$$

$$A_1 = \left(A_1^q + A_2^{q+1}\right)u^q, \tag{4.27}$$

$$A_1^qA_2 = (1 + A_1A_2^q)\,u^{q^2}. \tag{4.28}$$

Taking q-power of (4.26) and comparing with (4.27) we get

$$u^q = \frac{A_2^{q^2}}{A_2^q + A_1^{q^2+q}} = \frac{A_1}{A_1^q + A_2^{q+1}}$$

and hence

$$A_2^{q^2+q+1} + A_2^{q^2}A_1^q = A_2^qA_1 + A_1^{q^2+q+1}. \tag{4.29}$$

Putting $A_2^qA_1 = a$ we have $A_2^{q^2}A_1^q = a^q$ and using (4.29) we obtain that

$$a^q - a = A_1^{q^2+q+1} - A_2^{q^2+q+1} \in \mathbb{F}_q.$$

In particular the equation

$$y^q - y = A_1^{q^2+q+1} - A_2^{q^2+q+1}$$

in one variable $y \in \mathbb{F}_{q^3}$ is solvable. Using Hilbert's Theorem 90 (see [**12**, Theorem 2.25]) we conclude that

$$0 = \mathrm{Tr}_{\mathbb{F}_{q^3}/\mathbb{F}_q}\left(A_1^{q^2+q+1} - A_2^{q^2+q+1}\right) = 3\left(A_1^{q^2+q+1} - A_2^{q^2+q+1}\right)$$

and hence

$$a = A_2^qA_1 \in \mathbb{F}_q \quad \text{and} \quad A_1^{q^2+q+1} = A_2^{q^2+q+1}. \tag{4.30}$$

Combining (4.26) and (4.30) we get that

$$\text{(4.31)}\quad \frac{a}{A_1} = \left(\frac{a}{A_1^{q^2}} + A_1^{q+1}\right) u \text{ and equivalently } A_1^{q^2-1} = \frac{a + A_1^{q^2+q+1}}{a} u.$$

Combining (4.28) and (4.30) we get that

$$\text{(4.32)}\qquad A_1^q \frac{a}{A_1^{q^2}} = (1+a)u^{q^2} \text{ and equivalently } A_1^{q^2-1} = \frac{a+1}{a} u.$$

Comparing (4.31) and (4.32) we immediately obtain

$$\text{(4.33)}\qquad A_1^{q^2+q+1} = 1.$$

From (4.30) and (4.33) we have

$$\text{(4.34)}\qquad a^3 = (A_2^q A_1)^{q^2+q+1} = A_2^{q^2+q+1} A_1^{q^2+q+1} = 1.$$

Next we show that $a \neq 1$. Assume the contrary that $a = 1$. Recall that $x + A_1 x^q + A_2 x^{q^2} \in \mathbb{F}_{q^3}[x]$ is a permutation polynomial on $\mathbb{F}_{q^3}$. Using [**12**, Theorem 3.9 and page 362] we conclude that the matrix

$$\text{(4.35)}\qquad \begin{bmatrix} 1 & A_1 & A_2 \\ A_2^q & 1 & A_1^q \\ A_1^{q^2} & A_2^{q^2} & 1 \end{bmatrix}$$

must be nonsingular. By elementary row operations on (4.35) we obtain the matrix

$$\begin{bmatrix} 1 & A_1 & A_2 \\ 0 & 1 - A_1 A_2^q & A_1^q - A_2^{q+1} \\ 0 & A_2^{q^2} - A_1^{q^2+1} & 1 - A_1^{q^2} A_2 \end{bmatrix},$$

whose determinant is

$$\begin{aligned} &(1 - A_1 A_2^q)(1 - A_1^{q^2} A_2) - (A_1^q - A_2^{q+1})(A_2^{q^2} - A_1^{q^2+1}) \\ &= 1 - A_1^{q^2} A_2 - A_1 A_2^q + A_1^{q^2+1} A_2^{q+1} - (A_1^q A_2^{q^2} - A_1^{q^2+q+1} - A_2^{q^2+q+1} + A_1^{q^2+1} A_2^{q+1}) \\ &= 1 - a - a - (a - 1 - 1) = 3 - 3a = 0. \end{aligned}$$

Here we use the facts $A_1 A_2^q = a \in \mathbb{F}_q$, $A_1^{q^2+q+1} = A_2^{q^2+q+1} = 1$ and the assumption $a = 1$. This is contradiction to the nonsingularity of the matrix in (4.35) and we conclude that $a \neq 1$. Therefore using (4.32) and (4.34) we obtain that

$$A_1^{q^2-1} = \frac{a+1}{a} u \text{ and } a^2 + a + 1 = 0.$$

Then we have

$$\mathrm{Norm}_{\mathbb{F}_{q^3}/\mathbb{F}_q}\left(A_1^{q^2-1}\right) = \mathrm{Norm}_{\mathbb{F}_{q^3}/\mathbb{F}_q}\left(\frac{a+1}{a}\right) \mathrm{Norm}_{\mathbb{F}_{q^3}/\mathbb{F}_q}(u),$$

$$\mathrm{Norm}_{\mathbb{F}_{q^3}/\mathbb{F}_q}\left(A_1^{q^2-1}\right) = \mathrm{Norm}_{\mathbb{F}_{q^3}/\mathbb{F}_q}\left(\frac{A_1^{q^2}}{A_1}\right) = 1,$$

$$\mathrm{Norm}_{\mathbb{F}_{q^3}/\mathbb{F}_q}\left(\frac{a+1}{a}\right) = \mathrm{Norm}_{\mathbb{F}_{q^3}/\mathbb{F}_q}\left(\frac{-a^2}{a}\right) = \mathrm{Norm}_{\mathbb{F}_{q^3}/\mathbb{F}_q}(-1)a^3 = -1,$$

and hence $\mathrm{Norm}_{\mathbb{F}_{q^3}/\mathbb{F}_q}(u) = -1$. This completes the proof. □

Now we are ready to give the main result of our paper.

THEOREM 4.5. *Let q be a power of an odd prime and $u \in \mathbb{F}_{q^3}^*$. Let $F_u(x)$ be the polynomial $x^{q^2+q} + ux^2$ in $\mathbb{F}_{q^3}[x]$. If $q \equiv 1 \mod 3$, then let G and H be the subgroups of $\mathbb{F}_{q^3}^*$ with $|G| = q^2+q+1$ and $|H| = (q^2+q+1)/3$. Then the polynomial $F_u(x)$ is planar if and only if $q \equiv 1 \mod 3$ and*

$$u \in \left(-(G \setminus H) \cup \frac{1}{2}(G \setminus H)\right). \tag{4.36}$$

Moreover we have the followings: Assume that $q \equiv 1 \mod 3$.

- *The polynomial $F_u(x)$ is equivalent to x^2 if and only if $u \in \frac{1}{2}(G \setminus H)$.*
- *The polynomial $F_u(x)$ is equivalent to x^{q+1} if and only if $u \in -(G \setminus H)$.*

PROOF. We can assume that $q \equiv 1 \mod 3$ (see Corollary 3.4). Let $q = p^e$, where p is the characteristic of $\mathbb{F}_q$. Note that for $\alpha \in \mathbb{F}_{q^3}^*$ we have $\alpha \in H$ if and only if $\alpha^{(q^2+q+1)/3} = 1$. Then it is easy to observe that

$$\alpha \in H \Longleftrightarrow \alpha^p \in H. \tag{4.37}$$

Similarly we have that

$$\alpha \in G \Longleftrightarrow \alpha^p \in G. \tag{4.38}$$

Let i be an integer with $0 \leq i \leq 3e-1$ and v be the element of $\mathbb{F}_{q^3}^*$ such that $u = v^{p^i}$. As $1/8, -1 \in \mathbb{F}_p$ we also have that

$$\mathrm{Norm}_{\mathbb{F}_{q^3}/\mathbb{F}_q}(u) = \frac{1}{2^3} \Longleftrightarrow \mathrm{Norm}_{\mathbb{F}_{q^3}/\mathbb{F}_q}(v) = \frac{1}{2^3}, \tag{4.39}$$

and

$$\mathrm{Norm}_{\mathbb{F}_{q^3}/\mathbb{F}_q}(u) = -1 \Longleftrightarrow \mathrm{Norm}_{\mathbb{F}_{q^3}/\mathbb{F}_q}(v) = -1. \tag{4.40}$$

Using (4.37), (4.38), (4.39) and (4.40) we conclude that u satisfies the condition in (4.36) if and only if v satisfies the condition in (4.36). This implies that we can disregard the possible changes from u to v in Lemmas 4.1 and 4.2 and assume $v = u$ in them without loss of generality. Combining Theorem 2.5, Corollary 3.7, the arguments in this section before Lemma 4.1, Lemma 4.2 and Lemma 4.4 we obtain the results corresponding to the polynomial x^{q+1}. Similarly using instead Theorem 2.6, Lemma 4.1 and Lemma 4.4 we obtain the results corresponding to the polynomial x^2. □

The equivalences in Theorem 4.5 can be given explicitly. Assume that $q \equiv 1 \mod 3$ and $u \in \frac{1}{2}(G \setminus H)$. Extending the arguments in the proof of Lemma 4.3 we determine some linearized permutation polynomials $A_0x + A_1x^q + A_2x^{q^2}$, $B_0x + B_1x^q + B_2x^{q^2} \in \mathbb{F}_{q^3}[x]$ satisfying (4.21) as follows:

i) Choose $a \in \mathbb{F}_{q^3}^*$ with $a^2 + a + 1 = 0$ and $A_1 \in \mathbb{F}_{q^3}^*$ such that

$$A_1^{q^2+q+1} = 1 \quad \text{and} \quad u = \frac{1}{2a} A_1^{q^2-1}. \tag{4.41}$$

ii) Put $A_2 = \dfrac{a}{A_1^{q^2}}$, $A_0 = 1$, $B_0 = 2A_1A_2$, $B_1 = 2A_2$, $B_2 = 2A_1$.

Here we construct an a satisfying (4.41). First note that

$$\gcd\left(q^2-1, \frac{q^2+q+1}{3}\right) = 1.$$

Hence for $h \in H$ there exists a uniquely determined $A \in H$ satisfying $A^{q^2-1} = h$.

- If $u = \frac{1}{2}\epsilon h$, then we let $a = \epsilon^2$ and $A_1 \in H$ satisfying $A_1^{q^2-1} = h$. Then (4.41) is satisfied as $2au = A_1^{q^2-1}$ and $A^{q^2+q+1} = 1$ as $A_1 \in H \subseteq G$.

- If $u = \frac{1}{2}\epsilon^2 h$, then similarly we let $a = \epsilon$ and $A_1 \in H$ satisfying $A_1^{q^2-1} = h$.

Now assume that $q \equiv 1 \mod 3$ and $a \in -(G \setminus H)$. Extending the arguments in the proof of Lemma 4.24 we determine some linearized permutation polynomials $A_0x + A_1x^q + A_2x^{q^2}$, $B_0x + B_1x^q + B_2x^{q^2} \in \mathbb{F}_{q^3}[x]$ satisfying (4.24) as follows:

i) Choose $a \in \mathbb{F}_{q^3}^*$ with $a^2 + a + 1 = 0$ and $A_1 \in \mathbb{F}_{q^3}^*$ such that

$$A_1^{q^2+q+1} = 1 \quad \text{and} \quad u = \frac{a}{a+1}A_1^{q^2-1}. \tag{4.42}$$

ii) Put $A_2 = \dfrac{a}{A_1^{q^2}}$, $A_0 = 1$, $B_0 = A_2 + A_1^{q+1}$, $B_1 = A_1^q + A_2^{q+1}$, $B_2 = 1 + A_1 A_2^q$.

Here we construct an a satisfying (4.42).

- If $u = -\epsilon h$, then we let $a = \epsilon^2$ and $A_1 \in H$ satisfying $A_1^{q^2-1} = h$. Then (4.42) is satisfied as $\dfrac{a}{a+1}A_1^{q^2-1} = \dfrac{a}{-a^2}h = -\epsilon h = u$ and $A_1^{q^2+q+1} = 1$.

- If $u = -\epsilon^2 h$, then similarly we let $a = \epsilon$ and $A_1 \in H$ satisfying $A_1^{q^2-1} = h$.

Finally we present a corollary that we referred earlier in Section 3 before Remark 3.8. The following corollary gives certain bounds on $N^*(\mathbb{L}_u)$ when $q \equiv 1 \mod 3$ and $\mathrm{Norm}_{\mathbb{F}_{q^3}/\mathbb{F}_q}(u) \notin \{-1, 1/8\}$.

For $a \in \mathbb{F}_q^*$ we recall that $J_3(-a-2, -a)$ denotes the cardinality of the set $\{\nu \in \mathbb{F}_q : x^3 + (a+2)x^2 + \nu x + a \in \mathbb{F}_q[x] \text{ is irreducible}\}$ and we have $J_3(-a-2, -a) \geq 2$ when $a \neq 1$ and $a \neq -8$ (see Lemma 3.5 above).

COROLLARY 4.6. *Let* $q \equiv 1 \mod 3$, $\mathrm{Norm}_{\mathbb{F}_{q^3}/\mathbb{F}_q}(2u) \neq 1$ *and* $\mathrm{Norm}_{\mathbb{F}_{q^3}/\mathbb{F}_q}(u) \neq -1$. *Set* $a = \mathrm{Norm}_{\mathbb{F}_{q^3}/\mathbb{F}_q}(2u)$. *We have*

$$3(q-1)^2 \leq N^*(\mathbb{L}_u) \leq 9(q-1)^2 J_3(-a-2, -a) - 6(q-1)^2. \tag{4.43}$$

PROOF. Recall $N^*(\mathbb{L}_u)$ is the number of $(x, z) \in \mathbb{F}_{q^3}^* \times \mathbb{F}_{q^3}^*$ satisfying

$$x^{q^2-1} + x^{q-1} + 2uz^{3(q-1)} = 0. \tag{4.44}$$

Assume that $N^*(\mathbb{L}_u) \geq 1$ and let $(x_0, z_0) \in \mathbb{F}_{q^3}^* \times \mathbb{F}_{q^3}^*$ with

$$x_0^{q^2-1} + x_0^{q-1} + 2uz_0^{3(q-1)} = 0.$$

Let A and B be the subgroups of $\mathbb{F}_{q^3}^*$ of order $|A| = (q-1)$ and $|B| = 3(q-1)$. Here we use that $q \equiv 1 \mod 3$ for the existence of B. For $\alpha \in A$ and $\beta \in B$ we have $\alpha^{q-1} = \alpha^{q^2-1} = 1$ and $\beta^{3(q-1)} = 1$. If $\alpha \in A$ and $\beta \in B$, then we have

$$(\alpha x_0)^{q^2-1} + (\alpha x_0)^{q-1} + 2u(\beta z_0)^{3(q-1)} = 0.$$

Moreover if $\alpha_1, \alpha_2 \in A$ and $\beta_1, \beta_2 \in B$, then

$$(\alpha_1 x_0, \beta_1 z_0) = (\alpha_2 x_0, \beta_2 z_0) \iff \alpha_1 = \alpha_2 \text{ and } \beta_1 = \beta_2, \text{ as } x_0 \neq 0 \text{ and } z_0 \neq 0.$$

Hence the set $\left\{(\alpha x_0, \beta z_0) \in \mathbb{F}_{q^3}^* \times \mathbb{F}_{q^3}^* : \alpha \in A, \beta \in B\right\}$ has $|A||B| = 3(q-1)^2$ elements and all of its elements satisfy the equation in (4.44). This implies that $N^*(\mathbb{L}_u)$ is divisible by $3(q-1)^2$. Similarly $N^*(\mathbb{L}_{\epsilon u})$ and $N^*(\mathbb{L}_{\epsilon^2 u})$ are divisible by $3(q-1)^2$.

Using Theorem 4.5 we obtain that $N^*(\mathbb{L}_u)$, $N^*(\mathbb{L}_{\epsilon u})$ and $N^*(\mathbb{L}_{\epsilon^2 u})$ are nonzero. Combining this with the divisibility results above we conclude that

$$N^*(\mathbb{L}_u) \geq 3(q-1)^2, \quad N^*(\mathbb{L}_{\epsilon u}) \geq 3(q-1)^2, \quad N^*(\mathbb{L}_{\epsilon^2 u}) \geq 3(q-1)^2. \tag{4.45}$$

Combining Lemma 3.5 and Lemma 3.6 we get

$$3N^*(\mathbb{K}_u) = N^*(\mathbb{L}_u) + N^*(\mathbb{L}_{\epsilon u}) + N^*(\mathbb{L}_{\epsilon^2 u}) = 9(q-1)^2 J_3(-a-2, -a). \tag{4.46}$$

Combining (4.45) and (4.46) we obtain that $-N^*(\mathbb{L}_{\epsilon u}) \leq -3(q-1)^2$, $-N^*(\mathbb{L}_{\epsilon^2 u}) \leq -3(q-1)^2$ and

$$\begin{aligned} N^*(\mathbb{L}_u) &= 3N^*(\mathbb{K}_u) - N^*(\mathbb{L}_{\epsilon u}) - N^*(\mathbb{L}_{\epsilon^2 u}) \\ &= 9(q-1)^2 J_3(-a-2, -a) - N^*(\mathbb{L}_{\epsilon u}) - N^*(\mathbb{L}_{\epsilon^2 u}) \\ &\leq 9(q-1)^2 J_3(-a-2, -a) - 6(q-1)^2. \end{aligned}$$

This completes the proof of the inequality in the right hand side of (4.43). The inequality in the left hand side of (4.43) follows from (4.45). This completes the proof. □

Acknowledgments

The second named author thanks to Otto-von-Guericke University of Magdeburg for hospitality.

The authors would like to thank the anonymous reviewer for the useful suggestions which improved the paper.

References

[1] A. A. Albert, *Generalized twisted fields*, Pacific J. Mathematics, vol. 11, pp. 1-8, 1961. MR0122850 (23:A182)

[2] J. Bierbrauer, *New semifields, PN and APN functions*, Des. Codes Cryptography, vol. 54, pp. 189-200, 2010. MR2584973 (2011h:12018)

[3] J. Bierbrauer, *New commutative semifields and their nuclei*, in: Proceedings of AAECC 2009, LNCS 5527, pp. 179-185, 2009. MR2580866 (2011k:12010)

[4] C. Bracken, E. Bryne, N. Markin and G. McGuire, *New families of quadratic almost perfect nonlinear trinomials and multinomials*, Finite Fields Appl., vol. 14, pp. 703-714, 2008. MR2435056 (2009e:11224)

[5] R. S. Coulter and M. Henderson, *Commutative presemifields and semifields*, Adv. Math., vol. 217, pp. 282-304, 2008. MR2365198 (2009b:12019)

[6] P. Dembowski, *Finite Geometries*, Springer-Verlag, Berlin, 1997 (reprint of the 1968 edition). MR0233275 (38:1597)

[7] P. Dembowski and T. Ostrom, *Planes of order n with collineation group of order n^2*, Math. Z., vol. 103, pp. 239-258, 1968. MR0226486 (37:2075)

[8] Y. Edel, G. Kyureghyan and A. Pott, *A new APN function which is not equivalent to a power mapping*, IEEE Trans. Inform. Theory, vol. 52, no. 2, pp. 744-747, 2006. MR2236189 (2007e:94055)

[9] X.-D Hou and C. Sze, *On certain diagonal equations over finite fields*, Finite Fields Appl., vol. 15, pp. 633-643, 2009. MR2552272 (2010m:11075)

[10] W. M. Kantor, *Commutative semifields and sympletic spreads*, *J. Algebra*, vol. 270, pp.96-114, 2003. MR2015931 (2004k:51003)
[11] G. Leander and F. Rodier, *Bounds on the degree of APN polynomials: the case of* $x^{-1}+g(x)$, Des. Codes Cryptogr., vol. 59, pp. 207-222, 2011. MR2781610 (2012c:94085)
[12] R. Lidl, H. Niederreiter, *Finite Fields*, Cambridge University Press, Cambridge, 1997. MR1429394 (97i:11115)
[13] G. Menichetti, *On a Kaplansky conjecture concerning three-dimensional division algebras over a finite field*, J. Algebra vol. 47, pp. 400-410, 1977. MR0453823 (56:12076)
[14] K. Nyberg, *Differentially uniform mappings for cryptography*, in: Advances in Cryptography-Eurocrypt 93, LNCS 765, pp. 134-144, 1994. MR1290329 (95e:94039)
[15] J. Wolfmann, *The number of solutions of certain diagonal equations over finite fields*, J. Number Theory, vol. 42, pp. 247-257, 1992. MR1189504 (94j:11055)
[16] H. Stichtenoth, *Algebraic Function Fields and Codes*, Second edition, Springer-Verlag, Berlin, 2009. MR2464941 (2010d:14034)

Department of Mathematics, Otto-von-Guericke University of Magdeburg, Universitätsplatz 2, 39106 Magdeburg, Germany.
E-mail address: gohar.kyureghyan@ovgu.de

Department of Mathematics and Institute of Applied Mathematics, Middle East Technical University, Dumlupinar Bulvari No. 1, 06800, Ankara, Turkey.
E-mail address: ozbudak@metu.edu.tr

Department of Mathematics, Otto-von-Guericke University of Magdeburg, Universitätsplatz 2, 39106 Magdeburg, Germany.
E-mail address: alexander.pott@ovgu.de

Contemporary Mathematics
Volume **574**, 2012
http://dx.doi.org/10.1090/conm/574/11419

New families of APN functions in characteristic 3 or 5

Elodie Leducq

ABSTRACT. In this paper, we prove two conjectures of Dobbertin and al. in characteristic 3. We also give a new family of APN power mappings in characteristic 5.

1. Introduction

In the following, p is a prime number, n a positive integer, $q = p^n$ and $\mathbb{F}_q$ is a finite fields with q elements.

To resist to differential cryptanalysis, a function f from $\mathbb{F}_q$ to $\mathbb{F}_q$ used in a bloc cypher like DES has to have a low uniformity, that is to say for all $a \in \mathbb{F}_q^*$ and $b \in \mathbb{F}_q$, the equation

$$f(x+a) - f(x) = b \tag{1.1}$$

must have few solutions.

We denote by $N_f(a, b)$ the number of solutions of (1.1) and

$$\Delta_f := \max(N_f(a, b) : a, b \in \mathbb{F}_q, a \neq 0).$$

In characteristic 2, if $x \in \mathbb{F}_q$ is a solution of (1.1) then $x + a$ is also a solution of (1.1). So, Δ_f is at least 2. In odd characteristic, we can have $\Delta_f = 1$.

DEFINITION 1.1. Let $f : \mathbb{F}_q \to \mathbb{F}_q$; We say that f is almost perfectly nonlinear (APN) if $\Delta_f = 2$ and that f is perfectly nonlinear (PN) if $\Delta_f = 1$.

In the following, we consider power mappings. We denote $\Delta_{x \mapsto x^d}$ by Δ_d. Then, on $\mathbb{F}_q$, we have $\Delta_d = \max(N_{x \mapsto x^d}(1, b) : b \in \mathbb{F}_q)$ and $\Delta_d = \Delta_{dp^i}$ for all $1 \leq i \leq n-1$.

The only known PN power mappings are given in the following theorem

THEOREM 1.2. *Let $f : x \mapsto x^d$ a power mapping. Then f is PN over $\mathbb{F}_{p^n}$ for*

(1) $d = 2$
(2) $d = p^k + 1$ *where k is a positive integer such that $\frac{n}{\gcd(n,k)}$ is odd* [**CM97, DO68**].
(3) $d = \frac{3^k+1}{2}$ *where $p = 3$ and k is an odd integer such that* $\gcd(n, k) = 1$ [**CM97**].

In characteristic 2, the only known APN power mappings are given in the following theorem :

1991 *Mathematics Subject Classification.* Primary 12E10; Secondary 11T71.
Key words and phrases. Almost perfect nonlinear functions, odd characteristic.

THEOREM 1.3. *Let $f : x \mapsto x^d$ a power mapping. Then f is APN over $\mathbb{F}_{2^n}$ for*

(1) $d = 2^k + 1$ *where k is a positive integer such that* $\gcd(n,k) = 1$ [**Nyb94**].
(2) $d = 2^{2k} - 2^k + 1$ *where k is a positive integer such that* $\gcd(k,n) = 1$ [**JW93, Dob99c, Dob99b**].
(3) $d = 2^e - 2$ *if n is odd* [**Nyb94**].
(4) $d = 2^{4k} + 2^{3k} + 2^{2k} + 2^k - 1$ *if $n = 5k$* [**Dob01**].
(5) $d = 2^k + 3$ *if $n = 2k + 1$ is odd* [**Dob99b**].
(6) $d = 2^{2k} + 2^k - 1$ *where k is an positive integer such that* $4k + 1 \equiv 0 \mod n$ [**Dob99a**].

In odd characteristic, in [**HRS99**], after a computer search, the authors give a complete list of almost perfectly nonlinear (APN) power mappings on $\mathbb{F}_{p^n}$, for

$$p^n \in \{2^2, 2^3, \ldots, 2^{11}, 3^2, \ldots, 3^7, 5^2, \ldots, 5^5, 7^2, 7^3, 7^4, 11^2, 11^3\}.$$

Their article contains many theorems showing that some of these functions are members of infinite families of APN functions. In the following table, we give all cases which have not been included in infinite families in [**HRS99**]:

	p^n	d	$dp^i \mod p^n - 1$
I	3^5	134	(134,160,238,230,206)
II	3^5	152	(152,214,158,232,212)
III	3^7	40	(40,120,360,1080,1054,976,742)
IV	3^7	224	(224,672,2016,1676,656,1968,1532)
V	3^7	274	(274,822,280,840,334,1002,820)
VI	5^3	14	(14,70,102)
VII	5^5	843	(843,1091,2331,2283,2043)

We put $m = \frac{n+1}{2}$.

In [**DMM$^+$03**], Dobbertin, Mills, Muller, Pott and Willems make conjectures which give infinite families for cases I, II, III, V and VII of the table :

CONJECTURE 1.4. *For $n \geq 5$ an odd integer, the function $x \mapsto x^d$ is APN over $\mathbb{F}_{3^n}$ for*

$$d = \begin{cases} \frac{3^m-1}{2} & \textit{if } n \equiv 3 \mod 4 \\ \frac{3^m-1}{2} + \frac{3^n-1}{2} & \textit{if } n \equiv 1 \mod 4 \end{cases}.$$

CONJECTURE 1.5. *For $n \geq 5$ an odd integer, the function $x \mapsto x^d$ is APN over $\mathbb{F}_{3^n}$ for*

$$d = \begin{cases} \frac{3^{n+1}-1}{8} & \textit{si } n \equiv 3 \mod 4 \\ \frac{3^{n+1}-1}{8} + \frac{3^n-1}{2} & \textit{si } n \equiv 1 \mod 4 \end{cases}.$$

CONJECTURE 1.6. *Let n be an odd integer. The function $x \mapsto x^d$ is APN over $\mathbb{F}_{5^n}$ for*

$$d = \frac{5^n - 1}{4} + \frac{5^m - 1}{2}.$$

Now, we recall two theorems proved in [**ZW10**] (theorem 4.1 and 4.4) that explain cases IV and VII of the table. However, these theorems do not give new family to complete the table.

THEOREM 1.7 (Zha, Wang). *On $\mathbb{F}_{3^n}$, the function $f : x \mapsto x^d$ satisfies $\Delta_d \leq 2$ for d such that $(3^k+1)d - 2 = u(3^n - 1)$ where u is odd and $\gcd(n,k) = 1$. Furthermore, f is APN if $2k < n$.*

THEOREM 1.8 (Zha, Wang). *The function $x \mapsto x^d$ is APN over $\mathbb{F}_{5^n}$ for d such that $(5^k+1)d - 2 = u(5^n - 1)$ where $\gcd(n,k) = 1$, u is odd and k is even.*

Using theorem 1.8 Zha and Wang prove conjecture 1.6 (theorem 4.5).

In the next section, we prove conjectures 1.4 and 1.5. In section 3, we prove the following theorem, which gives a new infinite family of APN functions :

THEOREM 1.9. *Let $l \geq 2$ and n an integer such that $n \equiv -1 \mod 2^l$ then the function $x \mapsto x^d$ is APN over $\mathbb{F}_{5^n}$ for*

$$d = \frac{1}{2}\frac{5^{n+1}-1}{5^{\frac{n+1}{2^l}}+1} + \frac{5^n-1}{4}.$$

Finally, in the last section, we make some remarks about Zha and Wang theorems. In particular we give a new proof of conjecture 1.6 in the case where $n \equiv 3 \mod 4$ and we give an explicit family for the case IV of the table using theorem 1.7.

2. Proof of conjectures 1.4 and 1.5

In this part $p = 3$.

The following theorems are proved in [**DMM$^+$03**] (Theorem 2.1 and 2.2) :

THEOREM 2.1 (Dobbertin, Mills, Müller, Pott, Willems). *Let n be an odd integer. In $\mathbb{F}_{3^n}$, the function $x \mapsto x^d$ satisfies $\Delta_d \leq 2$ for*

$$d = \begin{cases} \frac{3^m-1}{2} & \text{if } n \equiv 3 \mod 4 \\ \frac{3^m-1}{2} + \frac{3^n-1}{2} & \text{if } n \equiv 1 \mod 4 \end{cases}.$$

THEOREM 2.2 (Dobbertin, Mills, Müller, Pott, Willems). *Let n be an odd integer. In $\mathbb{F}_{3^n}$, the function $x \mapsto x^d$ satisfies $\Delta_d \leq 2$ for*

$$d = \begin{cases} \frac{3^{n+1}-1}{8} & \text{si } n \equiv 3 \mod 4 \\ \frac{3^{n+1}-1}{8} + \frac{3^n-1}{2} & \text{si } n \equiv 1 \mod 4 \end{cases}.$$

So, we only have to show that $\Delta_d \neq 1$ which means that $(x+1)^d - x^d$ is not a permutation polynomial. We do that with Dickson and Hermite's criterion which gives a necessary and sufficient condition for a polynomial with coefficients in $\mathbb{F}_q$ to be a permutation polynomial over $\mathbb{F}_q$:

THEOREM 2.3 (Hermite an Dickson's criterion, see [**LN97**] p. 349). *The following propositions are equivalent :*

1. *$f(x) \in \mathbb{F}_q[x]$ is a permutation polynomial over $\mathbb{F}_q$;*
2. *$f(x)$ has exactly one root in $\mathbb{F}_q$ and*
   $\forall t$, $1 \leq t \leq q-2$, $t \not\equiv 0 \mod p$, the reduction $(f(x))^t \mod x^q - x$ has degree less than $q-1$.

We also need Lucas' theorem :

THEOREM 2.4 (Lucas see [**Luc78**], p. 230). *Let p be a prime number, n and r integers. We consider p-adic decomposition of n and r :*
$n = n_0 + n_1 p + \ldots + n_k p^k$ with $0 \leq n_i \leq p-1$,

$r = r_0 + r_1 p + \ldots + r_k p^k$ with $0 \le r_i \le p-1$.
Then

$$\binom{n}{r} \equiv \prod_{i=0}^{k} \binom{n_i}{r_i} \mod p.$$

Now we can prove the conjectures 1.4 and 1.5.

Since in each case $\gcd(d, 3^n - 1) = 2$, we have to find t in Hermite and Dickson's criterion such that the degree of $((x+1)^d - x^d)^t \mod x^q - x$ is $q-1$. Then,

$$\begin{aligned}
\left((x+1)^d - x^d\right)^t &= \sum_{k=0}^{t} \binom{t}{k} (x+1)^{dk} (-1)^{t-k} x^{d(t-k)} \\
&= \sum_{k=0}^{t} \binom{t}{k} (-1)^{t-k} \left(\sum_{j=0}^{dk} \binom{dk}{j} x^{dk-j} \right) x^{d(t-k)} \\
&= \sum_{k=0}^{t} \binom{t}{k} (-1)^{t-k} \sum_{j=0}^{dk} \binom{dk}{j} x^{dt-j} \\
&= \sum_{j=0}^{dt} x^{dt-j} \sum_{k=\lceil \frac{j}{d} \rceil}^{t} (-1)^{t-k} \binom{t}{k} \binom{dk}{j}.
\end{aligned}$$

The degree of $x^k \mod x^q - x$ is $q-1$ if and only if $k \equiv 0 \mod q-1$ and $k \neq 0$. For $v \in \mathbb{R}$ we denote by $\lceil v \rceil$ the ceilling of x and by $\lfloor v \rfloor$ its floor. Then the coefficient of x^{q-1} in $\left((x+1)^d - x^d\right)^t \mod x^q - x$ is

$$C = \sum_{i=1}^{\lfloor \frac{dt}{q-1} \rfloor} \sum_{k=\lceil \frac{dt - i(q-1)}{d} \rceil}^{t} (-1)^{t-k} \binom{t}{k} \binom{dk}{dt - i(q-1)}.$$

Case where $n \equiv 1 \mod 4$ in conjecture 1.4.

We have $d = \frac{3^m-1}{2} + \frac{3^n-1}{2}$. We choose $t = 2$ then, since $n \neq 1$, $\lfloor \frac{dt}{q-1} \rfloor = 1$ and $\lceil \frac{dt-(q-1)}{d} \rceil = 1$. So

$$C = \binom{3^n + 3^m - 2}{3^m - 1} + \binom{\frac{3^n-1}{2} + \frac{3^m-1}{2}}{3^m - 1}.$$

On one side, $3^n + 3^m - 2 = 3^n + \sum_{k=1}^{m-1} 2 \times 3^k + 1$ and $3^m - 1 = \sum_{k=0}^{m-1} 2 \times 3^k$ so by Lucas' theorem, $\binom{3^n+3^m-2}{3^m-1} \equiv 0 \mod 3$.

On the other side, $\frac{3^n-1}{2} + \frac{3^m-1}{2} = \sum_{k=0}^{n-1} 3^k + \sum_{k=0}^{m-1} 3^k = \sum_{k=m}^{n-1} 3^k + \sum_{k=0}^{m-1} 2 \times 3^k$. So $\binom{\frac{3^n-1}{2}+\frac{3^m-1}{2}}{3^m-1} \equiv 1 \mod 3$.

Case where $n \equiv 3 \mod 4$ in conjecture 1.4.

We have $d = \frac{3^m-1}{2}$. Let $s = 2 \times (3^{m-1} + 1)$, we choose $t = 2s + 4$. For $n > 3$, $0 < t < q-1$ and $t \not\equiv 0 \mod 3$. Furthermore, $\lfloor \frac{dt}{q-1} \rfloor = 2$, $\lceil \frac{dt-(q-1)}{d} \rceil = s + 6$ and

$\lceil \frac{dt-2(q-1)}{d} \rceil = 7$. So

$$C = \sum_{k=s+6}^{t} (-1)^k \binom{t}{k} \binom{dk}{dt-(q-1)} + \sum_{k=7}^{t} (-1)^k \binom{t}{k} \binom{dk}{dt-2(q-1)}.$$

We have $t = 4 \times 3^{m-1} + 8 = 3^m + 3^{m-1} + 2 \times 3 + 2$.
So by Lucas' theorem, $\binom{t}{k} \not\equiv 0 \mod 3$ if and only if $k = a3^m + b3^{m-1} + c3 + d$ where $a,\ b \in \{0,1\}$ and $c,\ d \in \{0,1,2\}$.
Then

$$dk = a3^n + (a+b) \sum_{j=m+1}^{n-1} 3^j + (a+b+c)3^m + (b+c+d)3^{m-1} + (c+d) \sum_{j=1}^{m-2} 3^j + d.$$

Furthermore, $dt - 2(q-1) = 3^{m+1} + \sum_{j=1}^{m-2} 2 \times 3^j + 1$ and

$dt - (q-1) = 3^n + 3^{m+1} + \sum_{j=1}^{m-2} 2 \times 3^j$.

So $\binom{dk}{dt-(q-1)} \not\equiv 0 \mod 3$ if $a = 1$ and $c + d = 2$.
Assume that $b = 1$, then $b+c+d = 3$ and $5 \geq a+b+c+1 \geq 3$. So the coefficient of 3^{m+1} in 3-adic decomposition of dk is 0 which means that $\binom{dk}{dt-2(q-1)} \equiv 0 \mod 3$. Hence the only k remaining in the first sum are $3^m + 2 \times 3$, $3^m + 3 + 1$ and $3^m + 2$. Now we consider the second sum : $\binom{dk}{dt-2(q-1)} \not\equiv 0 \mod 3$ if $d \in \{1,2\}$ and $c+d = 2$, namely if $c = 1$ and $d = 1$ or if $c = 0$ and $d = 2$. Since $k \geq 7$, a and b can't be both 0. Assume that $a = b = 1$ then $b + c + d = 3$ and $4 \geq a + c + b + 1 \geq 3$. So the coefficient of 3^{m+1} in 3-adic decomposition of dk is 0, which means that $\binom{dk}{dt-2(q-1)} \equiv 0 \mod 3$. Hence the only k remaining in this sum are $3^m + 3 + 1$, $3^m + 2$, $3^{m-1} + 3 + 1$ and $3^{m-1} + 2$.
Finally :

$$\begin{aligned} C = &-\binom{t}{3^m + 2 \times 3}\binom{d(3^m + 2 \times 3)}{dt-(q-1)} - \binom{t}{3^m+3+1}\binom{d(3^m+3+1)}{dt-(q-1)} \\ &- \binom{t}{3^m+2}\binom{d(3^m+2)}{dt-(q-1)} - \binom{t}{3^m+3+1}\binom{d(3^m+3+1)}{dt-2(q-1)} \\ &- \binom{t}{3^m+2}\binom{d(3^m+2)}{dt-2(q-1)} - \binom{t}{3^{m-1}+3+1}\binom{d(3^{m-1}+3+1)}{dt-2(q-1)} \\ &- \binom{t}{3^{m-1}+2}\binom{d(3^{m-1}+2)}{dt-2(q-1)}. \end{aligned}$$

Now, we have $t = 3^m + 3^{m-1} + 2 \times 3 + 2$, so by Lucas' theorem, $\binom{t}{3^m+2\times3} \equiv 1 \mod 3$. Moreover $d(3^m + 2 \times 3) = 3^n + \sum_{j=m+2}^{n-1} 3^j + 2 \times 3^{m+1} + 2 \times 3^{m-1} + 2 \sum_{j=1}^{m-2} 3^j$ and $dt - (q-1) = 3^n + 3^{m+1} + \sum_{j=1}^{m-2} 2 \times 3^j$, so $\binom{d(3^m+2\times3)}{dt-(q-1)} \equiv 2 \mod 3$.
We do the same for all binomials and we get $C \equiv 1 \mod 3$.

Case where $n \equiv 1 \mod 4$ in conjecture 1.5.

We have $d = \frac{3^{n+1}-1}{8} + \frac{3^n-1}{2}$. We choose $t = 2$ then $\lfloor \frac{dt}{q-1} \rfloor = 1$ and $\lceil \frac{dt-(q-1)}{d} \rceil = 1$. So

$$C = \binom{\frac{3^{n+1}-1}{4} + 3^n - 1}{\frac{3^{n+1}-1}{4}} + \binom{\frac{3^n-1}{2} + \frac{3^{n+1}-1}{8}}{\frac{3^{n+1}-1}{4}}.$$

On one side, we have $\frac{3^{n+1}-1}{4} + 3^n - 1 = \sum_{k=1}^{\frac{n-1}{2}} 2 \times 3^{2k} + 3^n + 1$ and $\frac{3^{n+1}-1}{4} = 2 \sum_{k=0}^{\frac{n-1}{2}} 3^{2k}$.

So by Lucas' theorem, $\binom{\frac{3^{n+1}-1}{4}+3^n-1}{\frac{3^{n+1}-1}{4}} \equiv 0 \mod 3$.

On the other side, $\frac{3^n-1}{2} + \frac{3^{n+1}-1}{8} = \sum_{k=0}^{\frac{n-1}{2}} 2 \times 3^{2k} + \sum_{k=0}^{\frac{n-1}{2}-1} 3^{2k+1}$.

So, $\binom{\frac{3^n-1}{2}+\frac{3^{n+1}-1}{8}}{\frac{3^{n+1}-1}{4}} \equiv 1 \mod 3$.

Case where $n \equiv 3 \mod 4$ in conjecture 1.5.

We have $d = \frac{3^{n+1}-1}{8}$. We choose $t = 26$, for $n \geq 5$, $1 \leq t \leq q-2$ and $t \not\equiv 0 \mod 3$. Furthermore

$$\frac{dt}{q-1} = \frac{13}{4}(2\frac{3^n}{3^n-1} + 1).$$

For $n \geq 5$, $\frac{243}{242} \geq \frac{3^n}{3^n-1} \geq 1$. So $\lfloor \frac{dt}{q-1} \rfloor = 9$.
We have $t = 2 + 2 \times 3 + 2 \times 9$ so by Lucas' theorem, $\binom{t}{k} \not\equiv 0 \mod 3$ if and only if $k = a + 3b + 9c$, a, b, $c \in \{0, 1, 2\}$.

In addition, $d = \sum_{k=0}^{\frac{n-1}{2}} 3^{2k}$, so $dk = a + \sum_{k=1}^{\frac{n-1}{2}} (a+c)3^{2k} + \sum_{k=0}^{\frac{n-1}{2}} b3^{2k+1} + c3^{n+1}$.

We have $dt - (q-1) = \sum_{k=1}^{\frac{n-1}{2}} 2 \times 3^{2k} + 2 \times 3^n + 2 \times 3^{n+1}$. So, if we write $j = \alpha + 3\beta$, α, $\beta \in \{0, 1, 2\}$,

$$dt - (j+1)(q-1) = \alpha + \beta 3 + \sum_{k=1}^{\frac{n-1}{2}} 2 \times 3^{2k} + (2-\alpha)3^n + (2-\beta)3^{n+1}.$$

So $\binom{dk}{dt-(j+1)(q-1)} \not\equiv 0 \mod 3$ if and only if $a \geq \alpha$, $b \geq \beta$, $b \geq 2-\alpha$, $c \geq 2-\beta$ and $a + c = 2$.
Finally,

$$\begin{aligned} C \equiv \binom{26}{24} - \binom{26}{16} + \binom{26}{24} - \binom{26}{13} + \binom{26}{16} + \binom{26}{8} + \binom{26}{16} \\ + \binom{26}{24} - \binom{26}{16} + \binom{26}{8} + \binom{26}{8} \mod 3 \\ \equiv 1 \mod 3. \end{aligned}$$

In all cases we have proved that $C \not\equiv 0 \mod 3$; so by Hermite and Dickson's criterion $\Delta_d \neq 1$.

3. Proof of theorem 1.9

We give first some preliminary results :

LEMMA 3.1 (see [**DMM**$^+$**03**] p. 97). *If* $\gcd(d, q-1) = 1$ *then* $\Delta_d = \Delta_{d^{-1}}$ *where* d^{-1} *is the inverse of* d *modulo* $q-1$.

PROPOSITION 3.2 (see [**HRS99**], corollary 1 p. 484). *Let* n *and* k *be integers such that* $\gcd(2n, k) = 1$. *Then* $x \mapsto x^d$ *is APN over* $\mathbb{F}_{5^n}$ *for* $d = \frac{5^k+1}{2}$.

COROLLARY 3.3. *Let* n *be a integer such that* $n \equiv 2^l - 1 \mod 2^{l+1}$ *then* $x \mapsto x^d$ *is APN over* $\mathbb{F}_{5^n}$ *for* $d = \frac{5^{\frac{n+1}{2^l}}+1}{2}$.

PROOF. If $n \equiv 2^l - 1 \mod 2^{l+1}$ then $\frac{n+1}{2^l} \equiv 1 \mod 2$. So $\gcd(2n, \frac{n+1}{2^l}) = 1$ and by proposition 3.2 we get the result. □

Now we are able to prove theorem 1.9.

First we consider the case where $n \equiv 2^l - 1 \mod 2^{l+1}$. We have

$$\begin{aligned}\frac{5^{\frac{n+1}{2^l}}+1}{2}d &= \frac{5^{n+1}-1}{4} + \frac{5^{\frac{n+1}{2^l}}+1}{2}\frac{5^n-1}{4} \\ &= 5^n + \frac{5^n-1}{4}\frac{5^{\frac{n+1}{2^l}}+3}{2}\end{aligned}$$

but if $n \equiv 2^l - 1 \mod 2^{l+1}$, $\frac{n+1}{2^l} \equiv 1 \mod 2$ and $5^{\frac{n+1}{2^l}} + 3 \equiv 0 \mod 8$. So

$$\frac{5^{\frac{n+1}{2^l}}+1}{2}d \equiv 1 \mod 5^n - 1.$$

Hence d is invertible modulo $5^n - 1$ and $d^{-1} = \frac{5^{\frac{n+1}{2^l}}+1}{2}$. By corollary 3.3, $x \mapsto x^{d^{-1}}$ is APN over $\mathbb{F}_{5^n}$, so by lemma 3.1, we get the result for $n \equiv 2^l - 1 \mod 2^{l+1}$.

If $n \equiv -1 \mod 2^{l+1}$ then $(2^l+1)n \equiv 2^l - 1 \mod 2^{l+1}$. We put

$$e = \frac{1}{2}\frac{5^{(2^l+1)n+1}-1}{5^{\frac{(2^l+1)n+1}{2^l}}+1} + \frac{5^{(2^l+1)n}-1}{4}.$$

By the first case $x \mapsto x^e$ is APN over $\mathbb{F}_{5^{(2^l+1)n}}$. Furthermore,

$$\begin{aligned}e &= \frac{1}{2}\sum_{k=0}^{2^l-1}(-1)^{k+1}(5^{\frac{(2^l+1)n+1}{2^l}})^k + \frac{5^{(2^l+1)n}-1}{4} \\ &= \frac{1}{2}\sum_{k=0}^{2^l-1}(-1)^{k+1}5^{nk}(5^{\frac{n+1}{2^l}})^k + 5^n\frac{5^{2^l n}-1}{4} + \frac{5^n-1}{4} \\ &= 5^n(5^n-1)\frac{1}{4}\sum_{j=0}^{2^l-1}5^{nj} + \frac{5^n-1}{4} + (5^n-1)\frac{1}{2}\sum_{k=1}^{2^l-1}(-1)^{k+1}\sum_{j=0}^{k-1}5^{nj}(5^{\frac{n+1}{2^l}})^k \\ &\quad + \frac{1}{2}\sum_{k=0}^{2^l-1}(-1)^{k+1}(5^{\frac{n+1}{2^l}})^k.\end{aligned}$$

Since $\sum_{k=0}^{2^l-1} 5^{nj} \equiv 0 \mod 4$ and $\sum_{k=1}^{2^l-1}(-1)^{k+1}\sum_{j=0}^{k-1} 5^{nj}(5^{\frac{n+1}{2^l}})^k \equiv 0 \mod 2$ for $l \geq 2$, we have :

$$e \equiv \frac{1}{2}\frac{5^{n+1}-1}{5^{\frac{n+1}{2^l}}+1} + \frac{5^n-1}{4} \mod 5^n - 1.$$

Since $x \mapsto x^e$ is APN over $\mathbb{F}_{5^{(2^l+1)n}}$, for all $b \in \mathbb{F}_{5^n}$ the equation $(x+1)^e - x^e = b$ has at most two solutions in $\mathbb{F}_{5^{(2^l+1)n}} \supset \mathbb{F}_{5^n}$. So, since $e \equiv d \mod 5^n - 1$, for all $b \in \mathbb{F}_{5^n}$, $(x+1)^d - x^d = b$ has at most two solutions in $\mathbb{F}_{5^n}$ and $\Delta_d \leq 2$ for $n \equiv -1 \mod 2^{l+1}$.

Furthermore since n is odd, $\frac{5^n-1}{4} \equiv 1 \mod 2$. In addition $\sum_{k=0}^{2^l-1}(-1)^{k+1}(5^{\frac{n+1}{2^l}})^k \equiv 0 \mod 4$, so $d \equiv 1 \mod 2$. Then both 0 and -1 are solutions of $(x+1)^d - x^d = 1$ and we get the result for $n \equiv -1 \mod 2^{l+1}$.

We notice that theorem 1.8 gives another proof of the case $n \equiv -1 \mod 2^{l+1}$ of theorem 1.9.

4. Some remarks about Zha and Wang theorems

4.1. A new proof of conjecture 1.6. We give another proof of conjecture 1.6. In the case where $n \equiv 1 \mod 4$, it is the same that in Zha and Wang's proof. But the proof of the case where $n \equiv 3 \mod 4$ doesn't use theorem 1.8.

First we recall the idea of the proof in case where $n \equiv 1 \mod 4$:

$$\frac{5^m+1}{2}d \equiv 1 \mod (5^n - 1).$$

So by lemma 3.1 and corollary 3.3 we get the result.

If $n \equiv 3 \mod 4$ then $3n \equiv 1 \mod 4$. So

$$\begin{aligned}\frac{5^{3n}-1}{4} + \frac{5^{\frac{3n+1}{2}}-1}{2} - \frac{5^n-1}{4} - \frac{5^m-1}{2} &= (5^n-1)\frac{2.5^m + 5^n(5^n+1)}{4} \\ &\equiv 0 \mod (5^n-1).\end{aligned}$$

Put $e = \frac{5^{3n}-1}{4} + \frac{5^{\frac{3n+1}{2}}-1}{2} \equiv \frac{5^n-1}{4} + \frac{5^m-1}{2} \mod 5^n - 1$ and, by the case where $n \equiv 1 \mod 4$, $x \mapsto x^e$ is APN over $\mathbb{F}_{5^{3n}}$ and $\Delta_d \leq 2$ for $n \equiv 3 \mod 4$. Furthermore d is odd, so $(0+1)^d - 0^d = 1 = (-1+1)^d - (-1)^d$ and $\Delta_d = 2$.

4.2. A remark on theorem 1.7. For $l \in \mathbb{N}$ and $n \equiv -1 \mod 2^l$, if we take $k = \frac{n+1}{2^l}$ and $u = 3$ in theorem 1.8, we get that $x \mapsto x^d$ is APN for

$$d = \frac{3^{n+1}-1}{3^{\frac{n+1}{2^l}}+1}.$$

For $l = 2$, this gives an explicit family to explain case IV of the table. Actually this family contains $656 = 224 \times 3^4 \mod 3^7 - 1$.

References

[CM97] Robert S. Coulter and Rex W. Matthews, *Planar functions and planes of Lenz-Barlotti class II*, Des. Codes Cryptogr. **10** (1997), no. 2, 167–184. MR1432296 (97j:51010)

[DMM+03] Hans Dobbertin, Donald Mills, Eva Nuria Müller, Alexander Pott, and Wolfgang Willems, *APN functions in odd characteristic*, Discrete Math. **267** (2003), no. 1-3, 95–112, Combinatorics 2000 (Gaeta). MR1991565 (2004j:12001)

[DO68] Peter Dembowski and T. G. Ostrom, *Planes of order n with collineation groups of order n^2*, Math. Z. **103** (1968), 239–258. MR0226486 (37:2075)

[Dob99a] Hans Dobbertin, *Almost perfect nonlinear power functions on $GF(2^n)$: the Niho case*, Inform. and Comput. **151** (1999), no. 1-2, 57–72. MR1692816 (2000d:94021)

[Dob99b] ———, *Almost perfect nonlinear power functions on $GF(2^n)$: the Welch case*, IEEE Trans. Inform. Theory **45** (1999), no. 4, 1271–1275. MR1686267 (2000b:94012)

[Dob99c] ———, *Another proof of Kasami's theorem*, Des. Codes Cryptogr. **17** (1999), no. 1-3, 177–180. MR1714379 (2000h:94025)

[Dob01] ———, *Almost perfect nonlinear power functions on $GF(2^n)$: a new case for n divisible by 5*, Finite fields and applications (Augsburg, 1999), Springer, Berlin, 2001, pp. 113–121. MR1849084 (2002i:11125)

[HRS99] Tor Helleseth, Chunming Rong, and Daniel Sandberg, *New families of almost perfect nonlinear power mappings*, IEEE Trans. Inform. Theory **45** (1999), no. 2, 474–485. MR1677012 (2000c:11201)

[JW93] H. Janwa and R. M. Wilson, *Hyperplane sections of Fermat varieties in $\mathbf{P}^3$ in char. 2 and some applications to cyclic codes*, Applied algebra, algebraic algorithms and error-correcting codes (San Juan, PR, 1993), Lecture Notes in Comput. Sci., vol. 673, Springer, Berlin, 1993, pp. 180–194. MR1251978 (95b:11119)

[LN97] Rudolf Lidl and Harald Niederreiter, *Finite fields*, second ed., Encyclopedia of Mathematics and its Applications, vol. 20, Cambridge University Press, Cambridge, 1997, With a foreword by P. M. Cohn. MR1429394 (97i:11115)

[Luc78] Edouard Lucas, *Theorie des Fonctions Numeriques Simplement Periodiques. [Continued]*, Amer. J. Math. **1** (1878), no. 3, 197–240. MR1505164

[Nyb94] Kaisa Nyberg, *Differentially uniform mappings for cryptography*, Advances in cryptology—EUROCRYPT '93 (Lofthus, 1993), Lecture Notes in Comput. Sci., vol. 765, Springer, Berlin, 1994, pp. 55–64. MR1290329 (95e:94039)

[ZW10] ZhengBang Zha and XueLi Wang, *Power functions with low uniformity on odd characteristic finite fields*, Sci. China Math. **53** (2010), no. 8, 1931–1940. MR2679075

INSTITUT DE MATHÉMATIQUES DE JUSSIEU, 175 RUE DU CHEVALERET, 75013 PARIS, FRANCE
E-mail address: `elodieleducq@math.jussieu.fr`

Contemporary Mathematics
Volume **574**, 2012
http://dx.doi.org/10.1090/conm/574/11421

Identities for Kloosterman sums and modular curves

Petr Lisoněk

ABSTRACT. We use the correspondence between Kloosterman sums and elliptic curves to prove identities for binary and ternary Kloosterman sums. The identities follow from rational parametrizations of modular curves $X_0(\ell)$. In the binary case we give the results from an unpublished manuscript by Mika R.S. Kojo and in the ternary case we give new results. We show that two identities published by other authors are consequences of identities proved in this paper.

1. Background

1.1. Kloosterman sums and elliptic curves. For a prime p and a positive integer m, let $q = p^m$. Let $\mathbb{F}_q$ denote the finite field of order q and $\mathbb{F}_q^* = \mathbb{F}_q \setminus \{0\}$. Let $\mathrm{Tr} : \mathbb{F}_q \to \mathbb{F}_p$ be the absolute trace.

The *Kloosterman sum* on $\mathbb{F}_q$ is the mapping $\mathcal{K} : \mathbb{F}_q \to \mathbb{R}$ defined for each $a \in \mathbb{F}_q$ by

$$\mathcal{K}(a) = 1 + \sum_{x \in \mathbb{F}_q^*} \zeta_p^{\mathrm{Tr}(x^{-1}+ax)}, \tag{1.1}$$

where $\zeta_p = e^{2\pi i/p}$ is a primitive p-th root of unity. In classical references the term 1 is not present in (1.1), however the divisibility properties of Kloosterman sums by powers of the field characteristic can be stated more compactly when the definition (1.1) is used, and we follow this recent trend. Since $\zeta_3 + \zeta_3^{-1} = -1$, it is easy to see that $\mathcal{K}(a)$ is an integer for $p \in \{2, 3\}$.

Kloosterman sums occur frequently in cryptography and in coding theory. Their applications include the construction of bent functions of high algebraic degree [**4, 6**], the assessment of the cryptographic security of the field inverse function [**3**], weight enumerators of cyclic codes [**12**] and construction of block designs from cyclic codes [**14**]. For the sake of brevity we have included only a small selection of references, however the available literature is much more extensive.

For an elliptic curve E defined over $\mathbb{F}_q$ we denote by $\#E(\mathbb{F}_q)$ the number of $\mathbb{F}_q$-rational points on E. We start by recalling two theorems that connect the value $\mathcal{K}(a)$ with the number of rational points on a corresponding elliptic curve.

2010 *Mathematics Subject Classification.* Primary 11T23; Secondary 11T71.

Research supported by the Natural Sciences and Engineering Research Council of Canada (NSERC).

THEOREM 1.1 ([**12**]). *Let $a \in \mathbb{F}_{2^m}^*$ and let E_2^a be the elliptic curve defined by*

$$E_2^a: \quad y^2 + xy = x^3 + a.$$

Then $\#E_2^a(\mathbb{F}_{2^m}) = 2^m + \mathcal{K}(a)$.

THEOREM 1.2 ([**10**]). *Let $a \in \mathbb{F}_{3^m}^*$ and let E_3^a be the elliptic curve defined by*

$$E_3^a: \quad y^2 + xy + ay = x^3.$$

Then $\#E_3^a(\mathbb{F}_{3^m}) = 3^m + \mathcal{K}(a)$.

1.2. Kloosterman sums and modular curves. By a classical result of Deuring, later generalized by Tate [**15**, Theorem 1], for two elliptic curves E and E' over $\mathbb{F}_q$ we have $\#E(\mathbb{F}_q) = \#E'(\mathbb{F}_q)$ if and only if E and E' are isogenous. We recall that ℓ-isogeny of elliptic curves is parametrized by the modular curve $X_0(\ell)$. This curve has a model $\Phi_\ell(x,y) \in \mathbb{Z}[x,y]$ (often called the classical modular polynomial of level ℓ, or the modular equation) and elliptic curves E, E' with j-invariants $j(E), j(E')$ are ℓ-isogenous if and only if $\Phi_\ell(j(E), j(E')) = 0$. The j-invariants of curves introduced in Theorems 1.1 and 1.2 are $j(E_2^a) = a^{-1}$ and $j(E_3^a) = a^{-3}$, respectively. Note that $a \mapsto j(E_2^a)$ is a bijection from $\mathbb{F}_{2^m}^*$ to $\mathbb{F}_{2^m}^*$ and $a \mapsto j(E_3^a)$ is a bijection from $\mathbb{F}_{3^m}^*$ to $\mathbb{F}_{3^m}^*$.

By applying Theorems 1.1 and 1.2 we obtain:

THEOREM 1.3 ([**11**]). *For $a, b \in \mathbb{F}_{2^m}^*$ we have $\mathcal{K}(a) = \mathcal{K}(b)$ if and only if there exists $\ell \in \mathbb{N}$ such that $\Phi_\ell(a^{-1}, b^{-1}) = 0$.*

THEOREM 1.4. *For $a, b \in \mathbb{F}_{3^m}^*$ we have $\mathcal{K}(a) = \mathcal{K}(b)$ if and only if there exists $\ell \in \mathbb{N}$ such that $\Phi_\ell(a^{-1}, b^{-1}) = 0$.*

In Theorem 1.4 we also applied the inverse Frobenius automorphism $z \mapsto z^{1/3}$ in order to reduce the exponents -3 to -1, thereby reducing the complexity of calculations that will follow.

Theorem 1.3 is implicitly used and proved in [**11**]. We will give more background on [**11**] in the next section.

For prime levels $\ell = 2, 3, 5, 7, 13$, the curve $X_0(\ell)$ has genus 0 and thus it has a rational parametrization. These parametrizations were first explicitly calculated by F. Klein in 1879; we will use the recent paper [**13**] as the reference. In Sections 2.1 and 2.2 we will use the parametrizations of $X_0(\ell)$ given in [**13**, Table 7] to derive identities for Kloosterman sums. We will comment on some identities obtained from composite values ℓ in Section 2.3.

By the Kronecker congruence relation, the level p modular polynomial Φ_p factors over $\mathbb{F}_p$ as $\Phi_p(X,Y) = (X^p - Y)(X - Y^p) \in \mathbb{F}_p[X,Y]$. For $p \in \{2,3\}$ the corresponding Kloosterman sum identity is simply $\mathcal{K}(a) = \mathcal{K}(a^p)$, which is well known and very easy to prove. In Sections 2.1 and 2.2 (where we deal with prime ℓ) we thus further restrict our attention to cases when ℓ is different from the field characteristic.

2. Kloosterman sum identities

For $p \in \{2,3\}$ we are interested in identities of the form $\mathcal{K}(f_1(b)) = \mathcal{K}(f_2(b))$ that hold for all $b \in \mathbb{F}_{p^m}$ and all $m \in \mathbb{N}$, where f_1 and f_2 are rational functions over $\mathbb{F}_p$. There also exist identities that only hold in fields of specific orders; we do not deal with such identities in this paper.

2.1. Binary identities at prime levels. In characteristic 2 the first Kloosterman sum identity that appeared in the literature is part (a) of Theorem 2.1 below, which was proved in [**14**] for odd m and in [**8**] for all m. This identity has an interesting combinatorial context, since it is used in the construction of certain block designs from codewords of the $\mathbb{Z}_4$-Goethals codes [**14**]. The identity listed as part (b) of Theorem 2.1 below was proved in [**7**]. Jyrki Lahtonen and Lauri Ojala found a proof for the identity in part (a) of Theorem 2.1 by constructing an explicit isogeny between the corresponding elliptic curves.

The Kloosterman sum identities proved in [**14**] and [**7**] and the proof by Lahtonen and Ojala were put into the context of modular curves by Mika R.S. Kojo [**11**]. Kojo realized that the identities follow from the parametrizations of modular curves $X_0(\ell)$. The draft manuscript [**11**] has never been published.

THEOREM 2.1. *Let $b \in \mathbb{F}_{2^m}$. We have:*

(a) [**14, 8**]

$$\mathcal{K}\left(b^3(1+b)\right) = \mathcal{K}\left(b(1+b)^3\right)$$

(b) [**7**]

$$\mathcal{K}\left(b^5(1+b)\right) = \mathcal{K}\left(b(1+b)^5\right)$$

(c) [**11**] *If $b^2+b+1 \neq 0$, then*

$$\mathcal{K}\left(\frac{b^7(1+b)}{(b^2+b+1)^4}\right) = \mathcal{K}\left(\frac{b(1+b)^7}{(b^2+b+1)^4}\right).$$

(d) [**11**] *If $b^2+b+1 \neq 0$, then*

$$\mathcal{K}\left(\frac{b^{13}(1+b)}{(b^2+b+1)^4}\right) = \mathcal{K}\left(\frac{b(1+b)^{13}}{(b^2+b+1)^4}\right).$$

PROOF. In [**11**] all four identities are proved using the well known parametrizations of $X_0(\ell)$ for $\ell = 3, 5, 7, 13$, respectively, combined with Theorem 1.3 above. These parametrizations can be found, for example, by reducing modulo 2 the parametrizations given in [**13**, Table 7] for characteristic 0. (The characteristic 0 parametrizations can be also found for example in [**5**].) Additionally Kojo applied the transformation $t = (1+b)/b$ (where t is the parametrization variable used by [**13**]) so that identities (a) and (b) match the forms given in [**8, 7**]; this additional transformation gives more compact formulas in all four cases. In the singular cases $b = 0, 1$ the Kloosterman sums in Theorem 2.1 are not associated with an elliptic curve; the identities hold trivially. □

2.2. Ternary identities at prime levels. There are no results for characteristic 3 in [**11**], however the proof technique works in the same way; the key to this is the similarity between Theorem 1.3 and Theorem 1.4. We obtain the following identities, none of which has appeared in the literature.

THEOREM 2.2. *Let $b \in \mathbb{F}_{3^m}$. We have:*

(a)

$$\mathcal{K}\left(b^2(1-b)\right) = \mathcal{K}\left(b(1-b)^2\right)$$

(b) *If $b^2+b-1 \neq 0$, then*

$$\mathcal{K}\left(\frac{b^5(b-1)}{(b^2+b-1)^3}\right) = \mathcal{K}\left(\frac{b(b-1)^5}{(b^2+b-1)^3}\right).$$

(c) *If* $b \neq -1$, *then*

$$\mathcal{K}\left(\frac{b^7(1-b)}{(b+1)^2}\right) = \mathcal{K}\left(\frac{b(1-b)^7}{(b+1)^2}\right).$$

(d) *If* $b \neq -1$ *and* $b^2 - b - 1 \neq 0$, *then*

$$\mathcal{K}\left(\frac{b^{13}(b-1)}{(b+1)^6(b^2-b-1)^3}\right) = \mathcal{K}\left(\frac{b(b-1)^{13}}{(b+1)^6(b^2-b-1)^3}\right).$$

PROOF. In analogy to the binary case all four identities are proved using the well known parametrizations of $X_0(\ell)$ for $\ell = 2, 5, 7, 13$, respectively, combined with Theorem 1.4 above. We use the parametrizations given in [**13**, Table 7] for characteristic 0, reduce them modulo 3, and additionally we apply the transformation $t = (1-c)/c$ (where t is the parametrization variable used by [**13**]). As in the binary case, the additional transformation gives more compact formulas in all four cases. Again in the singular cases $b = 0, 1$ the Kloosterman sums in Theorem 2.2 are not associated with an elliptic curve but the identities hold trivially. □

2.3. Identities at composite levels. For composite ℓ an ℓ-isogeny can be obtained by composing isogenies of prime levels. Thus a Kloosterman sum identity that is produced from $X_0(\ell)$ for composite ℓ can be seen as a combination of identities at prime levels. We now illustrate this phenomenon on two identities that have been published in the literature.

THEOREM 2.3 ([**9**, Theorem 1.1(iii)]). *For* $b \in \mathbb{F}_{2^m}$ *we have*

$$\mathcal{K}\left(b^8(b^4+b)\right) = \mathcal{K}\left((b^4+b)(1+b)^8\right). \tag{2.1}$$

PROOF. A quick calculation in Magma [**1**] (which contains a data base of $\Phi_\ell \in \mathbb{Z}[X,Y]$ for modest values of ℓ) reveals that (2.1) follows from Theorem 1.3 with $\ell = 9$. Since $9 = 3^2$, we see that (2.1) also can be proved by combining two binary level 3 identities as follows:

Let us rewrite (2.1) in the equivalent form (where we replaced b with c)

$$\mathcal{K}\left(c^9(c^3+1)\right) = \mathcal{K}\left(c(c+1)^9(c^2+c+1)\right). \tag{2.2}$$

By letting $b = c^3$ in Theorem 2.1(a) we get

$$\mathcal{K}\left(c^9(c^3+1)\right) = \mathcal{K}\left(c^3(c^3+1)^3\right), \tag{2.3}$$

and by letting $b = c(c^2+c+1)$ in Theorem 2.1(a) we get

$$\mathcal{K}\left(c^3(c^2+c+1)^3(c+1)^3\right) = \mathcal{K}\left(c(c^2+c+1)((c+1)^3)^3\right). \tag{2.4}$$

Equation (2.2) now follows by combining (2.3) and (2.4). □

THEOREM 2.4 ([**2**, Theorem 1.8]). *For* $b \in \mathbb{F}_{3^m}$ *we have*

$$\mathcal{K}\left(b^3(b-b^3)\right) = \mathcal{K}\left((b^3+1)(b-b^3)\right). \tag{2.5}$$

PROOF. A quick calculation in Magma reveals that (2.5) follows from Theorem 1.4 with $\ell = 4$. Since $4 = 2^2$, we see that (2.5) also can be proved by combining two ternary level 2 identities as follows:

Let us rewrite (2.5) in the equivalent form

$$\mathcal{K}\left(b^4(1+b)(1-b)\right) = \mathcal{K}\left(b(1+b)^4(1-b)\right). \tag{2.6}$$

By letting $a = (1+b)(1-b)$ in Theorem 2.2(a) we get

$$\mathcal{K}\left((1+b)(1-b)b^4\right) = \mathcal{K}\left(((1+b)(1-b))^2 b^2\right), \tag{2.7}$$

and by letting $a = (1+b)^2$ in Theorem 2.2(a) we get

$$\mathcal{K}\left((1+b)^2\left((-1+b)(-b)\right)^2\right) = \mathcal{K}\left((1+b)^4(-1+b)(-b)\right). \tag{2.8}$$

Equation (2.6) now follows by combining (2.7) and (2.8). □

Of course more identities for binary and ternary Kloosterman sums can be produced using composite levels ℓ for which $X_0(\ell)$ has genus 0; let us note that [**13**] lists the parametrizations also for composite ℓ. We chose to include in Section 2.3 only the two identities that have been already published elsewhere, because the fact that these identities are implied by simpler identities has been apparently overlooked in the original references. In particular we would like to point out that Theorem 2.4, which is the only ternary identity that has been published in the literature so far, is a consequence of a much simpler identity that we proved as part (a) in Theorem 2.2.

Acknowledgment

The author is indebted to Jyrki Lahtonen for providing a copy of [**11**].

References

1. W. Bosma, J. Cannon, C. Playoust, The Magma algebra system. I. The user language. J. Symbolic Comput. 24 (1997), no. 3-4, 235–265. MR1484478
2. X. Cao, H.D.L. Hollmann, Q. Xiang, New Kloosterman sum identities and equalities over finite fields. Finite Fields Appl. 14 (2008), no. 3, 823–833. MR2435065 (2009f:11153)
3. P. Charpin, T. Helleseth, V. Zinoviev, Propagation characteristics of $x \mapsto x^{-1}$ and Kloosterman sums. Finite Fields Appl. 13 (2007), no. 2, 366–381. MR2307134 (2008c:11110)
4. J.F. Dillon, Elementary Hadamard difference sets. Proceedings of the Sixth Southeastern Conference on Combinatorics, Graph Theory, and Computing (Florida Atlantic Univ., Boca Raton, Fla., 1975), pp. 237–249. Congressus Numerantium, No. XIV, Utilitas Math., Winnipeg, Man., 1975. MR0409221 (53:12981)
5. N.D. Elkies, Elliptic and modular curves over finite fields and related computational issues. Computational perspectives on number theory (Chicago, IL, 1995), 21–76, AMS/IP Stud. Adv. Math., 7, Amer. Math. Soc., Providence, RI, 1998. MR1486831 (99a:11078)
6. T. Helleseth, A. Kholosha, Monomial and quadratic bent functions over the finite fields of odd characteristic. IEEE Trans. Inform. Theory 52 (2006), no. 5, 2018–2032. MR2234462 (2007b:11190)
7. T. Helleseth, V. Zinoviev, New Kloosterman sums identities over F_{2^m} for all m. Finite Fields Appl. 9 (2003), no. 2, 187–193. MR1968029 (2003k:11183)
8. T. Helleseth, V. Zinoviev, On a new identity for Kloosterman sums and nonlinear system of equations over finite fields of characteristic 2. Discrete Math. 274 (2004), no. 1-3, 109–124. MR2025078 (2004h:11108)
9. H.D.L. Hollmann, Q. Xiang, Kloosterman sum identities over F_{2^m}. Discrete Math. 279 (2004), no. 1-3, 277–286. MR2059995 (2005c:11159)
10. N. Katz, R. Livné, Sommes de Kloosterman et courbes elliptiques universelles en caractéristiques 2 et 3. C. R. Acad. Sci. Paris Sér. I Math. 309 (1989), no. 11, 723–726. MR1054286 (91e:11066)
11. M.R.S. Kojo, Modular curves and identities of classical Kloosterman sums. Unpublished manuscript, 17 October 2002.
12. G. Lachaud, J. Wolfmann, The weights of the orthogonals of the extended quadratic binary Goppa codes. IEEE Trans. Inform. Theory 36 (1990), no. 3, 686–692. MR1053865 (92b:94040)

13. R.S. Maier, On rationally parametrized modular equations. J. Ramanujan Math. Soc. 24 (2009), no. 1, 1–73. MR2514149 (2010f:11060)
14. D. Shin, P.V. Kumar, T. Helleseth, 3-designs from the Z_4-Goethals codes via a new Kloosterman sum identity. Des. Codes Cryptogr. 28 (2003), no. 3, 247–263. MR1976960 (2004d:94044)
15. J. Tate, Endomorphisms of abelian varieties over finite fields. Invent. Math. 2 (1966) 134–144. MR0206004 (34:5829)

Department of Mathematics, Simon Fraser University, Burnaby, BC, Canada V5A 1S6

E-mail address: plisonek@sfu.ca

Contemporary Mathematics
Volume **574**, 2012
http://dx.doi.org/10.1090/conm/574/11426

Degree growth, linear independence and periods of a class of rational dynamical systems

Alina Ostafe and Igor Shparlinski

ABSTRACT. We introduce and study algebraic dynamical systems generated by triangular systems of rational functions. We obtain several results about the degree growth and linear independence of iterates as well as about possible lengths of trajectories generated by such dynamical systems over finite fields. Some of these results are generalisations of those known in the polynomial case, some are new even in this case.

1. Introduction

Let $\mathbb{F}$ be an arbitrary field and let $F_1, \dots, F_m \in \mathbb{F}(X_1, \dots, X_m)$ be m rational functions in m variables over $\mathbb{F}$. For each $i = 1, \dots, m$ we define the k-th iteration of the rational function F_i by the recurrence relation

$$F_i^{(0)} = X_i, \quad F_i^{(k)} = F_i\left(F_1^{(k-1)}, \dots, F_m^{(k-1)}\right), \quad k = 1, 2, \dots . \tag{1}$$

In this paper we consider dynamical systems generated by multivariate rational functions, we refer to [**1, 21, 22**] for a background on algebraic dynamical systems.

More precisely, we define the vectors $\mathbf{u}_n = (u_{n,1}, \dots, u_{n,m}) \in \mathbb{F}^m$ by the recurrence relation

$$u_{n+1,i} = F_i(u_{n,1}, \dots, u_{n,m}), \qquad n = 0, 1, \dots, \qquad i = 1, \dots, m, \tag{2}$$

with some *initial vector* $\mathbf{u}_0 = (u_{0,1}, \dots, u_{0,m}) \in \mathbb{F}^m$.

As we work with rational functions, we make the standard convention (see [**3, 9, 10**]) that

$$0^{-1} = 0. \tag{3}$$

Using the following vector notation

$$\mathbf{F} = (F_1(X_1, \dots, X_m), \dots, F_m(X_1, \dots, X_m)),$$

we have the recurrence relation

$$\mathbf{u}_{n+1} = \mathbf{F}(\mathbf{u}_n), \quad n = 0, 1, \dots . \tag{4}$$

2010 *Mathematics Subject Classification.* Primary 37P05; Secondary 11T06, 37P25, 65C10 .

During the preparation of this paper, A. O. was supported by the Swiss National Science Foundation Grant PBZHP2–133399 and I. S. by the Australian Research Council Grant DP1092835.

In particular, for any $n \ge 0$ and $i = 1, \dots, m$ we have

$$u_{n,i} = F_i^{(n)}(\mathbf{u}_0) = F_i^{(n)}(u_{0,1}, \dots, u_{0,m})$$

or

$$\mathbf{u}_n = \mathbf{F}^{(n)}(\mathbf{u}_0),$$

provided that $\mathbf{u}_n$ has been generated by (4) without using the convention (3) (that is, no poles have been encountered).

Clearly, if we work over a finite field of q elements, the above sequence (4) of vectors $\{\mathbf{u}_n\}$ is eventually periodic with some period $\tau \le q^m$.

One of the important characteristics of the dynamical system generated by $F_1, \dots, F_m \in \mathbb{F}(X_1, \dots, X_m)$ is the degree growth of the functions (1). It is of great interest for the theory of dynamical systems and has been studied in a number of works, see, for example, [**2, 25**] and references therein. It is also important for applications to pseudorandom number generators [**24**].

More precisely, although for a "typical" system an exponential degree growth is expected, there are several examples of systems where the degree grows much slower (which is highly beneficial for their applications), and such systems are of special interest.

For example, in [**14, 17**] several types of multivariate polynomial systems $\mathcal{F} = \{F_1, \dots, F_m\}$ of m polynomials in m variables over a finite field $\mathbb{F}_q$ have been constructed and studied, having the "triangular" form

$$\begin{aligned} F_1(X_1, \dots, X_m) &= X_1 G_1(X_2, \dots, X_m) + H_1(X_2, \dots, X_m), \\ &\dots \\ F_{m-1}(X_1, \dots, X_m) &= X_{m-1} G_{m-1}(X_m) + H_{m-1}(X_m), \\ F_m(X_1, \dots, X_m) &= g_m X_m + h_m, \end{aligned} \tag{5}$$

with $G_i, H_i \in \mathbb{F}_q[X_{i+1}, \dots, X_m]$, $i = 1, \dots, m-1$, and $g_m, h_m \in \mathbb{F}_q$, $g_m \ne 0$. These systems have been further investigated in [**13, 18, 19, 20**].

For the systems (5), in the case of constant polynomials $G_i \in \mathbb{F}_q^*$ in [**14**] and polynomials G_i with leading terms of special form in [**17, 18**], a series of results have been obtained about the distribution of the corresponding sequences given by (2) that are much stronger than those known for generic systems. Moreover, for these classes of polynomials, it has been shown in [**17**] that the degrees of the iterations of the polynomials F_i, $i = 1, \dots, m$, grow significantly slower than the exponential growth expected for the iterations of a "generic" system of m polynomials in m variables. In turn, this leads (see [**18**]) to much better estimates of exponential sums, and thus of discrepancy, for vectors generated by (5) than for those originated from arbitrary polynomial systems (see [**4, 5, 16**]).

We also note that the results obtained in [**17, 18**] regarding the degree growth of the iterations of the polynomials in (5) hold over any field $\mathbb{F}$.

In this paper we extend the class of rational dynamical systems with slow degree growth and present an analogue of the construction (5), but with rational functions defined by

$$\begin{aligned} F_1(X_1, \dots, X_m) &= X_1^{e_1} G_1(X_2, \dots, X_m) + H_1(X_2, \dots, X_m), \\ &\dots \\ F_{m-1}(X_1, \dots, X_m) &= X_{m-1}^{e_{m-1}} G_{m-1}(X_m) + H_{m-1}(X_m), \\ F_m(X_1, \dots, X_m) &= g_m X_m^{e_m} + h_m, \end{aligned} \tag{6}$$

with $e_1, \dots, e_m \in \{-1, 1\}$, $G_i, H_i \in \mathbb{F}[X_{i+1}, \dots, X_m]$, $i = 1, \dots, m-1$, and $g_m, h_m \in \mathbb{F}$, $g_m \neq 0$.

We note that for $m = 1$ and $e = 1$ we obtain the classical linear congruential generator which have been successully used for decades in the theory of Quasi Monte Carlo methods, see [**7, 8**], and for $m = 1$ and $e = -1$, the classical inversive generator, see [**9, 10, 11, 12**].

For the above class of multivariate rational functions, we study the degree growth under iterations and, using an approach similar to that of [**17**, Lemma 1], we show in Section 3 that under certain additional conditions imposed on the systems of rational functions (6), the degree grows polynomially in the number of iterations.

Moreover, for applications to pseudorandom number generators, following the standard technique almost identical to that of [**17**], one almost immediately obtains bounds on the exponential sums with elements of the sequence (4) generated by the system (6) (satisfying the conditions outlined in Section 3), that in turn leads to estimates on the uniformity of distribution of the vectors (4). However, one has also to prove that for any $k \neq l$ and nonzero vector $\mathbf{a} = (a_1, \dots, a_{m-1}) \in \mathbb{F}^{m-1}$, the linear combination

$$Q_{k,l,\mathbf{a}} = \sum_{i=1}^{m-1} a_i (F_i^{(k)} - F_i^{(l)}) \tag{7}$$

is a non-constant rational function. We note that in the case of rational functions this does not follow directly from the degree argument as in the case of the polynomial systems (5) in [**17**], but we give such a result in Section 4.

We note that we do not include the polynomials $F_m^{(k)}$ and $F_m^{(l)}$ in $Q_{k,l,\mathbf{a}}$ as generally speaking, due to the linearity of F_m (as being a linear or inversive function), such a linear combination may vanish even for nontrivial coefficients (note that it is possible that $F_m^{(k)} = F_m^{(l)}$ for $k \neq l$).

Since the derivation of such bounds of exponential sums for our systems does not bring anything new to the area, we do not do this here but rather concentrate on the study of the degree and linear independence of iterates, which is also of interest for the general area of algebraic dynamics.

Furthermore, we consider a related question about the length of trajectories generated by iterations (4) over a finite field $\mathbb{F}_q$. We remark that in this case a trajectory falls into a cycle if $\mathbf{u}_t = \mathbf{u}_s$ for some integers $t > s \geq 0$. In particular, we show that under some rather broad conditions for any fixed $\varepsilon > 0$, for all but $o(q^m)$ initial vectors $\mathbf{u}_0 \in \mathbb{F}_q^m$, the trajectory length t of the iterations (4) is at least $q^{1/3-\varepsilon}$.

We note that Silverman [**23**] has considered a question about periods of general polynomial systems but in somewhat dual situation when the initial value is fixed and the iterations are considered over a family of finite fields. The results of [**23**] apply to very general systems, however the estimates are only logarithmic rather than a power of the field size.

Moreover, we give necessary and sufficient conditions for the systems (6) to generate sequences of maximal period. We note that for the case $e_i = 1$ for all $i = 1, \dots, m$, the maximal period length of the sequence generated by the system (5) is achieved whenever the conditions of [**15**, Theorem 6] are satisfied. Our result is a generalisation of that of [**15**].

2. Structure of the Iterations

As in [**17**], we can describe explicitly the iterations of the rational functions F_i, $i = 1, \dots, m$, defined by (6), as follows.

Let us define the sets

$$I_+ = \{1 \le i \le m \ : \ e_i = 1\} \quad \text{and} \quad I_- = \{1 \le i \le m \ : \ e_i = -1\}. \tag{8}$$

We also define

$$\begin{aligned} G_i^{(\ell)}(X_{i+1}, \dots, X_m) &= G_i\left(F_{i+1}^{(\ell-1)}, \dots, F_m^{(\ell-1)}\right), \\ H_i^{(\ell)}(X_{i+1}, \dots, X_m) &= H_i\left(F_{i+1}^{(\ell-1)}, \dots, F_m^{(\ell-1)}\right). \end{aligned}$$

LEMMA 1. *Let $F_1, \dots, F_m$ be rational functions defined by (6). Then for $i = 1, \dots, m-1$ and $k = 0, 1, \dots,$ for the rational functions $F_i^{(k)}$ given by (1),*

(1) *for every $i \in I_+$, $i < m$, we have*

$$F_i^{(k)} = X_i G_{i,k} + H_{i,k}, \tag{9}$$

where $G_{i,k}, H_{i,k} \in \mathbb{F}(X_{i+1}, \dots, X_m)$ are defined by

$$\begin{aligned} G_{i,k} &= G_i G_i^{(2)} \dots G_i^{(k)}, \\ H_{i,k} &= H_i G_i^{(2)} \dots G_i^{(k)} + H_i^{(2)} G_i^{(3)} \dots G_i^{(k)} + \dots + H_i^{(k-1)} G_i^{(k)} + H_i^{(k)}; \end{aligned} \tag{10}$$

(2) *for every $i \in I_-$, $i < m$, we have:*

$$F_i^{(k)} = \frac{X_i R_{i,k} + S_{i,k}}{X_i R_{i,k-1} + S_{i,k-1}}, \tag{11}$$

where $R_{i,k}, S_{i,k}$ are defined by the recurrence relations

$$\begin{aligned} R_{i,k} &= G_i^{(k)} R_{i,k-2} + H_i^{(k)} R_{i,k-1}, \\ S_{i,k} &= G_i^{(k)} S_{i,k-2} + H_i^{(k)} S_{i,k-1}, \end{aligned} \tag{12}$$

for $k \ge 1$, with the initial rational functions

$$R_{i,0} = 1, \quad S_{i,0} = 0, \quad R_{i,1} = H_i, \quad S_{i,1} = G_i;$$

(3) *if $m \in I_+$, then*

$$F_m^{(k)} = g_m^k X_m + (g_m^{k-1} + \dots + g_m + 1) h_m;$$

(4) *if $m \in I_-$, then*

$$F_m^{(k)} = \frac{(A^k)_{1,1} X_m + (A^k)_{1,2}}{(A^k)_{2,1} X_m + (A^k)_{2,2}},$$

where

$$A^k = \begin{pmatrix} h_m & g_m \\ 1 & 0 \end{pmatrix}^k = \begin{pmatrix} (A^k)_{1,1} & (A^k)_{1,2} \\ (A^k)_{2,1} & (A^k)2,2 \end{pmatrix}.$$

PROOF. The case $e_i = 1$, $i = 1, \dots, m$, is given by [**18**, Lemma 1]. We consider now that $e_i = -1$ and prove the result by induction on the number of iterations k. For $k = 1$ it is clear from the definition of the system, so we consider the statement true for the first $k-1$ iterations and we prove it for the k-th iteration. For $i = 1, \ldots, m-1$, we have

$$\begin{aligned} F_i^{(k)} &= F_i(F_i^{(k-1)}, F_{i+1}^{(k-1)}, \ldots, F_m^{(k-1)}) \\ &= \frac{F_i^{(k-1)} H_i^{(k)} + G_i^{(k)}}{F_i^{(k-1)}} = \frac{\frac{X_i R_{i,k-1}+S_{i,k-1}}{X_i R_{i,k-2}+S_{i,k-2}} H_i^{(k)} + G_i^{(k)}}{\frac{X_i R_{i,k-1}+S_{i,k-1}}{X_i R_{i,k-2}+S_{i,k-2}}} \\ &= \frac{X_i(G_i^{(k)} R_{i,k-2} + H_i^{(k)} R_{i,k-1}) + G_i^{(k)} S_{i,k-2} + H_i^{(k)} S_{i,k-1}}{X_i R_{i,k-1} + S_{i,k-1}} \end{aligned}$$

and thus we conclude this case. When $e_m = -1$, it is also clear as the k-th iteration of

$$F_m = \frac{h_m X_m + g_m}{X_m}$$

is given by A^k as simple calculations show. □

We want to describe the degree growth of the iterations of the rational functions defined by (6), and in particular to prove that we have the same effect of slow degree growth as for the polynomial systems (5) described in [**17**, Lemma 1]. To be able to give an explicit formula for the degree growth we need to impose some further conditions on the degrees of the polynomials G_i and H_i, $i = 1, \ldots, m-1$.

Let $F_1, \ldots, F_m$ be rational functions defined by (6). From now on we consider the system (6) satisfying the following conditions for F_i for any $i = 1, \ldots, m-1$:

(1) if $e_i = 1$, as in [**17, 18**], we assume that the polynomial G_i has a unique leading monomial $X_{i+1}^{s_{i,i+1}} \ldots X_m^{s_{i,m}}$, that is

$$G_i = g_i X_{i+1}^{s_{i,i+1}} \ldots X_m^{s_{i,m}} + \widetilde{G}_i,$$

where $g_i \in \mathbb{F}^*$ and $\widetilde{G}_i \in \mathbb{F}[X_{i+1}, \ldots, X_m]$ with

$$\deg_{X_j} \widetilde{G}_i < s_{i,j}, \quad \deg_{X_j} H_i < s_{i,j}, \quad j = i+1, \ldots, m; \tag{13}$$

(2) if $e_i = -1$, we assume that the polynomial H_i has a unique leading monomial $X_{i+1}^{s_{i,i+1}} \ldots X_m^{s_{i,m}}$, that is

$$H_i = h_i X_{i+1}^{s_{i,i+1}} \ldots X_m^{s_{i,m}} + \widetilde{H}_i,$$

where $h_i \in \mathbb{F}^*$ and $\widetilde{H}_i \in \mathbb{F}[X_{i+1}, \ldots, X_m]$ with

$$\deg_{X_j} \widetilde{H}_i < s_{i,j}, \quad \deg_{X_j} G_i < 2s_{i,j}, \quad j = i+1, \ldots, m. \tag{14}$$

Having these conditions, we prove the following formula for the degree growth which coincides with [**17**, Lemma 1].

3. Degree Growth

THEOREM 2. *Let $F_1, \ldots, F_m$ be rational functions defined by (6) satisfying the conditions (13) and (14) and such that $s_{i,i+1} \neq 0$, $i = 1, \ldots, m-1$. Then, the degrees of the iterations of $F_1, \ldots, F_m$ grow as follows*

$$\begin{aligned} \deg F_i^{(k)} &= \frac{1}{(m-i)!} k^{m-i} s_{i,i+1} \ldots s_{m-1,m} + \psi_i(k), \qquad i = 1, \ldots, m-1, \\ \deg F_m^{(k)} &= 1, \end{aligned}$$

where $\psi_i(T) \in \mathbb{Q}[T]$ is a polynomial of degree $\deg \psi_i < m-i$.

PROOF. The proof is based on Lemma 1. The case $e_i = 1$, using (9) and (11), follows exactly the same as in [**17**, Lemma 1].

We prove now the case when $e_i = -1$. Using the conditions (13) and (14) and the recurrence relation (12), it is easy to see that

$$
\begin{aligned}
\deg F_i^{(k)} &= \deg R_{i,k} + 1 = \deg H_i^{(k)} R_{i,k-1} + 1 \\
&= \deg H_i H_i^{(2)} \dots H_i^{(k)} + 1 = \sum_{j=1}^{k} \deg H_i^{(j)} + 1.
\end{aligned} \tag{15}
$$

As in [**17**, Lemma 1], we use induction on the number of variables m. For $m = 2$ we easily see that $\deg H_1^{(j)} = \deg H_1 = s_{1,2}$, and thus, by (15), we have that $\deg F_1^{(k)} = ks_{1,2} + 1$. We assume now that the theorem is true for $m - 1$ variables and we prove it for m. For any $i = 1, \dots, m-1$, by the induction hypothesis, we have

$$
\begin{aligned}
\deg F_i^{(k)} &= \sum_{j=1}^{k} \deg H_i^{(j)} + 1 = \sum_{j=1}^{k} \deg H_i(F_{i+1}^{(j-1)}, \dots, F_m^{(j-1)}) + 1 \\
&= \sum_{j=1}^{k} \deg \left((F_{i+1}^{(j-1)})^{s_{i,i+1}} \dots (F_m^{(j-1)})^{s_{i,m}} \right) + 1 \\
&= \sum_{j=1}^{k} \left(\sum_{\ell=i+1}^{m} \frac{1}{(m-\ell)!} (j-1)^{m-\ell} s_{i,\ell} s_{\ell,\ell+1} \dots s_{m-1,m} \right) + 1.
\end{aligned}
$$

As

$$
\sum_{j=1}^{k} j^{m-1-i} = \frac{1}{m-i} (B_{m-i}(k+1) - B_{m-i}(0)),
$$

where B_{m-i} is the Bernoulli polynomial of degree $m - i$ (which has the leading coefficient equal to 1), we finally obtain the desired result. □

4. Linear Independence

THEOREM 3. *Let $F_1, \dots, F_m$ be rational functions defined by (6) satisfying the conditions (13) and (14) and such that $s_{i,i+1} \neq 0$, $i = 1, \dots, m-1$. Then, for $k \neq l$ and a nonzero vector $\mathbf{a} \in \mathbb{F}^{m-1}$, $Q_{k,l,\mathbf{a}}$ is a non-constant rational function.*

PROOF. The proof reduces to proving that $\deg_{X_s}(F_s^{(k)} - F_s^{(l)}) > 1$, where $s \leq m - 1$ is the smallest index such that $a_s \neq 0$, as the variable X_s does not appear in the polynomial

$$
Q_{k,l,\mathbf{a}} - a_s(F_s^{(k)} - F_s^{(l)}) = \sum_{i=s+1}^{m-1} a_i (F_i^{(k)} - F_i^{(l)}).
$$

If $e_s = 1$, it is clear, as from Theorem 2, for $k > l$ we have $\deg G_{s,k} > \deg G_{s,l}$. If $e_s = -1$, by (11), we have

$$
F_s^{(k)} - F_s^{(l)} = \frac{X_s R_{s,k} + S_{s,k}}{X_s R_{s,k-1} + S_{s,k-1}} - \frac{X_s R_{s,l} + S_{s,l}}{X_s R_{s,l-1} + S_{s,l-1}} = \frac{U_{k,l,s}}{V_{k,l,s}},
$$

where

$$\begin{aligned} U_{k,l,s} = X_s^2(R_{s,k}R_{s,l-1} - R_{s,k-1}R_{s,l}) \\ + X_s(R_{s,k}S_{s,l-1} + S_{s,k}R_{s,l-1} - R_{s,k-1}S_{s,l} - S_{s,k-1}R_{s,l}) \\ + S_{s,k}S_{s,l-1} - S_{s,k-1}S_{s,l} \end{aligned}$$

and

$$V_{k,l,s} = X_s^2 R_{s,k-1}R_{s,l-1} + X_s(R_{s,k-1}S_{s,l-1} + R_{s,l-1}S_{s,k-1}) + S_{s,k-1}S_{s,l-1}.$$

Without loss of generality we may assume that $k > l$. Using (12), we obtain

$$\begin{aligned} R_{s,k}R_{s,l-1} - R_{s,k-1}R_{s,l} = (G_s^{(k)}R_{s,k-2} + H_s^{(k)}R_{s,k-1})R_{s,l-1} \\ - R_{s,k-1}(G_s^{(l)}R_{s,l-2} + H_s^{(l)}R_{s,l-1}), \end{aligned}$$

and thus, using Lemma 1 and Theorem 2, we derive

$$\begin{aligned} \deg(R_{s,k}R_{s,l-1} - R_{s,k-1}R_{s,l}) &= \deg(H_s^{(k)} - H_s^{(l)})R_{s,k-1}R_{s,l-1} \\ &= \deg H_s^{(k)}R_{s,k-1}R_{s,l-1} > \deg R_{s,k-1}R_{s,l-1} > 1 \end{aligned} \tag{16}$$

for $k > l$, which concludes the proof. □

Note that, as in [**18**], we can include m-term linear combinations

$$\overline{Q}_{k,l,\mathbf{a}} = \sum_{i=1}^{m} a_i(F_i^{(k)} - F_i^{(l)})$$

with $\mathbf{a} \in \mathbb{F}^m$, but in the case of $a_1 = \ldots = a_{m-1} = 0$, $a_m \neq 0$, the nontriviality also depends on the divisibility of $k - l$ by the multiplicative order of g_m.

5. Trajectory Lengths

In this section we work over a finite field $\mathbb{F}_q$.

THEOREM 4. *Let $F_1, \ldots, F_m$ be rational functions defined by (6) satisfying the conditions (13) and (14) and such that $s_{i,i+1} \neq 0$, $i = 1, \ldots, m-1$. Then, for any $T \geq 1$ for all but $O(T^3 q^{m-1})$ initial vectors $\mathbf{u}_0 \in \mathbb{F}_q^m$, the trajectory length of the iterations (4) exceeds T.*

PROOF. Let $\mathcal{U}$ be the set of $\mathbf{u} = (u_1, \ldots, u_m) \in \mathbb{F}_q^m$ such that

$$G_i(u_{i+1}, \ldots, u_m) = 0$$

for some $i = 1, \ldots, m-1$. Clearly $\#\mathcal{U} = O(q^{m-1})$.

We now build a sequence of sets $\mathcal{U}_k$, $k = 0, 1, \ldots$, recursively.

We put $\mathcal{U}_0 = \mathcal{U}$.

Assume that $\mathcal{U}_0, \ldots, \mathcal{U}_k$ are defined and let

$$\mathcal{W}_k = \mathcal{U}_0 \cup \ldots \cup \mathcal{U}_k.$$

Then, we let $\mathcal{U}_{k+1}$ be the set of the initial values

$$\mathbf{u}_0 \in \mathbb{F}_q^m \setminus \mathcal{W}_k$$

such that for the corresponding sequence of vectors (4) we have $\mathbf{u}_{k+1} \in \mathcal{U}$. Inspecting (6), we now see that, by our construction, for any $\mathbf{v} \in \mathbb{F}_q^m$ with $\mathbf{v} = \mathbf{F}(\mathbf{u})$, for $\mathbf{u} \in \mathbb{F}_q^m \setminus \mathcal{U}$, there is a unique preimage under the map given by (6) (that is, the vector $\mathbf{u}$ itself). In turn, we see that for any $\mathbf{v} \in \mathbb{F}_q^m$ there is at most one initial value $\mathbf{u}_0 \in \mathbb{F}_q^m \setminus \mathcal{W}_{k-1}$ with $\mathbf{u}_k = \mathbf{v}$. Thus $\#\mathcal{U}_k \leq \#\mathcal{U}$.

Since there are $O(q^{m-1})$ vectors $\mathbf{u} \in \mathbb{F}_q^m$ that contain a zero in at least one component, we see that the set $\mathcal{E}_T$ of initial values for which, for some integer $t \le T$, the vector $\mathbf{u}_t$ has a zero component, satisfies

$$\mathcal{E}_T = O(q^{m-1} + T\#\mathcal{U}) = O(Tq^{m-1}). \tag{17}$$

We see that if a vector $\mathbf{u}_0 \in \mathbb{F}_q^m \setminus \mathcal{E}_T$ generates a trajectory of length $t \le T$, then $\mathbf{u}_t = \mathbf{u}_s$ for some nonnegative integer $s < t$.

Now, if $e_{m-1} = 1$, then we remove the set $\mathcal{F}_T$ of initial vectors $\mathbf{u}_0 \in \mathbb{F}_q^m$ such that

$$G_{m-1,t}(\mathbf{u}_0) = G_{m-1,s}(\mathbf{u}_0)$$

for some integers s and t with $T \ge t > s \ge 0$. By Lemma 1 we see that $G_{m-1,t} - G_{m-1,s}$ is a nontrivial polynomial of degree $O(t)$. Hence,

$$\#\mathcal{F}_T = O\left(\sum_{0 \le s < t \le T} tq^{m-1}\right) = O\left(T^3 q^{m-1}\right). \tag{18}$$

Furthermore, if $e_{m-1} = -1$, then we remove the set $\mathcal{F}_T$ of initial vectors $\mathbf{u}_0 \in \mathbb{F}_q^m$ such that

$$R_{m-1,t}(\mathbf{u}_0)R_{m-1,s-1}(\mathbf{u}_0) = R_{m-1,t-1}(\mathbf{u}_0)R_{m-1,s}(\mathbf{u}_0)$$

for some integers s and t with $T \ge t > s \ge 0$. As in the proof of Theorem 3 (in particular, see (16)) we note that Lemma 1 implies that $R_{m-1,t}R_{m-1,s-1} - R_{m-1,t-1}R_{m-1,s}$ is a nontrivial polynomial of degree $O(t)$. Hence, again we obtain the bound (18).

We remark that for $T \ge t > s \ge 0$, for any solution $\mathbf{u}_0 = (u_{0,1}, \dots, u_{0,m}) \in \mathbb{F}_q^m \setminus \mathcal{F}_T$ to the equation

$$\mathbf{F}^{(t)}(\mathbf{u}_0) = \mathbf{F}^{(s)}(\mathbf{u}_0),$$

the component $u_{0,m-1}$ is uniquely defined by $u_{0,m}$. So there at most q^{m-1} such solutions for every fixed t and s with $T \ge t > s \ge 0$ and thus at most T^2q^{m-1} for such t and s. Combining this bound with (17) and (18) we conclude the proof. □

Clearly if the map $\mathbf{u} \mapsto \mathbf{F}(\mathbf{u})$ is a permutation, as for example, in [**13**], then all trajectories are purely periodic. So we always have $s = 0$ in the argument of the proof of Theorem 4. This leads to a better estimate $O(T^2q^{m-1})$ on the number of initial values generating trajectories of length at most T.

6. Maximal Periods

In this section we show that the periods of the rational function systems over $\mathbb{F}_q$ defined by (6) with $e_i = -1$ for all $i = 1, \dots, m$ are given by the orbit lengths of certain linear fractional transformations, also called Möbius transformations. In particular, we describe the case when the systems (6) achieve maximal periods in their orbits. We also note that in [**15**, Theorem 6] there are given necessary and sufficient conditions for the system (6) to achieve maximal period in the case $e_i = 1$ for all $i = 1, \dots, m$.

We denote

$$\widetilde{\mathbf{u}}_{0,i} = (u_{0,i+1}, \dots, u_{0,m}) \in \mathbb{F}_q^{m-i}, \qquad i = 1, \dots, m-1.$$

LEMMA 5. *Let $F_1, \dots, F_m \in \mathbb{F}_q[X_1, \dots, X_m]$ be as in (6) with $e_i = -1$ for some $i = 1, \dots, m$. Assume that the sequence generated by the lower $m-i$ rational functions $F_{i+1}, \dots, F_m$ in $\mathbb{F}_q^{m-i}$ is purely periodic with period τ_{i+1} for some $i = 1, \dots, m-1$. Then we have the following description for the $k\tau_{i+1}$-th iteration of F_i on any initial vector $\mathbf{u}_0 \in \mathbb{F}_q^m$:*

$$F_i^{(k\tau_{i+1})}(\mathbf{u}_0) = f_i^{(k)}(u_{0,i}), \qquad k \geq 1, \tag{19}$$

where $R_{i,\tau_{i+1}}$ and $S_{i,\tau_{i+1}}$ are defined by (12) and f_i is the Möbius transformation in the variable Y,

$$f_i(Y) = \frac{YR_{i,\tau_{i+1}}(\widetilde{\mathbf{u}}_{0,i}) + S_{i,\tau_{i+1}}(\widetilde{\mathbf{u}}_{0,i})}{YR_{i,\tau_{i+1}-1}(\widetilde{\mathbf{u}}_{0,i}) + S_{i,\tau_{i+1}-1}(\widetilde{\mathbf{u}}_{0,i})}.$$

In particular, the orbit length of F_i in $\mathbf{u}_0$ is given by the orbit length of f_i in $u_{0,i}$.

PROOF. We first note that the orbit length of F_i is a multiple of τ_{i+1}. Indeed, let τ_i be the orbit length of the system $F_i, \dots, F_m$ in the initial vector $\mathbf{u}_0$. Then $\tau_i = \operatorname{lcm}(\tau_{i+1}, \eta_i)$, where η_i is the period of the sequence $\{u_{n,i}\}$ defined by the iterations of the polynomial F_i, and thus τ_i is a multiple of τ_{i+1}. This shows that, in order to describe the period of $F_i, \dots, F_m$ on the initial vector $\mathbf{u}_0$, it is enough to consider only $k\tau_{i+1}$-th iterations of F_i.

By (11) we have

$$F_i^{(\tau_{i+1})}(\mathbf{u}_0) = \frac{u_{0,i}R_{i,\tau_{i+1}}(\widetilde{\mathbf{u}}_{0,i}) + S_{i,\tau_{i+1}}(\widetilde{\mathbf{u}}_{0,i})}{u_{0,i}R_{i,\tau_{i+1}-1}(\widetilde{\mathbf{u}}_{0,i}) + S_{i,\tau_{i+1}-1}(\widetilde{\mathbf{u}}_{0,i})} = f_i(u_{0,i}). \tag{20}$$

Now,

$$\begin{aligned} &F_i^{(k\tau_{i+1})}(\mathbf{u}_0) \\ &= F_i^{(\tau_{i+1})}\left(F_i^{((k-1)\tau_{i+1})}(\mathbf{u}_0), F_{i+1}^{((k-1)\tau_{i+1})}(\mathbf{u}_0), \dots, F_m^{((k-1)\tau_{i+1})}(\mathbf{u}_0)\right) \\ &= F_i^{(\tau_{i+1})}\left(F_i^{((k-1)\tau_{i+1})}(\mathbf{u}_0), u_{0,i+1}, \dots, u_{0,m}\right) \\ &= \frac{F_i^{((k-1)\tau_{i+1})}(\mathbf{u}_0)R_{i,\tau_{i+1}}(\widetilde{\mathbf{u}}_{0,i}) + S_{i,\tau_{i+1}}(\widetilde{\mathbf{u}}_{0,i})}{F_i^{((k-1)\tau_{i+1})}(\mathbf{u}_0)R_{i,\tau_{i+1}-1}(\widetilde{\mathbf{u}}_{0,i}) + S_{i,\tau_{i+1}-1}(\widetilde{\mathbf{u}}_{0,i})}. \end{aligned} \tag{21}$$

To prove (19) we use induction over k. For $k = 1$ it is clear.

We now assume that the statement is true for $k-1$ and we prove it for k. Using (20), (21) and the induction hypothesis, we derive

$$F_i^{(k\tau_{i+1})}(\mathbf{u}_0) = \frac{f_i^{(k-1)}(u_{0,i})R_{i,\tau_{i+1}}(\widetilde{\mathbf{u}}_{0,i}) + S_{i,\tau_{i+1}}(\widetilde{\mathbf{u}}_{0,i})}{f_i^{(k-1)}(u_{0,i})R_{i,\tau_{i+1}-1}(\widetilde{\mathbf{u}}_{0,i}) + S_{i,\tau_{i+1}-1}(\widetilde{\mathbf{u}}_{0,i})} = f_i^{(k)}(u_{0,i}),$$

which concludes the proof. □

LEMMA 6. *Let $\mathcal{F} = \{F_1, \dots, F_m\}$ be a system of polynomials over $\mathbb{F}_q$ defined by (6). Let $i = 1, \dots, m$ such that $e_i = -1$ in the system (6). Then we have*

$$R_{i,k}S_{i,k-1} - R_{i,k-1}S_{i,k} = (-1)^k G_i G_i^{(2)} \dots G_i^{(k)},$$

where $R_{i,k}, S_{i,k}$ are defined by (12).

PROOF. We use induction on k. If $k = 1$, by (12), we get

$$R_{i,1}S_{i,0} - R_{i,0}S_{i,1} = -G_i.$$

We assume the statement true for k and we prove it for $k+1$. By (12) and the induction hypothesis we have

$$\begin{aligned} R_{i,k+1}&S_{i,k} - R_{i,k}S_{i,k+1} \\ &=(G_i^{(k+1)}R_{i,k-1} + H_i^{(k+1)}R_{i,k})S_{i,k} - R_{i,k}(G_i^{(k+1)}S_{i,k-1} + H_i^{(k+1)}S_{i,k}) \\ &= - G_i^{(k+1)}(R_{i,k}S_{i,k-1} - R_{i,k-1}S_{i,k}) \\ &=(-1)^{k+1}G_iG_i^{(2)}\dots G_i^{(k)}G_i^{(k+1)} \end{aligned}$$

and thus we conclude the proof. □

As usual, we say that a polynomial $f \in \mathbb{F}_q[X]$ of degree $d \geq 1$ is *primitive* if it is the minimal polynomial over $\mathbb{F}_q$ of a primitive element of $\mathbb{F}_{q^d}$ (that is, an element of multiplicative order $q^d - 1$), see [**6**].

Next, we present necessary and sufficient conditions for the system (6) to achieve maximal period over the prime field $\mathbb{F}_p$.

Using [**15**, Lemma 2] which holds for the functions F_i in the system (6) for which $e_i = 1$, we have the following analogue of [**15**, Lemma 5] (with an almost identical proof which we do not present here).

LEMMA 7. *Let $\mathcal{F} = \{F_1, \dots, F_m\}$ be a system of polynomials over $\mathbb{F}_p$ defined by (6). Let the index $1 \leq i \leq m$ such that $e_i = 1$ and assume that the period of the sequence generated by the lower $m-i$ polynomials $F_{i+1}, \dots, F_m$ in $\mathbb{F}_p^{m-i}$ is p^{m-i} and that $G_{i,p^{m-i}}(\widetilde{\mathbf{u}}_{0,i}) = 1$. Then, for the rational functions $H_{i,p^{m-i}}$ defined by (10), we have*

$$H_{i,p^{m-i}}(\widetilde{\mathbf{u}}_{0,i}) = \sum_{\mathbf{v}\in\mathbb{F}_p^{m-i}} L_i(\mathbf{v}),$$

where

$$L_i = H_iG_i^{(2)}\dots G_i^{(p^{m-i})}. \tag{22}$$

Now, using Lemma 7 and [**15**, Theorem 6] for F_i with $e_i = 1$ in the system (6), we have the following result.

We recall that the sets I_+ and I_- are given by (8).

THEOREM 8. *Let $\mathcal{F} = \{F_1, \dots, F_m\}$ be a system of polynomials over $\mathbb{F}_p$ defined by (6). Then the sequence $\{\mathbf{u}_n\}$ generated by (4) is purely periodic with period $\tau = p^m$ if and only if the following conditions are satisfied*

(1) *for every $i \in I_+$, $i < m$, we have*

$$\prod_{\mathbf{v}\in\mathbb{F}_p^{m-i}} G_i(\mathbf{v}) = 1 \quad \textit{and} \quad \sum_{\mathbf{v}\in\mathbb{F}_p^{m-i}} L_i(\mathbf{v}) \neq 0,$$

where L_i is defined by (22);

(2) *for every $i \in I_-$, $i < m$, we have:*

(a) *if $R_{i,p^{m-i}-1}(\mathbf{u}_0) = 0$, then*

$$R_{i,p^{m-i}}(\mathbf{u}_0) = S_{i,p^{m-i}-1}(\mathbf{u}_0) \quad \textit{and} \quad S_{i,p^{m-i}}(\mathbf{u}_0)S_{i,p^{m-i}-1}(\mathbf{u}_0) \neq 0;$$

(b) *if* $R_{i,p^{m-i}-1}(\mathbf{u}_0) \neq 0$, *then*

$$X^2 - \frac{R_{i,p^{m-i}}(\mathbf{u}_0)}{R_{i,p^{m-i}-1}(\mathbf{u}_0)}X - \frac{\prod_{\mathbf{v}\in\mathbb{F}_p^{m-i}} G_i(\mathbf{v})}{R_{i,p^{m-i}-1}(\mathbf{u}_0)}$$

is a primitive polynomial over $\mathbb{F}_p$;

(3) *if* $m \in I_+$, *then* $g_m = 1$ *and* $h_m \in \mathbb{F}_p^*$;

(4) *if* $m \in I_-$, *then* $X^2 - h_m X - g_m$ *is a primitive polynomial over* $\mathbb{F}_p$.

PROOF. We prove the result by induction on m. For $m = 1$, if $m \in I_+$, then it's clear that the period p is achieved if and only if $g_m = 1$ and $h_m \in \mathbb{F}_p^*$. If $m \in I_-$, we have

$$F_1 = g_1 X_1^{-1} + h_1,$$

which, by [**3**, Theorem 1], has maximal period p if and only if the polynomial $X^2 - h_1 X - g_1$ is a primitive polynomial over $\mathbb{F}_p$.

We assume the statement true for the first $m-1$ variables and we want to prove it for m. Let the sequence $\{\widetilde{\mathbf{u}}_{n,1}\} = \{(u_{n,2}, \dots, u_{n,m})\}$ be defined by the last $m-1$ components of the vectors in the sequence $\{\mathbf{u}_n\}$. By the induction hypothesis we know that the period $\widetilde{\tau}$ of the sequence $\{\widetilde{\mathbf{u}}_{n,1}\}$ is p^{m-1}, and taking into account the first remark in the proof of Lemma 5, we see that the period of $\{\widetilde{\mathbf{u}}_{n,1}\}$ is of the form kp^{m-1}, for some $1 \leq k \leq p$. Thus, proving the maximality of the period of $\{\mathbf{u}_n\}$ reduces to proving that $k = p$. We note that the representation of $F_1^{(k)}$ given by Lemma 1 does not depend how we choose $e_2, \dots, e_m$ in the functions $F_2, \dots, F_m$, but only the functions $G_{1,k}, H_{1,k}$ or $G_1^{(k)}, H_1^{(k)}$ if $1 \in I_+$ or $1 \in I_-$, respectively. Thus, the representation of $F_1^{(k)}$ given by Lemma 1 is the same regardless if $i_j \in I_+$ or $i_j \in I_-$ for $2 \leq j \leq m$.

Thus, the case $1 \in I_+$ follows identically as in the proof of [**15**, Theorem], and we do not repeat it here.

We consider now the case $1 \in I_-$. By Lemma 5 we have

$$F_1^{(kp^{m-1})}(\mathbf{u}_0) = f_1^{(k)}(u_{0,1}), \qquad k \geq 1,$$

where

$$f_1(Y) = \frac{Y R_{1,p^{m-1}}(\widetilde{\mathbf{u}}_{0,1}) + S_{1,p^{m-1}}(\widetilde{\mathbf{u}}_{0,1})}{Y R_{1,p^{m-1}-1}(\widetilde{\mathbf{u}}_{0,1}) + S_{1,p^{m-1}-1}(\widetilde{\mathbf{u}}_{0,1})},$$

and thus the maximal period of the sequence generated by the iterations of F_1 is given by the case when f_1 achieves maximal orbit length in $u_{0,1}$.

We distinguish now two cases. If $R_{1,p^{m-1}-1}(\widetilde{\mathbf{u}}_{0,1}) = 0$, then we note that $S_{1,p^{m-1}-1}(\widetilde{\mathbf{u}}_{0,1}) \neq 0$, as otherwise $f_1 = 0$. Thus, we have the case of linear generator

$$f_1(Y) = \frac{R_{1,p^{m-1}}(\widetilde{\mathbf{u}}_{0,1})}{S_{1,p^{m-1}-1}(\widetilde{\mathbf{u}}_{0,1})}Y + \frac{S_{1,p^{m-1}}(\widetilde{\mathbf{u}}_{0,1})}{S_{1,p^{m-1}-1}(\widetilde{\mathbf{u}}_{0,1})}.$$

The maximal period is achieved in this case if and only if

$$\frac{R_{1,p^{m-1}}(\widetilde{\mathbf{u}}_{0,1})}{S_{1,p^{m-1}-1}(\widetilde{\mathbf{u}}_{0,1})} = 1, \qquad \frac{S_{1,p^{m-1}}(\widetilde{\mathbf{u}}_{0,1})}{S_{1,p^{m-1}-1}(\widetilde{\mathbf{u}}_{0,1})} \neq 0,$$

which is equivalent to

$$R_{1,p^{m-1}}(\widetilde{\mathbf{u}}_{0,1}) = S_{1,p^{m-1}-1}(\widetilde{\mathbf{u}}_{0,1}), \quad S_{1,p^{m-1}}(\widetilde{\mathbf{u}}_{0,1}) \neq 0.$$

We consider now the case of $R_{1,p^{m-1}-1}(\widetilde{\mathbf{u}}_{0,1}) \neq 0$. We note that in this case f_1 achieves maximal period if and only if, under a linear transformation, it has the same property. Taking into account that $R_{1,p^{m-1}-1}(\widetilde{\mathbf{u}}_{0,1}) \neq 0$, we can make the linear transformation

$$Y \to R_{1,p^{m-1}-1}(\widetilde{\mathbf{u}}_{0,1})^{-1}Y - R_{1,p^{m-1}-1}(\widetilde{\mathbf{u}}_{0,1})^{-1}S_{1,p^{m-1}-1}(\widetilde{\mathbf{u}}_{0,1})$$

and obtain the following inversive generator which, by a slightly abuse of notation, we denote also by f_1,

$$f_1(Y) = \frac{-R_{1,p^{m-1}}(\widetilde{\mathbf{u}}_{0,1})S_{1,p^{m-1}-1}(\widetilde{\mathbf{u}}_{0,1}) + R_{1,p^{m-1}-1}(\widetilde{\mathbf{u}}_{0,1})S_{1,p^{m-1}}(\widetilde{\mathbf{u}}_{0,1})}{R_{1,p^{m-1}-1}(\widetilde{\mathbf{u}}_{0,1})}Y^{-1} + \frac{R_{1,p^{m-1}}(\widetilde{\mathbf{u}}_{0,1})}{R_{1,p^{m-1}-1}(\widetilde{\mathbf{u}}_{0,1})}.$$

Applying now Lemma 6 we have

$$R_{1,p^{m-1}}(\widetilde{\mathbf{u}}_{0,1})S_{1,p^{m-1}-1}(\widetilde{\mathbf{u}}_{0,1}) - R_{1,p^{m-1}-1}(\widetilde{\mathbf{u}}_{0,1})S_{1,p^{m-1}}(\widetilde{\mathbf{u}}_{0,1}) = -G_1(\widetilde{\mathbf{u}}_{0,1})G_1^{(2)}(\widetilde{\mathbf{u}}_{0,1})\dots G_1^{(p^{m-1})}(\widetilde{\mathbf{u}}_{0,1}).$$

Let $\widetilde{\mathcal{F}} = \{F_2, \dots, F_m\}$. Now, as the period induced by $\widetilde{\mathcal{F}}$ is p^{m-1}, the elements

$$\widetilde{\mathbf{u}}_{0,1}, \widetilde{\mathcal{F}}(\widetilde{\mathbf{u}}_{0,1}), \dots, \widetilde{\mathcal{F}}^{(p^{m-1}-1)}(\widetilde{\mathbf{u}}_{0,1})$$

are all the distinct elements of $\mathbb{F}_p^{m-1}$, and thus we obtain

$$G_1(\widetilde{\mathbf{u}}_{0,1})G_1^{(2)}(\widetilde{\mathbf{u}}_{0,1})\dots G_1^{(p^{m-1})}(\widetilde{\mathbf{u}}_{0,1}) = G_1(\widetilde{\mathbf{u}}_{0,1})G_1(\widetilde{\mathcal{F}}(\widetilde{\mathbf{u}}_{0,1}))\dots G_1(\widetilde{\mathcal{F}}^{(p^{m-1}-1)}(\widetilde{\mathbf{u}}_{0,1})) = \prod_{\mathbf{v}\in\mathbb{F}_p^{m-1}} G_1(\mathbf{v}).$$

This implies that

$$f_1(Y) = \frac{\prod_{\mathbf{v}\in\mathbb{F}_p^{m-1}} G_1(\mathbf{v})}{R_{1,p^{m-1}-1}(\widetilde{\mathbf{u}}_{0,1})}Y^{-1} + \frac{R_{1,p^{m-1}}(\widetilde{\mathbf{u}}_{0,1})}{R_{1,p^{m-1}-1}(\widetilde{\mathbf{u}}_{0,1})}.$$

Applying now [**3**, Theorem 1], we know that f_1 achieves maximal period p if and only if the polynomial

$$X^2 - \frac{R_{1,p^{m-1}}(\widetilde{\mathbf{u}}_{0,1})}{R_{1,p^{m-1}-1}(\widetilde{\mathbf{u}}_{0,1})}X - \frac{\prod_{\mathbf{v}\in\mathbb{F}_p^{m-1}} G_1(\mathbf{v})}{R_{1,p^{m-1}-1}(\widetilde{\mathbf{u}}_{0,1})}$$

is a primitive polynomial over $\mathbb{F}_p$. □

References

[1] V. Anashin and A. Khrennikov, 'Applied algebraic dynamics', Walter de Gruyter, Berlin. 2009. MR2533085 (2011d:37002)

[2] E. Bedford and T. T. Truong, 'Degree complexity of birational maps related to matrix inversion', *Comm. Math. Phys.*, **298** (2010), 357–368. MR2669440 (2011g:37125)

[3] M. Flahive and H. Niederreiter, 'On inversive congruential generators for pseudorandom numbers', *Finite Fields, Coding Theory, and Advances in Commun. and Comp.*, Marcel Dekker, New York (1992), 75–80. MR1199823 (94a:11117)

[4] F. Griffin, H. Niederreiter and I. E. Shparlinski, 'On the distribution of nonlinear recursive congruential pseudorandom numbers of higher orders', *Lect. Notes in Comp. Sci.*, Springer-Verlag, Berlin, **1719** (1999), 87–93. MR1846485 (2002j:94038)

[5] J. Gutierrez and D. Gomez-Perez, 'Iterations of multivariate polynomials and discrepancy of pseudorandom numbers', *Lect. Notes in Comp. Sci.*, Springer-Verlag, Berlin, **2227** (2001), 192–199. MR1913465

[6] R. Lidl and H. Niederreiter, *Finite fields*, Cambridge Univ. Press, Cambridge, 1997. MR1429394 (97i:11115)

[7] H. Niederreiter, 'Quasi-Monte Carlo methods and pseudo-random numbers', *Bull. Amer. Math. Soc.*, **84** (1978), 957–1041. MR508447 (80d:65016)

[8] H. Niederreiter, *Random number generation and Quasi–Monte Carlo methods*, SIAM Press, 1992. MR1172997 (93h:65008)

[9] H. Niederreiter and I. E. Shparlinski, 'On the distribution of inversive congruential pseudorandom numbers in parts of the period', *Math. Comp.*, **70** (2001), 1569–1574. MR1836919 (2002e:11104)

[10] H. Niederreiter and I. E. Shparlinski, 'On the average distribution of inversive pseudorandom numbers', *Finite Fields and Their Appl.*, **8** (2002), 491–503. MR1933620 (2003g:11085)

[11] H. Niederreiter and I. E. Shparlinski, 'Dynamical systems generated by rational functions', *Lect. Notes in Comp. Sci.*, Springer-Verlag, Berlin, **2643** (2003), 6–17. MR2042407 (2005a:94047)

[12] H. Niederreiter and A. Winterhof, 'Exponential sums for nonlinear recurring sequences', *Finite Fields Appl.*, **14** (2008), 59–64. MR2381476 (2008m:11166)

[13] A. Ostafe, 'Multivariate permutation polynomial systems and pseudorandom number generators', *Finite Fields and Their Appl.*, **16** (2010), 144–154. MR2610705 (2011d:65014)

[14] A. Ostafe, 'Pseudorandom vector sequences derived from triangular polynomial systems with constant multipliers', *Lect. Notes in Comp. Sci.*, WAIFI 2010, Springer-Verlag, Berlin, (2010), 62–72. MR2674215 (2012a:65013)

[15] A. Ostafe, 'Pseudorandom vector sequences of maximal period generated by polynomial dynamical systems', *Designs, Codes and Cryptography*, **63** (2012), 59–72.

[16] A. Ostafe, E. Pelican and I. E. Shparlinski, 'On pseudorandom numbers from multivariate polynomial systems', *Finite Fields and Their Appl.*, **16** (2010), 320–328. MR2678620 (2011h:11085)

[17] A. Ostafe and I. E. Shparlinski, 'On the degree growth in some polynomial dynamical systems and nonlinear pseudorandom number generators', *Math. Comp.*, **79** (2010), 501–511. MR2552237 (2010i:11120)

[18] A. Ostafe and I. E. Shparlinski, 'Pseudorandom numbers and hash functions from iterations of multivariate polynomials', *Cryptography and Communications*, **2** (2010), 49–67. MR2592430 (2011a:11144)

[19] A. Ostafe, I. E. Shparlinski and A. Winterhof, 'On the generalized joint linear complexity profile of a class of nonlinear pseudorandom multisequences', *Adv.Math.Comm.*, **4** (2010), 369–379. MR2677868 (2011i:65009)

[20] A. Ostafe, I. E. Shparlinski and A. Winterhof, 'Multiplicative character sums of a class of nonlinear recurrence vector sequences', *Intern. J. Number Theory*, **7** (2011), 1557–1571. MR2835810

[21] K. Schmidt, *Dynamical systems of algebraic origin*, Progress in Math., v.128. Birkhäuser Verlag, Basel, 1995. MR1345152 (97c:28041)

[22] J. H. Silverman, *The arithmetic of dynamical systems*, Springer, New York, 2007. MR2316407 (2008c:11002)

[23] J. H. Silverman, 'Variation of periods modulo p in arithmetic dynamics', *New York J. Math.*, **14** (2008), 601–616. MR2448661 (2009h:11100)

[24] A. Topuzoğlu and A. Winterhof, 'Pseudorandom sequences', *Topics in Geometry, Coding Theory and Cryptography*, Springer-Verlag, Berlin, 2006, 135–166. MR2278037 (2007m:11106)

[25] C.-M. Viallet, 'Algebraic dynamics and algebraic entropy', *Int. J. Geom. Methods Mod. Phys.*, **5** (2008), 1373–1391. MR2484559 (2009m:37011)

Department of Computing, Macquarie University, Sydney, NSW 2109, Australia
E-mail address: alina.ostafe@mq.edu.au

Department of Computing, Macquarie University, Sydney, NSW 2109, Australia
E-mail address: igor.shparlinski@mq.edu.au

Contemporary Mathematics
Volume **574**, 2012
http://dx.doi.org/10.1090/conm/574/11422

Computer search for curves with many points among abelian covers of genus 2 curves

Karl Rökaeus

Abstract. Using class field theory one associates to each curve C over a finite field, and each subgroup G of its divisor class group, unramified abelian covers of C whose genus is determined by the index of G. By listing class groups of curves of small genus one may get examples of curves with many points; we do this for all curves of genus 2 over the fields of cardinality 5,7,9,11,13 and 16, giving new entries for the tables of curves with many points.

Background

Let q be a prime power and let g be a non-negative integer. We are interested in how many rational points there can be on a projective, smooth and geometrically irreducible curve of genus g, defined over the finite field $\mathbf{F}_q$; equivalently, how many rational places there can be in a global function field of genus g with full constant field $\mathbf{F}_q$. Denote by $\mathrm{N}_q(g)$ the maximum possible number of rational places in such a field. The intervals in which $\mathrm{N}_q(g)$ in known to lie for $g \leq 50$, together with references, are given in the tables [**6**]. In this report we improve upon the lower bounds of these intervals, by constructing curves with many points for some genera over the fields with $q = 5, 7, 9, 11, 13$ and 16 elements. We do this by going through unramified abelian covers of all genus 2 curves over these fields, using class field theory. The method of looking at such covers has been used before by various authors, see [**2**]. Still, by doing the systematic search we found some new curves.

The search was done using Magma [**1**], first in terms of curves, using its arithmetic geometry part; then everything was checked also using its function field part. In this report we have formulated everything in terms of function fields.

Notation. See [**15**] for an introduction to function fields in one variable. By a function field F/k we will always mean a global function field with full constant field k, i.e., k is algebraically closed in F. We write $\mathbf{P}_F$ for the set of places of F and $\mathrm{Cl}(F)$ for its divisor class group of degree zero divisors.

Description of the method

Class field theory for global function fields: In this section we state the theorems that we use for the construction of the curves; see [**11**] for details. Let F/k

2000 *Mathematics Subject Classification.* Primary 11-04; Secondary 11-06, 11G20, 11R37.
The author is supported by The Wenner-Gren Foundations Postdoctoral Grant.

be a global function field. Let O be a rational place of F and set $S = \mathbf{P}_F \setminus \{O\}$. We use F_O to denote the S-Hilbert class field of F, i.e., the maximal unramified abelian extension of F (inside a fixed separable closure) in which O splits completely. There is an isomorphism

$$\psi\colon \mathrm{Cl}(F) \to \mathrm{Gal}(F_O/F).$$

under which $[P - \deg(P)O]$ maps to $\left[\frac{F_O/F}{P}\right]$, the Artin symbol of P, for any $P \in \mathbf{P}_F$. Fix a subgroup G of $\mathrm{Cl}(F)$ of index d and write F_O^G for $F_O^{\psi(G)}$, the subfield of F_O fixed by $\psi(G)$. Then F_O^G is a Galois extension of F with Galois group isomorphic to $\mathrm{Cl}(F)/G$; in particular $[F_O^G : F] = d$. A rational place $P \in \mathbf{P}_F$ splits completely in F_O^G if and only if $\left[\frac{F_O/F}{P}\right] \in \psi(G)$, i.e., if and only if $[P - O] \in G$.

Therefore, if F/k is a global function field of genus g with rational places $\{P_1, \ldots, P_m\}$ and if $G \subset \mathrm{Cl}(F)$ is a subgroup of index d then, for every rational place O, we get an unramified extension F_O^G/F of degree d, hence of genus $d(g-1)+1$, with $d \cdot |G \cap \{[P_i - O]\}_{i=1}^m|$ rational places. Also, k is the full constant field of F_O^G.

Note that, for fixed G, two different rational places O and O' need not give isomorphic extensions. Although F_O^G/F and $F_{O'}^G/F$ have the same genus and isomorphic Galois groups they may have a different number of rational places.

We give an example that can be done by hand, a construction of a genus 6 curve over $\mathbf{F}_2$ with 10 rational points.

Example. Let F be the degree 2 extension of $\mathbf{F}_2(x)$ given by

$$y^2 + xy = x^5 + x^3 + x^2 + x.$$

It is of genus 2 and has 4 rational places; one containing x and one containing $1/x$, P_0 and P_∞; and two containing $x+1$, $P_{1,0}$ and $P_{1,1}$. Moreover, F has two places of degree 2 so its class number is $h = 10$. This already gives the existence of four covers of degree 10, namely the S-Hilbert class fields F_O, where $S = \mathbf{P}_F \setminus \{O\}$ and O is any rational place in F. These are all of genus 11, with 10 rational places.

Let $a = [P_0 - P_\infty] \in \mathrm{Cl}(F)$. Then $a \neq 0$ (because F is not rational), while $2a = 0$ because $(x) = 2P_0 - 2P_\infty$. Let $G = \{0, a\} \subset \mathrm{Cl}(F)$. If we choose $O = P_\infty$ then $[P_\infty - O] = 0$ and $[P_0 - O] = a$ lie in G. The extension F_O^G/F therefore has degree $10/2 = 5$. It has genus 6 and contains $2 \cdot 5 = 10$ rational places.

Similarly, if we choose $O = P_0$ we also get a genus 6 extension with 10 rational places. However, if we instead choose O to be $P_{1,0}$ or $P_{1,1}$ then F_O^G is a genus 6 extension with only 5 rational places.

Organization of the search. Let k be a finite field and let $g > 1$ be an integer. Let $\mathcal{F}_g$ be a list of global function fields over k of genus g. (In this report we will only consider the case of $\mathcal{F}_2$, the fields of genus 2.) For each function field $F \in \mathcal{F}_g$, do the following:

- Compute the class group $\mathrm{Cl}(F)$ of F.
- For each rational place O compute the image I_O of the rational places in $\mathrm{Cl}(F)$ under the map $P \mapsto [P - \deg(P)O]\colon \mathbf{P}_F \to \mathrm{Cl}(F)$.
- List all subgroups of $\mathrm{Cl}(F)$. For each of these, G, compute its index d_G and the cardinality $m_{G,O}$ of its intersection with I_O. This gives the existence of an unramified abelian extension of F of degree d_G, of genus $d_G(g-1)+1$ and with $d_G m_{G,O}$ rational places.

Implementation and results

Using Magma [**1**] we implemented the search with all genus 2 curves as base, over $k = \mathbf{F}_q$ for $q \leq 16$. For q smaller than 9 the listing of genus 2 curves could be done in the straightforward way; for the bigger values of q we used the Magma package *G2Twists* [**5**]. The curves are then represented by Magma as hyperelliptic curves. For each such curve we computed its Jacobian and a list of its subgroups, using the arithmetic geometry part of Magma. All the records we found were then checked using the function field part of Magma, this is also how we describe them in this report.

Below we have recorded all the new curves we found by this method, together with the details required for their construction: First a genus 2 function field F. Then a set of rational places M that generate the subgroup of the class group that defines the extension. More precisely, the extension is F_O^G where O is any place in M and G is the subgroup of $\mathrm{Cl}(F)$ generated by $\{[P - O]\}_{P \in M}$. (It can also be constructed as the S-Hilbert class field of F, with $S = \mathbf{P}_F \setminus M$.)

The genus 2 fields are always given as a degree 2 extension $F = k(x,y)/k(x)$ defined by some $f \in k[x][y]$. We use the following notation for the rational places of F: For $\alpha \in k$ let Q_α be the place of $k(x)$ containing $x - \alpha$. If Q_α ramifies then there is a unique place of F containing $x - \alpha$ which we denote by P_α. If Q_α splits there are two such places, one for each $\beta \in k$ such that $f(\alpha, \beta) = 0$; it is characterized by containing $y - \beta$, we denote it $P_{\alpha,\beta}$. If there is a unique place of F containing $1/x$ we denote it P_∞.

Curves in characteristic 2. This is the case where most of the work of constructing curves has been done, in particular using the present method, and for $q = 2, 4$ and 8 we didn't find any improvements.

Over $\mathbf{F}_{16}$ there are 8470 different genus 2 curves. Among their unramified abelian covers we found 2 new records:

- $\underline{g = 8,\ N = 63.}$ F: $y^2 + (x^2 + x)y = \alpha^6 x^5 + \alpha^{12} x^4 + x^3 + \alpha^3 x^2 + \alpha^9 x$, where α is a primitive element of $\mathbf{F}_{16}$ satisfying $\alpha^4 + \alpha + 1 = 0$. Places: three ramified, P_∞, P_0, P_1 and the 3 pairs $P_{\alpha^3,\alpha^4}, P_{\alpha^3,\alpha^{10}}$, $P_{\alpha^6,\alpha^6}, P_{\alpha^6,\alpha^{12}}$ and $P_{\alpha^9,\alpha^3}, P_{\alpha^9,\alpha^9}$. Old interval $[62, 75]$.
- $\underline{g = 49,\ N = 240.}$ F: $y^2 + (x^2 + x)y = x^5 + x^3 + x^2 + x$. Places: With O any of P_∞, P_0, P_1, there are two such extensions. One in which these three places and P_{α^5,α^2} and P_{α^5,α^8} splits; and one in which these three places and $P_{\alpha^{10},\alpha}$ and P_{α^{10},α^4} splits. Old interval $[213, 286]$.

Curves in characteristic 3. For $q = 3$ we didn't find any new curves. For $q = 9$ there had been much previous activity, *e.g.*, [**2**], [**3**], [**7**], [**14**]. By doing a systematic search we found 3 small improvements of the old table entries. Let α be a primitive element of $\mathbf{F}_9$ satisfying $\alpha^2 - \alpha - 1 = 0$.

- $\underline{g = 18,\ N = 68.}$ F: $y^2 = x^5 + \alpha^6 x^3 + \alpha^6 x^2 + \alpha^3 x$. Places: P_∞, P_0, P_1 and P_α. Old interval $[67, 84]$.
- $\underline{g = 32,\ N = 93.}$ F: $y^2 = x^5 + \alpha^6 x^4 + \alpha^7 x^3 + 2x^2 + \alpha^5 x + \alpha^2$. Places: P_∞, $P_{0,\alpha}$ and $P_{0,-\alpha}$. Old interval $[92, 130]$.
- $\underline{g = 38,\ N = 111.}$ F: $y^2 = x^5 + \alpha^7 x^3 + 2x^2 + \alpha^2 x$. Places: P_∞, P_0 and $P_{-\alpha}$. Old interval $[105, 149]$.

Curves over $\mathbf{F}_5$. For $q = 5$ there had been some previous work, *e.g.*, [**8**], [**9**] and [**10**], in particular using the present method. By doing the systematic search we found 3 improvements of the old table entries:

- $g = 7$, $N = 24$. F: $y^2 = x^5 + x^2 - x$. Places: P_∞, P_0, $P_{4,1}$ and $P_{4,4}$. Old interval $[22, 26]$.
- $g = 9$, $N = 32$. F: $y^2 = x^5 - x^3 + x$. Places: P_∞, P_0 and either the two places containing $x - 1$ or the two places containing $x - 4$. Old interval $[26, 32]$.
- $g = 12$, $N = 33$. F: $y^2 = x^5 + x^4 + 3x^2 + 1$. Places: P_∞, $P_{0,1}, P_{0,4}$. Old interval $[30, 38]$.

We give some details about the construction of the genus 9 curve that we found which, together with the Oesterlé bound, show that $\mathrm{N}_5(9) = 32$.

EXAMPLE. Let $k = \mathbf{F}_5$ and let F/k be the hyperelliptic field of genus 2 given by

$$y^2 = x^5 - x^3 + x.$$

Take O to be P_0, the unique place containing x. Magma gives an isomorphism from $\mathrm{Cl}(F)$ to $\mathbf{Z}/8 \oplus \mathbf{Z}/8$ under which the divisor classes $[P_\infty - O], [P_{4,3} - O]$ and $[P_{4,2} - O]$ map to $(0, 4), (0, 5)$ and $(0, 3)$. They are hence contained in a subgroup of index 8. Therefore, F has an unramified extension of degree 8 in which these places and O split completely. This extension has genus $1 + 8 \cdot (2 - 1) = 9$ and $8 \cdot 4 = 32$ rational places.

We mention that Magma can give explicit equations for the class fields: The degree 8 extension of F in this example has minimal polynomial

$$\begin{aligned}&T^8 + ((4x + 4)y + 3x^4 + 3x^3 + 3x^2 + 3x + 3)T^4 + \\ &\quad + (x^5 + 2x^4 + 2x^3 + 2x^2 + 2x + 1)y + x^8 + 2x^6 + x^5 + 2x^4 + x^3 + 2x^2 + 1.\end{aligned}$$

Curves over $\mathbf{F}_7, \mathbf{F}_{11}$ and $\mathbf{F}_{13}$. For $q = 7, 11$ and 13 there where no lower bounds in the tables for genus greater than 4, so all we had to do to make it to the tables was to produce curves with more than $b(g)/\sqrt{2}$ points, where $b(g)$ is the best known upper bound for $\mathrm{N}_q(g)$. (This is the criteria used to decide if a curve is considered to have many points, see [**4**].) The results are given in the table below; we give integers g and N such that there exists a genus g curve with N points, and then the interval $[b(g)/\sqrt{2}, b(g)]$. The details of the construction are given in [**12**] and [**13**]. There, for each entry given in this table, a function field of genus 2 together with all the rest of the information required to prove the existence of the the extension, are given.

New entries for the tables over $\mathbf{F}_7, \mathbf{F}_{11}$ and $\mathbf{F}_{13}$.

$q=7$			$q=11$			$q=13$		
g	N	$[\frac{b(g)}{\sqrt{2}}, b(g)]$	g	N	$[\frac{b(g)}{\sqrt{2}}, b(g)]$	g	N	$[\frac{b(g)}{\sqrt{2}}, b(g)]$
5	24	[20,28]	5	36	[27,38]	5	40	[32,44]
6	25	[23,32]	6	40	[32,45]	6	50	[37,52]
7	30	[26,36]	7	42	[36,50]	7	48	[42,58]
8	35	[27,38]	8	42	[39,55]	8	56	[45,63]
9	32	[29,41]	9	48	[42,59]	9	56	[49,68]
10	36	[32,45]	10	54	[45,63]	10	63	[51,72]
11	40	[34,48]	11	60	[48,67]	11	70	[55,77]
12	44	[37,51]	12	66	[51,72]	12	66	[58,82]
13	48	[39,54]	13	60	[55,77]	13	72	[62,87]
14			14			14	65	[65,91]
15			15	70	[61,85]	15	84	[68,96]
16	45	[45,63]	16			16	90	[72,101]
17	64	[47,66]	17	80	[66,93]	17	96	[75,105]
18	51	[49,68]	18	85	[70,98]	18	85	[78,110]
19	54	[51,71]	19	90	[73,102]	19	90	[82,115]
20			20	76	[75,106]	20	95	[85,119]
21	60	[55,77]	21	80	[78,110]	21	100	[88,124]
22	63	[56,79]	22			22	105	[92,129]
23	66	[58,82]	23	88	[85,119]	...		
24			24	92	[87,123]	29	140	[114,161]
25	72	[62,87]	25	96	[90,127]	32	124	[124,175]
26			26			33	128	[127,179]
27	78	[66,93]	27			35	136	[133,187]
			28	108	[98,138]	36	140	[136,191]
			...			37	144	[138,195]
			33	128	[111,156]	41	160	[150,211]
						43	168	[155,219]

ACKNOWLEDGMENTS. The author thanks the Wenner-Gren Foundations for financial support; the Korteweg-de Vries Institute at the University of Amsterdam for hospitality; and Gerard van der Geer for his hospitality and for helpful conversations and comments.

References

1. W. Bosma, J. Cannon, C. Playoust, *The Magma algebra system I: The user language*, J. Symbolic Comput., 24 (1997), 235–265. MR1484478
2. G. van der Geer, *Hunting for curves with many points*, In: C. Xing et al. (Eds.): IWCC 2009, LNCS 5557, pp. 82–96, 2009. Available electronically: arXiv:0902.3882 (2009)
3. G. van der Geer, M. van der Vlugt *How to construct curves over finite fields with many points* In: Arithmetic Geometry, (Cortona 1994), F. Catanese Ed., Cambridge Univ. Press, Cambridge, 1997, p. 169-189. MR1472497 (98h:11077)
4. G. van der Geer and M. van der Vlugt, *Tables of curves with many points*, Math. Comp 69, (2000), no 230 pp. 797–810 MR1654002 (2000i:11097)
5. R. Lercier and C. Ritzenthaler, *G2Twists v1.1 : Reconstruction of genus 2 curves from their invariants and twists over a finite field*, Magma package, incorporated in the official magma distribution, since version 2.15, April 2009.

6. *manYPoints.org*, Tables of upper and lower bounds for the maximum number of points on a genus g curve over different finite fields. Moderators: G. van der Geer, E. Howe, K. Lauter, C. Ritzenthaler
7. H. Niederreiter; C. P. Xing A general method of constructing global function fields with many rational places In: Algorithmic Number Theory (Portland 1998), Lecture Notes in Comp. Science 1423, Springer, Berlin, 1998, p. 555-566. MR1726101 (2000i:11176)
8. H. Niederreiter; C. P. Xing, *Cyclotomic function fields, Hilbert class fields and global function fields with many rational places* Acta Arithm. 79 (1997), p. 59-76. MR1438117 (97m:11141)
9. H. Niederreiter; C. P. Xing, *Global function fields with many rational places over the quinary field* Demonstratio Math. 30 (1997), no. 4, 919–930 MR1617283 (99a:11133)
10. H. Niederreiter; C. P. Xing, *Global function fields with many rational places over the quinary field. II* Acta Arith. 86 (1998), no. 3, 277–288 MR1655986 (99i:11105)
11. H. Niederreiter; C. P. Xing, *Rational Points on Curves over Finite Fields*, LMS Lecture Note Series 285, 2001
12. K. Rökaeus, *Computer search for curves with many points among unramified covers of genus 2 curves over* $\mathbf{F}_5$, $\mathbf{F}_7$, $\mathbf{F}_9$ *and* $\mathbf{F}_{11}$ Report, available under *manYPoints.org*, http://manypoints.org/upload/1369548137.pdf
13. K. Rökaeus, *Computer search for curves with many points among abelian covers of genus 2 curves over* $\mathbf{F}_{13}$ Report, available under *manYPoints.org*, http://manypoints.org/upload/1225383491.pdf
14. V. Shabat, *Curves with many points*, Thesis, University of Amsterdam, 2001.
15. H. Stichtenoth, *Algebraic function fields and codes*, Graduate Texts in Mathematics 254, Springer-Verlag, 2009 MR2464941 (2010d:14034)

Korteweg de Vries Instituut voor Wiskunde, Universiteit van Amsterdam, P.O. Box 94248, 1090 GE Amsterdam, The Netherlands

E-mail address: S.K.F.Rokaeus@uva.nl

Contemporary Mathematics
Volume **574**, 2012
http://dx.doi.org/10.1090/conm/574/11424

The groups of points on abelian surfaces over finite fields

Sergey Rybakov

ABSTRACT. Let A be an abelian surface over a finite field k. The k-isogeny class of A is uniquely determined by a Weil polynomial f_A of degree 4. We give a classification of the groups of k-rational points on varieties from this class in terms of f_A.

1. Introduction

Classical results of Tate and Honda give us an explicit description for the set of isogeny classes of abelian varieties over a finite field k. Namely, an isogeny class is uniquely determined by a so called Weil polynomial of any variety from this class. It looks natural to consider classification problems concerning abelian varieties inside a given isogeny class. In this paper we classify groups of k-rational points on abelian surfaces. Xing partly classified groups of points on supersingular surfaces in [**Xi94**] and in [**Xi96**]. In the paper [**Ry10**] we show that one could use the language of Hodge polygons to describe groups of points. Moreover, we classify this groups for abelian varieties with commutative endomorphism algebra. An abelian variety has commutative endomorphism algebra if and only if its Weil polynomial has no multiple roots. We could say that this is the general case. From the classification of Weil polynomials of abelian surfaces due to Rück, Xing, Maisner and Nart [**MN02**] we get that there are only three cases more. Namely,

- the Weil polynomial is a square of a polynomial of degree 2 without multiple roots;
- the Weil polynomial is of the form $P(t)(t \pm \sqrt{q})^2$, where $P(t)$ has no multiple roots, and $\sqrt{q} \in \mathbb{Z}$;
- the Weil polynomial equals $(t \pm \sqrt{q})^4$, and $\sqrt{q} \in \mathbb{Z}$.

The paper is devoted to a classification of groups of points for these three cases. The author is grateful to referee for providing useful corrections on the paper.

2010 *Mathematics Subject Classification.* Primary 14K99, 14G05, 14G15.

Key words and phrases. Abelian variety, the group of rational points, finite field, Newton polygon, Hodge polygon.

Supported in part by RFBR grants no. 11-01-00393-a, 11-01-12072-ofi-m and 10-01-93110-CNRSLa.

2. Main result

Throughout this paper k is a finite field $\mathbb{F}_q$ of characteristic p. Let A be an abelian variety of dimension g over k, and let $\bar{k}$ be an algebraic closure of k. Fix a prime number ℓ. For a natural number m denote by A_m the kernel of multiplication by ℓ^m in $A(\bar{k})$. Let $T_\ell(A) = \varprojlim A_m$ be the Tate module, and $V_\ell(A) = T_\ell(A) \otimes_{\mathbb{Z}_\ell} \mathbb{Q}_\ell$ be the corresponding vector space over $\mathbb{Q}_\ell$. If $\ell \neq p$, then $T_\ell(A)$ is a free $\mathbb{Z}_\ell$-module of rank $2g$. The Frobenius endomorphism F of A acts on the Tate module by a semisimple linear operator, which we also denote by $F : T_\ell(A) \to T_\ell(A)$. The characteristic polynomial

$$f_A(t) = \det(t - F|T_\ell(A))$$

is called a *Weil polynomial of* A. It is a monic polynomial of degree $2g$ with rational integer coefficients independent of the choice of prime $\ell \neq p$. It is well known that for isogenous varieties A and B we have $f_A(t) = f_B(t)$. Moreover, Tate proved that the isogeny class of abelian variety is determined by its characteristic polynomial, that is $f_A(t) = f_B(t)$ implies that A is isogenous to B (see [**WM69**]). If $\ell = p$, then $T_p(A)$ is called a *physical Tate module.* In this case, $f_A(t) = f_1(t)f_2(t)$, where $f_1, f_2 \in \mathbb{Z}_p[t]$, and $f_1(t) = \det(t - F|T_p(A))$. Moreover $d = \deg f_1 \leq g$, and $f_2(t) \equiv t^{2g-d} \mod p$ (see [**De78**]).

Recall some definitions and results from [**Ry10**]. For an abelian group G we denote by G_ℓ the ℓ-primary component of G. The group $A(k)$ is a kernel of $1 - F : A \to A$, and the ℓ-component

$$A(k)_\ell \cong T_\ell(A)/(1 - F)T_\ell(A)$$

(see [**Ry10**, Proposition 3.1]). The proof of the following lemma is essentially the proof of [**Ry10**, Theorem 1.1].

Lemma 2.1. *Let A be an abelian variety over k, and let G be a finite abelian group of order $f_A(1)$. Suppose that for any prime number ℓ dividing order of G we have an F-invariant sublattice $T_\ell \subset T_\ell(A)$ such that $G_\ell \cong T_\ell/(1 - F)T_\ell$. Then there exists an abelian variety B over k, and an isogeny $\varphi : B \to A$ such that $T_\ell(\varphi)$ induces an isomorphism $T_\ell(B) \cong T_\ell$ for any ℓ. In particular, $B(k) \cong G$.*

We need a modification of this statement. Suppose we are looking for an abelian variety B such that $B(k) \cong G$, and A and B are isogenous. It is enough to find free $\mathbb{Z}_\ell$ module T_ℓ for any ℓ with semisimple action $F : T_\ell \to T_\ell$ such that $f_A(t) = \det(t - F)$, and $G_\ell \cong T_\ell/(1 - F)T_\ell$. Indeed, since the Frobenius action on the vector space V_ℓ is semisimple, it is determined up to isomorphism by the Weil polynomial f_A. We can reconstruct this action as follows. Put $V'_\ell = T_\ell \otimes_{\mathbb{Z}_\ell} \mathbb{Q}_\ell$, and choose any inclusion $T_\ell(A) \to V'_\ell$ such that the image of $T_\ell(A)$ contains T_ℓ. Clearly, this inclusion induces an isomorphism of F-vector spaces $V_\ell \to V'_\ell$.

Let $Q(t) = \sum_i Q_i t^i$ be a polynomial of degree d over $\mathbb{Q}_\ell$, and let $Q(0) = Q_0 \neq 0$. Take the lower convex hull of the points $(i, \mathrm{ord}_\ell(Q_i))$ for $0 \leq i \leq d$ in $\mathbb{R}^2$. The boundary of this region without vertical lines is called *the Newton polygon* $\mathrm{Np}_\ell(Q)$ *of* Q. Its vertices have integer coefficients, and $(0, \mathrm{ord}_\ell(Q_0))$ and $(d, \mathrm{ord}_\ell(Q_d))$ are its endpoints. One can associate to $\mathrm{Np}(Q)$ the set of its slopes, and each slope has a multiplicity. The Newton polygon of Q is uniquely determined by this data. Suppose we have two polynomials Q_1 and Q_2. Then the slope set for $\mathrm{Np}(Q_1Q_2)$

is the union of slope sets for $\mathrm{Np}(Q_1)$ and $\mathrm{Np}(Q_2)$, and multiplicities of slopes of $\mathrm{Np}(Q_1Q_2)$ are sums of multiplicities of corresponding slopes of $\mathrm{Np}(Q_1)$ and $\mathrm{Np}(Q_2)$.

We associate to $A(k)_\ell$ a polygon of special type.

DEFINITION 2.2. Let $0 \le m_1 \le m_2 \le \cdots \le m_r$ be nonnegative integers, and let $H = \oplus_{i=1}^r \mathbb{Z}/\ell^{m_i}\mathbb{Z}$ be an abelian group of order ℓ^m. The *Hodge polygon* $\mathrm{Hp}_\ell(H,r)$ *of a group* H is the convex polygon with vertices $(i, \sum_{j=1}^{r-i} m_j)$ for $0 \le i \le r$. It has $(0,m)$ and $(r,0)$ as its endpoints, and its slopes are $-m_r, \dots, -m_1$. We write $\mathrm{Hp}(H) = (m_1, \dots, m_r)$.

Note that some of the numbers m_i could be zero, in other words, the Hodge polygon is allowed to have zero slopes. The isomorphism class of H depends only on $\mathrm{Hp}_\ell(H,r)$. When we work with groups of points on abelian surfaces we often write $\mathrm{Hp}_\ell(H) = \mathrm{Hp}_\ell(H,4)$. We need the following well-known result (see [**Ke10**, 4.3.8] or [**BO**, 8.40]).

THEOREM 2.3. *Let E be an injective endomorphism of a free $\mathbb{Z}_\ell$-module T of finite rank. Let $f(t) = \det(E-t)$ be its characteristic polynomial. Then $\mathrm{Np}_\ell(f)$ lies on or above the Hodge polygon of T/ET, and these polygons have same endpoints.*

Weil polynomials of abelian surfaces were classified by Rück, Xing, Maisner and Nart [**MN02**]. We use a simplified version of this classification. Namely, we use that for the Weil polynomial only four cases of the main theorem below occur.

THEOREM 2.4. *Let A be an abelian surface over a finite field with Weil polynomial f_A. Let G be an abelian group of order $f_A(1)$. Then G is a group of points on some variety in the isogeny class of A if and only if for any prime number ℓ*

(1) *if f_A has no multiple roots, then $\mathrm{Np}_\ell(f_A(1-t))$ lies on or above $\mathrm{Hp}_\ell(G_\ell)$;*
(2) *if $f_A = P_A^2$, and P_A has no multiple roots, then $G_\ell \cong G_\ell^{(1)} \oplus G_\ell^{(2)}$, where $G_\ell^{(1)}$ and $G_\ell^{(2)}$ are ℓ-primary abelian groups with one or two generators such that $\mathrm{Np}_\ell(P_A(1-t))$ lies on or above $\mathrm{Hp}_\ell(G_\ell^{(1)},2)$ and $\mathrm{Hp}_\ell(G_\ell^{(2)},2)$.*
(3) *Suppose $f_A = (t^2 - bt + q)(t \pm \sqrt{q})^2$, and $\sqrt{q} \in \mathbb{Z}$, where $f(t) = t^2 - bt + q$ has no multiple roots; let*
    - $\mathrm{Hp}_\ell(G_\ell) = (m_1, m_2, m_3, m_4)$;
    - $m = \mathrm{ord}_\ell(f(1))$;
    - $m_q = \mathrm{ord}_\ell(1 \pm \sqrt{q})$;
    - $m_b = m_1 + m_3 - m_q$.

    Then
    (a) $0 \le m_b \le \mathrm{ord}_\ell(b-2)$;
    (b) $\min(m_b, m_q) \ge m_1$;
    (c) $\min(m - m_b, m_q) \ge m_2$.
(4) *Finally, if $f_A(t) = (t \pm \sqrt{q})^4$, and $\sqrt{q} \in \mathbb{Z}$, then $G \cong (\mathbb{Z}/(1 \pm \sqrt{q})\mathbb{Z})^4$.*

PROOF. Case (1) follows from [**Ry10**]. Let us prove that conditions of the case (2) are necessary. First, suppose that a prime ℓ divides $f_A(1)$, but $P_A(1-t) \not\equiv t^2$ mod ℓ. It is equivalent to say that $P_A(1-t) = (t-x_1)(t-x_2)$, where $x_1, x_2 \in \mathbb{Z}_\ell$, and ℓ divides x_1, but not divides x_2. Thus $T_\ell(A)$ is a direct sum of two modules T_1 and T_2 of $\mathbb{Z}_\ell$-rank two such that $1-F$ acts on T_i as multiplication by x_i. It follows that

$$G_\ell \cong T_1/(1-F)T_1 \cong (\mathbb{Z}_\ell/x_1\mathbb{Z}_\ell)^2,$$

and $G_\ell^{(1)} \cong G_\ell^{(2)} \cong \mathbb{Z}_\ell/x_1\mathbb{Z}_\ell$. The rest follows from Theorem 3.4 below. To prove that the conditions of (2) are sufficient we have to find an abelian variety B with a given group of points G. By Lemma 2.1 it is enough to construct a Tate module $T_\ell(B)$ for any prime ℓ dividing $f_A(1)$. By [**Ry10**, Theorem 3.2] there exist F-invariant $\mathbb{Z}_\ell$-modules T_1' and T_2' such that $T_i'/(1-F)T_i' \cong G_\ell^{(i)}$, and $T_i' \otimes_{\mathbb{Z}_\ell} \mathbb{Q}_\ell \cong \mathbb{Q}_\ell[t]/P_A(t)\mathbb{Q}_\ell[t]$ for $i \in \{1,2\}$. Now we let $T_\ell(B) \cong T_1' \oplus T_2'$.

We prove the case (3) in the last section. The case (4) is obvious since F acts as multiplication by $\mp\sqrt{q}$. □

3. Matrix factorizations

In this section we finish the proof of case (2). We assume that ℓ divides $P_A(1)$ and that

$$P_A(1-t) \equiv t^2 \mod \ell.$$

In this case $\ell \neq p$. Indeed, let $P_A(t) = t^2 - bt + c$, where $c = \pm q$. If $\ell = p$ satisfies our assumptions, we get that $2-b$ and $1-b$ are both divisible by p.

In [**Ry**] we show that the Tate module $T_\ell(A)$ corresponds to a matrix factorization. Fix a pair of polynomials $f, f_1 \in \mathbb{Z}_\ell[t]$ and a positive integer r. Put $R = \mathbb{Z}_\ell[t]/f_1\mathbb{Z}_\ell[t]$. Denote by $x \in R$ the image of t under the natural projection from $\mathbb{Z}_\ell[t]$.

Definition 3.1. A matrix factorization (X, Y) is a pair of $r \times r$ matrices with coefficients in $\mathbb{Z}_\ell[t]$ such that $\det X = f$ and $YX = f_1 \cdot I_r$, where I_r is the identity matrix.

Suppose we are given a matrix factorization (X, Y). The matrix X defines a map of free $\mathbb{Z}_\ell[t]$ modules:

$$\mathbb{Z}_\ell[t]^r \xrightarrow{X} \mathbb{Z}_\ell[t]^r.$$

Its cokernel T is annihilated by f_1. It is equivalent to say that T is an R-module. We see that the matrix factorization (X, Y) corresponds to a finitely generated R-module T given by the presentation:

$$\mathbb{Z}_\ell[t]^r \xrightarrow{X} \mathbb{Z}_\ell[t]^r \to T \to 0. \tag{3.1}$$

Proposition 3.2. [**Ry**, Proposition 5.2] *Suppose* $f_1 \equiv t^{d_1} \mod \ell$, *and* $\deg f_1 = d_1$. *The module T is free of finite rank over $\mathbb{Z}_\ell$, and characteristic polynomial of the action of x on T is equal to f.*

The following proposition shows that modules over R give rise to matrix factorizations.

Proposition 3.3. *Let T be an R-module which is free of finite rank over $\mathbb{Z}_\ell$. Suppose that T can be generated over R by r elements, and that* $\mathrm{Hp}(T/xT) = (m_1, \dots, m_r)$. *Then there exists a matrix factorization (X, Y) such that T has presentation* (3.1), *and*

$$X \equiv \mathrm{diag}(\ell^{m_1}, \dots, \ell^{m_r}) \mod t\mathbb{Z}_\ell[t].$$

Proof. By [**Ry**, Proposition 5.3] there exists a matrix factorization (X_1, Y_1) such that T has presentation

$$\mathbb{Z}_\ell[t]^r \xrightarrow{X_1} \mathbb{Z}_\ell[t]^r \to T \to 0. \tag{3.2}$$

Take the cokernel of the multiplication by t of the presentation(3.2); we get

$$\mathbb{Z}_\ell^r \xrightarrow{\bar{X}_1} \mathbb{Z}_\ell^r \to T/xT \to 0.$$

There exist matrices M_1 and M_2 over $\mathbb{Z}_\ell$ such that $\det M_1 = \det M_2 = 1$ and $\bar{X} = M_1\bar{X}_1M_2$ is a diagonal matrix. Since

$$\mathbb{Z}_\ell^r/\bar{X}\mathbb{Z}_\ell^r \cong T/xT,$$

we get that $\bar{X}=\operatorname{diag}(\ell^{m_1},\dots,\ell^{m_r})$. Now take $X=M_1X_1M_2$ and $Y=M_2^{-1}Y_1M_1^{-1}$. □

THEOREM 3.4. *Assume that* $f_A = P_A^2$, *and* $P_A(1-t) \equiv t^2 \mod \ell$. *Suppose that* P_A *has no multiple roots. Then* $A(k)_\ell \cong G_\ell^{(1)} \oplus G_\ell^{(2)}$, *where* $G_\ell^{(1)}$ *and* $G_\ell^{(2)}$ *are* ℓ*-primary abelian groups with one or two generators such that* $\mathrm{Np}_\ell(P_A(1-t))$ *lies on or above* $\mathrm{Hp}_\ell(G_\ell^{(1)},2)$ *and* $\mathrm{Hp}_\ell(G_\ell^{(2)},2)$.

PROOF. Apply Theorem 2.3 to $T = T_\ell(A)$ and $E = 1-F$. The slopes of $\mathrm{Np}(P_A(1-t))$ are the slopes of $\mathrm{Np}(P_A(1-t)^2)$, but multiplicities are doubled. We get that m_1 is not greater than the smallest slope of $\mathrm{Np}(P_A(1-t))$, i.e. that ℓ^{m_1} divides $b-2$. It is enough to prove that $m := \mathrm{ord}_\ell(P_A(1)) = m_1 + m_4$. Indeed, in this case $\mathrm{Np}_\ell(P_A(1-t))$ lies on or above $\mathrm{Hp}_\ell(G_\ell^{(1)},2) = (m_1,m_4)$. Moreover, $m = m_2+m_3$, and by Lemma 3.5 below $\mathrm{Np}_\ell(P_A(1-t))$ lies on or above $\mathrm{Hp}_\ell(G_\ell^{(2)},2) = (m_2,m_3)$.

Note that m_4 is a minimal number such that $\frac{\ell^{m_4}}{F-1} \in \mathrm{End}\, A$, and m_1 is a maximal number such that ℓ^{m_1} divides $F-1$. From the equality $(F-1)^2-(b-2)(F-1)-P_A(1)=0$ we get that

$$\frac{P_A(1)}{(F-1)\ell^{m_1}} = \frac{F-1}{\ell^{m_1}} - \frac{b-2}{\ell^{m_1}} \in \mathrm{End}\, A.$$

It follows that $m \geq m_1+m_4$.

The Frobenius action on the Tate module $T_\ell(A)$ is determined up to isomorphism by a module structure over the ring $R = \mathbb{Z}_\ell[t]/P_A(1-t)\mathbb{Z}_\ell[t]$ with t acting as $1-F$. Suppose we have an R-module T such that $\mathrm{Hp}(T/xT) = (m_1,m_2,m_3,m_4)$. By Proposition 3.3 there exists a matrix factorization (X,Y) such that $\det X = f_A(1-t)$ and

$$X \equiv \operatorname{diag}(\ell^{m_1},\ell^{m_2},\ell^{m_3},\ell^{m_4}) \mod t\mathbb{Z}_\ell[t].$$

The matrix factorization (Y,X) corresponds to a module T' over R, which is generated by 4 elements and the characteristic polynomial of x is equal to $\det Y = P_A^4(1-t)/f_A(1-t) = f_A(1-t)$. Moreover,

$$Y \equiv \operatorname{diag}(\ell^{m-m_1},\ell^{m-m_2},\ell^{m-m_3},\ell^{m-m_4}) \mod t\mathbb{Z}_\ell[t],$$

i.e. $\mathrm{Hp}(T'/xT') = (m-m_1, m-m_2, m-m_3, m-m_4)$. By Lemma 2.1 there exists an abelian surface B such that $T_\ell(B) \cong T'$. From the first part of the proof applied to $T_\ell(B)$ it follows that $(m-m_1)+(m-m_4) \leq m$. Thus $m = m_1+m_4$. □

LEMMA 3.5. $m_2 \leq \mathrm{ord}_\ell(b-2)$.

PROOF. Let L be a splitting field of P_A over $\mathbb{Q}_\ell$, and let S be its ring of integers. We have $P_A(1-t) = (t-a)(t-c)$, where $a,c \in S$, and $\mathrm{ord}_\ell(a) \leq \mathrm{ord}_\ell(b-2)$. (Here $\mathrm{ord}_\ell(a) \in \frac{1}{2}\mathbb{Z}$.) Consider the map

$$E = F+a-1 : T_\ell(A)\otimes_{\mathbb{Z}_\ell} S \to T_\ell(A)\otimes_{\mathbb{Z}_\ell} S.$$

Then $1-F$ acts on $T_1 = \ker E$ as multiplication by a. We get an exact sequence:

$$0 \to T_1 \to T_\ell(A) \otimes_{\mathbb{Z}_\ell} S \to T_2 = \operatorname{Im} E \to 0,$$

where $\operatorname{rk}_S T_1 = \operatorname{rk}_S T_2 = 2$. Act by $1-F$ on this sequence. By snake lemma we get an exact triple of abelian groups:

$$0 \to (S/aS)^2 \to A(k)_\ell \otimes_{\mathbb{Z}_\ell} S \to G' \to 0.$$

If $S = \mathbb{Z}_\ell$ then $A(k)_\ell$ contains $(\mathbb{Z}_\ell/a\mathbb{Z}_\ell)^2$, and $G' = T_2/(1-F)T_2$ is generated by 2 elements. It follows that $m_2 \leq \operatorname{ord}_\ell(a)$. If $S \neq \mathbb{Z}_\ell$, then $S \cong \mathbb{Z}_\ell^2$ as $\mathbb{Z}_\ell$-module and we multiply everything by two: $A(k)_\ell^2$ contains

$$(S/aS)^2 \cong (\mathbb{Z}/\lfloor \operatorname{ord}_\ell(a) \rfloor \mathbb{Z})^2 \oplus (\mathbb{Z}/\lceil \operatorname{ord}_\ell(a) \rceil \mathbb{Z})^2,$$

and $G' = T_2/(1-F)T_2$ is generated by 4 elements. Thus $m_2 \leq \lfloor \operatorname{ord}_\ell(a) \rfloor \leq \operatorname{ord}_\ell(b-2)$. □

4. Proof of case 3

The conditions (b) and (c) are equivalent to the following inequalities:

(1) $m_1 \leq m_b$, which is equivalent to $m_3 \geq m_q$;
(2) $m_1 \leq m_q$;
(3) $m_2 \leq m_q$;
(4) $m_2 \leq m - m_b$, which is equivalent to $m_4 \geq m_q$, since $f_A(1) = m_1 + m_2 + m_3 + m_4 = m + 2m_q$.

Note that inequalities $m_b \geq 0$ and (4) follow from (1), and (2) follows from (3). We have to prove (3), (1) and the second part of (a).

Let $\alpha = 1 \pm \sqrt{q}$, and let $G_\ell = A(k)_\ell$. Consider the map $E = F + \alpha - 1 : T_\ell(A) \to T_\ell(A)$. Then $1-F$ acts on $T_1 = \ker E$ by multiplication by α, and $1-F$ acts on $T_2 = \operatorname{Im} E$ with characteristic polynomial $f(1-t)$. We get an exact sequence:

$$0 \to T_1 \to T_\ell(A) \to T_2 \to 0,$$

and by the snake lemma

$$0 \to (\mathbb{Z}/\ell^{m_q}\mathbb{Z})^2 \to G_\ell \to G' \to 0.$$

The group G_ℓ contains $(\mathbb{Z}/\ell^{m_q}\mathbb{Z})^2$ only if $m_3 \geq m_q$, and G' has two generators only if (3) holds.

We now prove (a). Since $\operatorname{Hp}_\ell(T_\ell(A)/(F-1)T_\ell(A)) = (m_1, m_2, m_3, m_4)$ there exists a basis $v_1, v_2, v_3, v_4 \in T_\ell(A)$ such that

$$(F-1)v_1 \in \ell^{m_1} T_\ell(A),\ (F-1)v_2 \in \ell^{m_2} T_\ell(A),$$

$$(F-1)v_3 \in \ell^{m_3} T_\ell(A), \text{ and } (F-1)v_4 \in \ell^{m_4} T_\ell(A). \qquad (*)$$

The vectors Ev_4 and Ev_3 generate a lattice $T_3 \subset T_2$ of rank 2, thus $1-F$ acts on T_3 with characteristic polynomial $f(1-t)$. By Theorem 2.3 the Newton polygon $\operatorname{Np}_\ell(f(1-t))$ lies on or above $\operatorname{Hp}(T_3/(F-1)T_3, 2)$. Thus, we can find a linear combination v_3' of v_4 and v_3 such that v_1, v_2, v_3', v_4 is a basis and $(*)$ holds, but $(F-1)Ev_3'$ is not divisible by $(b-2)\ell$ in T_3. On the other hand, $F - 1 = E - \alpha$, and since $m_3 \geq m_q$, it follows that Ev_3' is not divisible by $\alpha\ell$ in $T_\ell(A)$. We proved that $(F-1)Ev_3'$ is not divisible by $\alpha(b-2)\ell$ in $T_\ell(A)$. By definition, $(F-1)v_3'$ is divisible by ℓ^{m_3} in $T_\ell(A)$. By the first inequality, $m_q \geq m_1$; thus for any vector

$v \in T_\ell(A)$ the vector Ev is divisible by ℓ^{m_1} in $T_\ell(A)$. We finally get that $(F-1)Ev_3'$ is divisible by $\ell^{m_1+m_3}$, but not divisible by $\ell\alpha(b-2)$ in $T_\ell(A)$. Thus

$$m_1 + m_3 < m_q + \operatorname{ord}_\ell(2-b) + 1.$$

Let us construct an abelian variety with a given group of points. We give an explicit construction of the Tate module T_ℓ such that $T_\ell/(F-1)T_\ell \cong G_\ell$, and apply Lemma 2.1. By assumption, we have an isomorphism of F-vector spaces:

$$V_\ell(A) \cong \mathbb{Q}_\ell[t]/P_A(1-t) \oplus (\mathbb{Q}_\ell[t]/(\alpha-t))^2,$$

where t acts as $1-F$ on the right-hand side. Thus there exists a basis v_1, v_2, v_3, v_4 of $V_\ell(A)$ such that

$$(1-F)v_1 = \ell^{m_b}v_2 + (b-2)v_1, \quad (1-F)v_2 = -\ell^{-m_b}f(1)v_1,$$
$$(1-F)v_3 = \alpha v_3, \qquad (1-F)v_4 = \alpha v_4.$$

Let T_ℓ be the $\mathbb{Z}_\ell$-submodule of $V_\ell(A)$ generated by

$$u_1 = v_1 + v_3, \qquad u_3 = v_2 + v_4,$$
$$u_2 = \frac{(1-F)u_1 - (b-2+\alpha)u_1 - \ell^{m_b}u_3}{\ell^{m_1}}, \text{ and}$$
$$u_4 = \frac{(1-F)u_3 - \alpha u_3 + \frac{f(1)}{\ell^{m_b}}u_1}{\ell^{m_2}}.$$

A straightforward calculation shows that

$$(1-F)u_1 = \ell^{m_1}u_2 + (b-2+\alpha)u_1 + \ell^{m_b}u_3;$$
$$(1-F)u_3 = \ell^{m_2}u_4 + \alpha u_3 - \frac{f(1)}{\ell^{m_b}}u_1;$$
$$(1-F)u_2 = \frac{-(b-2)\alpha u_1 - \ell^{m_b}\alpha u_3}{\ell^{m_1}};$$
$$(1-F)u_4 = \frac{f(1)\alpha u_1}{\ell^{m_2+m_b}} \in \ell^{m_4}T_\ell.$$

By (a) and (1), the vector $(1-F)u_1 \in \ell^{m_1}T_\ell$; by (3) and (4), the vector $(1-F)u_3 \in \ell^{m_2}T_\ell$; by (a) the vector $(1-F)u_2 \in \ell^{m_3}T_\ell$; by definition of m_b, the vector $(1-F)u_4 \in \ell^{m_4}T_\ell$. We get that the natural surjective map $T_\ell \to G_\ell$ factors through a surjective map $T_\ell/(1-F)T_\ell \to G_\ell$. We conclude that $T_\ell/(1-F)T_\ell \cong G_\ell$, since orders of both groups coincide.

References

[BO] Berthelot P., Ogus A., Notes on crystalline cohomology, Princeton University Press, Princeton, N.J.; University of Tokyo Press, Tokyo, 1978. MR0491705 (58:10908)

[De78] Demazure M., Lectures on p-divisible groups, Lecture notes in mathematics 302, Springer-Verlag, Berlin, 1972. MR883960 (88a:14049)

[Ke10] Kedlaya K., p-adic differential equations, Cambridge University Press, 2010. MR2663480 (2011m:12016)

[MN02] D. Maisner, E. Nart. *Abelian surfaces over finite fields as Jacobians.* With an appendix by Everett W. Howe. Experiment. Math. 11 (2002), no. 3, 321–337. MR1959745 (2003k:14057)

[Ry10] S. Rybakov. *The groups of points on abelian varieties over finite fields.* Cent. Eur. J. Math. 8(2), 2010, 282-288. arXiv:0903.0106v4 MR2610753 (2011c:14067)

[Ry] S. Rybakov. *The finite group subschemes of abelian varieties over finite fields.* arXiv:1006.5959v1

[Wa69] Waterhouse W., Abelian varieties over finite fields, Ann. scient. Éc. Norm. Sup., 1969, 4 serie 2, 521–560. MR0265369 (42:279)

[WM69] Waterhouse W., Milne J., Abelian varieties over finite fields, Proc. Sympos. Pure Math., Vol. XX, State Univ. New York, Stony Brook, N.Y., 1969, 53–64. MR0314847 (47:3397)

[Xi94] Xing Ch., The structure of the rational point groups of simple abelian varieties of dimension two over finite fields, Arch. Math., 1994,63, 427–430. MR1300737 (95k:11082)

[Xi96] Xing Ch., On supersingular abelian varieties of dimension two over finite fields, Finite Fields Appl., 1996, 2, no. 4, 407–421. MR1409453 (97g:11058)

Poncelet laboratory (UMI 2615 of CNRS and Independent University of Moscow)

Institute for information transmission problems of the Russian Academy of Sciences

Laboratory of Algebraic Geometry, GU-HSE, 7 Vavilova Str., Moscow, Russia, 117312

E-mail address: rybakov@mccme.ru, rybakov.sergey@gmail.com

Contemporary Mathematics
Volume **574**, 2012
http://dx.doi.org/10.1090/conm/574/11418

Computing low-degree isogenies in genus 2 with the Dolgachev–Lehavi method

Benjamin Smith

Abstract. Let ℓ be a prime, and $\mathcal{H}$ a curve of genus 2 over a field $\Bbbk$ of characteristic neither 2 nor ℓ. If S is a maximal Weil-isotropic subgroup of $\mathcal{J}_{\mathcal{H}}[\ell]$, then $\mathcal{J}_{\mathcal{H}}/S$ is isomorphic to the Jacobian $\mathcal{J}_{\mathcal{X}}$ of some (possibly reducible) curve $\mathcal{X}$. We investigate the Dolgachev–Lehavi method for constructing the curve $\mathcal{X}$, simplifying their approach and making it more explicit. The result, at least for $\ell = 3$, is an efficient and easily programmable algorithm suitable for number-theoretic calculations.

1. Introduction

Let $\ell \geq 3$ be prime, and let $\mathcal{H}$ be a curve of genus 2 over a perfect field $\Bbbk$ of characteristic neither 2 nor ℓ. Let $\mathcal{J}_{\mathcal{H}}$ be the Jacobian of $\mathcal{H}$, and let S be a maximal ℓ-Weil isotropic subgroup of $\mathcal{J}_H[\ell]$; since ℓ is prime, $S \cong (\mathbb{Z}/\ell\mathbb{Z})^2$. The quotient $\mathcal{J}_{\mathcal{H}}/S$ is isomorphic (as a principally polarized abelian variety) to a Jacobian $\mathcal{J}_{\mathcal{X}}$, where $\mathcal{X}$ is some curve of genus 2 (see [**14**]); hence, there exists an isogeny

$$\phi : \mathcal{J}_{\mathcal{H}} \to \mathcal{J}_{\mathcal{X}}$$

with kernel S (note that $\mathcal{X}$ may be reducible, in which case $\mathcal{J}_{\mathcal{X}}$ is a product of elliptic curves). Our aim is to compute an explicit form for $\mathcal{X}$ given $\mathcal{H}$ and S.

In the case $\ell = 2$, the problem is resolved by the well-known Richelot construction (see [**3**] and [**5**, Chapter 9]). For $\ell \geq 3$, if $\Bbbk$ is finite, then we can apply the explicit theta function-based algorithms of Lubicz and Robert [**10**], as implemented in the freely-available `avIsogenies` package [**1**].

Alternatively, there is the algebraic-geometric approach described by Dolgachev and Lehavi [**7**], which computes the image of the theta divisor on $\mathcal{J}_{\mathcal{H}}$ in the Kummer surface of $\mathcal{J}_{\mathcal{X}}$. As presented in [**7**], this approach has two drawbacks:

(1) it is not effective for $\ell \neq 3$, and
(2) for $\ell = 3$, where theta structures are involved, it assumes $\Bbbk \subset \mathbb{C}$.

In this work we render the kernel of the Dolgachev–Lehavi method completely explicit, with a view to computations in number theory. Our intention is to provide a sort of "user's guide" to this part of the algorithm and its implementation. For $\ell = 3$ we obtain a simple, efficient, and easily-programmable algorithm (that does not require $\Bbbk \subset \mathbb{C}$). Our algorithm retains the pleasing geometric flavour of the original, but is better-suited to everyday calculations.

2010 *Mathematics Subject Classification.* Primary 11Y99; Secondary 14Q05,14H45.

2. An overview of the Dolgachev–Lehavi construction

We begin by briefly recalling the Dolgachev–Lehavi construction, before treating it in detail in the following sections. Suppose $\mathcal{H}/\mathbb{k}$, S, ϕ, and $\mathcal{X}$ are as above; we assume we are given an explicit form for $\mathcal{H}$ and S, and we want to compute an explicit form for $\mathcal{X}$. Dolgachev and Lehavi observe that if $\Theta_{\mathcal{H}}$ and $\Theta_{\mathcal{X}}$ are theta divisors on $\mathcal{J}_{\mathcal{H}}$ and $\mathcal{J}_{\mathcal{X}}$, respectively, then $\phi(\Theta_{\mathcal{H}})$ is in $|\ell\Theta_{\mathcal{X}}|$ (see [**7**, Proposition 2.4]); hence, the image of $\phi(\Theta_{\mathcal{H}})$ in the Kummer surface $\mathcal{K}_{\mathcal{X}} = \mathcal{J}_{\mathcal{X}}/\langle\pm 1\rangle$ is a degree-2ℓ rational curve[1] in $\mathbb{P}^3$ of arithmetic genus $(\ell^2-1)/2$ and with $(\ell^2-1)/2$ ordinary double points, which correspond to the nonzero elements of S, up to sign [**7**, Proposition 3.1]. We can compute this curve *without* knowing ϕ by expressing the map $\Phi : \mathcal{H} \cong \Theta_{\mathcal{H}} \subset \mathcal{J}_{\mathcal{H}} \to \mathcal{J}_{\mathcal{X}} \to \mathcal{K}_{\mathcal{X}} \subset \mathbb{P}^3$ as a composition $\pi \circ \rho_{2\ell}$, where $\rho_{2\ell} : \mathcal{H} \to \mathbb{P}^{2\ell}$ is a double cover of a rational normal curve and $\pi : \mathbb{P}^{2\ell} \to \mathbb{P}^3$ is a projection whose centre depends on certain secants corresponding to the nonzero elements of S, up to sign. The images of the Weierstrass points of $\mathcal{H}$ under Φ lie on a conic $\mathcal{Q}$ contained in a hyperplane of $\mathbb{P}^3$; that is, a trope of $\mathcal{K}_{\mathcal{X}}$. The double cover of $\mathcal{Q}$ ramified over the Weierstrass point images is then (a quadratic twist of) $\mathcal{X}$.

3. The domain curve

We suppose that $\mathcal{H}/\mathbb{k}$ is presented as a nonsingular projective model

$$\mathcal{H} : Y^2 = F(X,Z) = \sum_{i=0}^{6} F_i X^i Z^{6-i} \subset \mathbb{P}(1,3,1), \tag{1}$$

where F is a squarefree sextic form over $\mathbb{k}$ (such a model always exists when $\mathbb{k}$ is perfect and has characteristic not 2: see [**5**, §1.3]). The hyperelliptic involution of $\mathcal{H}$ is

$$\iota_{\mathcal{H}} : (X : Y : Z) \longmapsto (X : -Y : Z).$$

The divisor at infinity on $\mathcal{H}$ is

$$D_\infty = \left(1 : \sqrt{F_6} : 0\right) + \left(1 : -\sqrt{F_6} : 0\right);$$

observe that D_∞ is defined over $\mathbb{k}$, fixed by $\iota_{\mathcal{H}}$, and equal to $2(1:0:0)$ if $F_6 = 0$.

The six Weierstrass points of $\mathcal{H}$ are the fixed points of $\iota_{\mathcal{H}}$; they correspond to the projective roots of the sextic F. The Weierstrass divisor $W_{\mathcal{H}}$ of $\mathcal{H}$ is the effective divisor cut out by $Y = 0$; if $F(X,Z) = \prod_{i=1}^{6}(z_i X - x_i Z)$ over $\overline{\mathbb{k}}$, then

$$W_{\mathcal{H}} = (x_1 : 0 : z_1) + \cdots + (x_6 : 0 : z_6).$$

Note that $W_{\mathcal{H}}$ is defined over $\mathbb{k}$. Finally, we fix a canonical divisor on $\mathcal{H}$, defining

$$K_{\mathcal{H}} = W_{\mathcal{H}} - 2D_\infty.$$

4. The kernel of the isogeny

When defining their method for $\ell = 3$, Dolgachev and Lehavi state "unfortunately, we do not know how to input explicitly the pair $(\mathcal{H}, S)$. Instead we consider $\mathcal{H}$ with an odd theta structure." We will take a rather more middlebrow approach to the problem: we suppose that $\mathcal{H}$ is presented in the form (1), and that S is given as a collection of divisor classes on $\mathcal{H}$ expressed using an extended Mumford representation (detailed below).

[1] By "rational curve" we mean a curve of genus 0. In all other contexts, "rational" means "defined over $\mathbb{k}$".

Our motivation for this choice is simple: this is precisely how one computes with hyperelliptic Jacobians in computational algebra systems such as Magma [**11, 2**] and Sage [**13**]. This choice also radically simplifies the algorithm: we can omit the theta structure calculations and pass directly to the secant computations (short-circuiting the first four steps of the algorithm in [**7**, §5.1]).

Points on $\mathcal{J}_{\mathcal{H}}$ correspond to divisor classes of degree zero on $\mathcal{H}$. The Riemann–Roch theorem tells us that every nontrivial degree-0 class has a unique representative in the form $P + Q - D_\infty$ (this representation fails to be unique for the trivial class, because $[P + \iota_{\mathcal{H}}(P) - D_\infty] = 0$ for every P in $\mathcal{H}(\overline{\Bbbk})$).

So let e be a point of $\mathcal{J}_{\mathcal{H}}$, corresponding to a divisor class $[P + Q - D_\infty]$. The effective divisor $P + Q$ is cut out by an ideal in the form $(A(X, Z), Y - B(X, Z))$, where B is a cubic form and A is a form of degree $d \leq 2$. The triple

$$\langle a(x), b(x), d\rangle := \langle A(x, 1), B(x, 1), d\rangle$$

then encodes the point e (with the convention that A is chosen such that a is monic). Note that if $\Bbbk'$ is an extension of $\Bbbk$, then $e = \langle a, b, d\rangle$ is in $\mathcal{J}_{\mathcal{H}}(\Bbbk')$ if and only if a and b have coefficients in $\Bbbk'$.

Conversely, given a triple $\langle a, b, d\rangle$, we recover the corresponding point of $\mathcal{J}_{\mathcal{H}}$ by computing the effective divisor cut out by $(A(X, Z), Y - B(X, Z))$, where B is the degree-3 homogenization of b and A is the degree-d homogenization of a, and then subtracting $(d/2)D_\infty$. If $\mathcal{H}$ has two points at infinity (that is, if $F_6 \neq 0$) then d must be either 2 or 0. In the case where $\mathcal{H}$ has a single point at infinity (that is, when $F_6 = 0$) we always have $d = \deg a$, and the pair $\langle a, b \bmod a\rangle$ is the standard Mumford representation. The advantage of the extended representation above is that it gracefully handles the general case where there are two points at infinity.

EXAMPLE 4.1. Consider the following points on the Jacobian of the genus 2 curve $\mathcal{H} : Y^2 = X^6 - Z^6$.

- 0 is represented by $\langle 1, 0, 0\rangle$.
- $[(1 : 0 : 1) + (-1 : 0 : 1) - D_\infty]$ is represented by $\langle x^2 - 1, 0, 2\rangle$.
- $[(1 : 1 : 0) - (1 : -1 : 0)] = [2(1 : 1 : 0) - D_\infty]$ is represented by $\langle 1, x^3, 2\rangle$.

In this article, we will assume that the points of S are all $\Bbbk$-rational: this simplifies both the exposition and the computations. However, all of our calculations are symmetric in the elements of S, and the algorithm should therefore be easily adapted to the case where S is rational but its elements are not.

5. The rational normal curve

The Riemann–Roch space $L(2\ell K_{\mathcal{H}})$ is a direct sum of subspaces

$$L(2\ell K_{\mathcal{H}}) = L(2\ell K_{\mathcal{H}})^+ \oplus L(2\ell K_{\mathcal{H}})^-,$$

where $\iota_{\mathcal{H}}$ acts as $+1$ on the elements of $L(2\ell K_{\mathcal{H}})^+$ and -1 on the elements of $L(2\ell K_{\mathcal{H}})^-$. Writing $x = X/Z$ and $y = Y/Z^3$, we have

$$L(2\ell K_{\mathcal{H}})^+ = \left\langle x^i/y^{2\ell}\right\rangle_{i=0}^{2\ell} \quad \text{and} \quad L(2\ell K_{\mathcal{H}})^- = \left\langle x^i/y^{2\ell-1}\right\rangle_{i=0}^{2\ell-3}.$$

The space $L(2\ell K_{\mathcal{H}})^+$ corresponds to the linear system $|2\ell K_{\mathcal{H}}|^{\langle\iota_{\mathcal{H}}\rangle}$; we see immediately that it is $(2\ell+1)$-dimensional, and therefore defines a map

$$\rho_{2\ell} : \mathcal{H} \longrightarrow \mathcal{R}_{2\ell} \subset \mathbb{P}^{2\ell}$$

onto a curve $\mathcal{R}_{2\ell}$ in $\mathbb{P}^{2\ell}$. Fixing coordinates on $\mathbb{P}^{2\ell}$, we take $\rho_{2\ell}$ to be defined by

$$\rho_{2\ell} : (X : Y : Z) \longmapsto (U_0 : \cdots : U_{2\ell}) = (X^0 Z^{2\ell} : X Z^{2\ell-1} : \cdots : X^{2\ell-1} Z : X^{2\ell}).$$

We see that $\mathcal{R}_{2\ell}$ is a rational normal curve of degree 2ℓ in $\mathbb{P}^{2\ell}$, and $\rho_{2\ell}$ is a double cover:

$$\rho_{2\ell}(P) = \rho_{2\ell}(Q) \iff \big(P = Q \quad \text{or} \quad P = \iota_{\mathcal{H}}(Q)\big). \tag{2}$$

(Essentially, $\rho_{2\ell}$ is the canonical map of $\mathcal{H}$ composed with an ℓ-uple embedding.)

6. The secant lines

We adopt the following convention: if S is a set of points in some projective space $\mathbb{P}^n$, then $\langle S \rangle$ denotes the linear subspace of $\mathbb{P}^n$ generated by S.

For any pair of points P and Q on $\mathcal{H}$, we define $\mathcal{L}_{P,Q}$ to be the line in $\mathbb{P}^{2\ell}$ intersecting $\mathcal{R}_{2\ell}$ in $\rho_{2\ell}(P) + \rho_{2\ell}(Q)$; that is,

$$\mathcal{L}_{P,Q} := \begin{cases} \langle \rho_{2\ell}(P), \rho_{2\ell}(Q) \rangle & \text{if } P \notin \{Q, \iota_{\mathcal{H}}(Q)\} \\ T_{\rho_{2\ell}(P)}(\mathcal{R}_{2\ell}) & \text{otherwise.} \end{cases}$$

We can also define secant lines corresponding to nonzero Jacobian elements: if e is a nonzero point on $\mathcal{J}_{\mathcal{H}}$, then we define

$$\mathcal{L}_e := \mathcal{L}_{P,Q} \quad \text{where } e = [P + Q - D_\infty].$$

Observe that $\mathcal{L}_{P,Q} = \mathcal{L}_{P,\iota_{\mathcal{H}}(Q)} = \mathcal{L}_{\iota_{\mathcal{H}}(P),Q} = \mathcal{L}_{\iota_{\mathcal{H}}(P),\iota_{\mathcal{H}}(Q)}$ for all P and Q on $\mathcal{H}$, so

$$\mathcal{L}_e = \mathcal{L}_{-e}$$

for all e in $\mathcal{J}_{\mathcal{H}} \setminus \{0\}$.

REMARK 6.1. Dolgachev and Lehavi define secants $l_e = \langle \rho_{2\ell}(P_1), \rho_{2\ell}(P_2) \rangle$ for each nontrivial point $e = [P_1 - P_2]$ in $\mathcal{J}_{\mathcal{H}}$ (see [**7**, Theorem 1.1]). Our $\mathcal{L}_e$ is equal to l_e, because $[P_1 - P_2] = [P_1 + \iota_{\mathcal{H}}(P_2) - D_\infty]$ and $\mathcal{L}_{P_1,P_2} = \mathcal{L}_{P_1,\iota_{\mathcal{H}}(P_2)}$.

The following lemma gives explicit and rational formulæ for the secants $\mathcal{L}_e$ and their intersections with arbitrary hyperplanes in $\mathbb{P}^2$. These formulæ are central to the explicit Dolgachev–Lehavi method.

LEMMA 6.2. *Let $e = \langle a, b, d \rangle$ be a nonzero point of $\mathcal{J}_{\mathcal{H}}$. Let $H : \sum_{i=0}^{2\ell} H_i U_i = 0$ be a hyperplane in $\mathbb{P}^{2\ell}$, and write $h(x) := \sum_{i=0}^{2\ell} H_i x^i$.*

(1) *If $a = 1$ and $d = 2$, then*

$$\mathcal{L}_e = \big\langle (0 : \cdots : 0 : 1 : 0), (0 : \cdots : 0 : 0 : 1) \big\rangle.$$

 (a) *If $H_{2\ell} = H_{2\ell-1} = 0$, then $\mathcal{L}_e \subset H$.*
 (b) *Otherwise, $\mathcal{L}_e \cap H = (0 : \cdots : 0 : H_{2\ell} : -H_{2\ell-1})$.*

(2) *If $a(x) = x - \alpha$, then*

$$\mathcal{L}_e = \big\langle (0 : \cdots : 0 : 1), (1 : \cdots : \alpha^{2\ell}) \big\rangle.$$

 (a) *If $h(\alpha) = 0$ and $H_{2\ell} = 0$, then $\mathcal{L}_e \subset H$.*
 (b) *Otherwise, $\mathcal{L}_e \cap H = \big(H_{2\ell} : H_{2\ell}\alpha : \cdots : H_{2\ell}\alpha^{2\ell-1} : H_{2\ell}\alpha^{2\ell} - h(\alpha)\big)$.*

(3) *If $a(x) = x^2 + a_1 x + a_0$ with $a_1^2 \neq 4a_0$, then*

$$\mathcal{L}_e = \big\langle (\pi_0 : \cdots : \pi_{2\ell}), (-a_1 : a_2\pi_0 : a_2\pi_1 : \cdots : a_2\pi_{2\ell-1}) \big\rangle$$

where $\pi_0 = 2$, $\pi_1 = -a_1$, and $\pi_i = -a_1\pi_{i-1} - a_2\pi_{i-2}$ for $i > 1$.

 (a) *If $a(x)$ divides $h(x)$, then $\mathcal{L}_e \subset H$.*

(b) *Otherwise,* $\mathcal{L}_e \cap H = (\gamma_0 : \cdots : \gamma_{2\ell})$ *where*

$$\gamma_i = \sum_{0 \le j \le 2\ell} H_j (a_2^j \sigma_{i-j} - a_2^i \sigma_{j-i})$$

with $\sigma_k = 0$ *for* $k < 1$, $\sigma_1 = 1$, *and* $\sigma_k = -a_1\sigma_{k-1} - a_2\sigma_{k-2}$ *for* $k > 1$.

(4) *If* $a(x) = x^2 + a_1 x + a_0$ *with* $a_1^2 = 4a_0$, *then writing* α *for* $-a_1/2$, *we have*

$$\mathcal{L}_e = \langle (1 : \alpha : \cdots : \alpha^{2\ell}), (0 : 1 : 2\alpha : \cdots : 2\ell\alpha^{2\ell-1}) \rangle.$$

(a) *If* $a(x)$ *divides* $h(x)$, *then* $\mathcal{L}_e \subset H$.
(b) *Otherwise,* $\mathcal{L}_e \cap H = (\gamma_0 : \cdots : \gamma_{2\ell})$ *where* $\gamma_i = i\alpha^{i-1}h(\alpha) - \alpha^i h'(\alpha)$.

PROOF. In general, given points $\alpha = (\alpha_0 : \cdots : \alpha_{2\ell})$ and $\beta = (\beta_0 : \cdots : \beta_{2\ell})$ in $\mathbb{P}^{2\ell}$, we have

$$H \cap \mathcal{L}_{\alpha,\beta} = (A\beta_0 - B\alpha_0 : \cdots : A\beta_{2\ell} - B\alpha_{2\ell})$$

where $A = \sum_{i=0}^{2\ell} H_i\alpha_i$ and $B = \sum_{i=0}^{2\ell} H_i\beta_i$; if $A = B = 0$, then $\mathcal{L}_{\alpha,\beta} \subset H$ (and the point above is not defined). In the following, we suppose $e = [P + Q - D_\infty]$; we have $e \neq 0$, so we can suppose $P \neq \iota_{\mathcal{H}}(Q)$.

In case (1) both P and Q are at infinity, so $\rho_{2\ell}(P) = \rho_{2\ell}(Q) = (0 : \cdots : 0 : 1)$; our expression for $\mathcal{L}_e$ gives generators for the tangent to $\mathcal{R}_{2\ell}$ at $(0 : \cdots : 0 : 1)$. The intersection formula follows immediately.

In case (2), we have $P = (1 : 0 : 0)$ and $Q = (\alpha : \pm\sqrt{F(\alpha,1)} : 1)$, so $\rho_{2\ell}(P) = (0 : \cdots : 0 : 1)$ and $\rho_{2\ell}(Q) = (1 : \alpha : \cdots : \alpha^{2\ell})$. The intersection formula follows immediately.

In case (3), we have $P = (\alpha : \pm\sqrt{F(\alpha,1)} : 1)$ and $Q = (\beta : \pm\sqrt{F(\beta,1)} : 1)$ with $\alpha \neq \beta$, $\alpha + \beta = -a_1$, and $\alpha\beta = a_2$; so $\rho_{2\ell}(P) = (1 : \alpha : \cdots : \alpha^{2\ell})$ and $\rho_{2\ell}(Q) = (1 : \beta : \cdots : \beta^{2\ell})$. If we take

$$T = (2 : \alpha + \beta : \cdots : \alpha^{2\ell} + \beta^{2\ell}) \quad \text{and} \quad S = (\alpha + \beta : 2\beta\alpha : \cdots : \alpha\beta^{2\ell} + \beta\alpha^{2\ell}),$$

then we easily verify that $\mathcal{L}_{P,Q} = \mathcal{L}_{T,S}$; it is a straightforward exercise with symmetric polynomials to show that $\alpha^i + \beta^i = \pi_i$ for $0 \le i \le 2\ell$ and $\alpha\beta^i + \beta\alpha^i = a_2\pi_{i-1}$ for $i > 0$, whence our formula for $\mathcal{L}_e$. The intersection $H \cap \mathcal{L}_e$ is

$$H \cap \mathcal{L}_{P,Q} = (h(\alpha) - h(\beta) : h(\alpha)\beta - h(\beta)\alpha : \cdots : h(\alpha)\beta^{2\ell} - h(\beta)\alpha^{2\ell});$$

it is another straightforward exercise to show that

$$\alpha^j\beta^i - \beta^j\alpha^i = (\beta - \alpha)(a_2^j\sigma_{i-j} - a_2^i\sigma_{j-i}),$$

so $h(\alpha)\beta^i - h(\beta)\alpha^i = \sum_{j=0}^{2\ell} H_j(\beta - \alpha)(a_2^j\sigma_{i-j} - a_2^i\sigma_{j-i}) = (\beta - \alpha)\gamma_i$ for $0 \le i \le 2\ell$, and thus $H \cap \mathcal{L}_e = (\gamma_0 : \cdots : \gamma_{2\ell})$.

In case (4), we have $P = Q = (\alpha : \pm\sqrt{F(\alpha,1)} : 1)$; our expression for $\mathcal{L}_e$ gives generators for the tangent to $\mathcal{R}_{2\ell}$ at $\rho_{2\ell}(P) = (1 : \alpha : \cdots : \alpha^{2\ell})$. The intersection formula follows. □

7. The Weierstrass subspace

Since $\mathcal{R}_{2\ell}$ is a rational normal curve of degree 2ℓ, any $2\ell + 1$ distinct points on $\mathcal{R}_{2\ell}$ are linearly independent. In particular, the images of the six Weierstrass points of $\mathcal{H}$ under $\rho_{2\ell}$ are linearly independent because $\ell \geq 3$. In view of (2) the images are distinct, so the subspace

$$W := \langle \rho_{2\ell}(W_{\mathcal{H}}) \rangle \subset \mathbb{P}^{2\ell}$$

is five-dimensional.

For each $0 \leq i \leq 2\ell - 6$, we define a linear form

$$W_i := \sum_{j=0}^{6} F_j U_{i+j}.$$

LEMMA 7.1. *The space W is*

$$W = \bigcap_{i=0}^{2\ell-6} V(W_i) = V(\{W_i : 0 \leq i \leq 2\ell - 6\}).$$

PROOF. Each $V(W_i)$ contains W, since $W_i \circ \rho_{2\ell} = X^i Z^{2\ell-6-i} F(X,Z)$. But the W_i are linearly independent, so the intersection $\bigcap_{i=0}^{2\ell-6} V(W_i)$ is 5-dimensional, and hence equal to W. □

8. The theorem of Dolgachev and Lehavi

We are now ready to state the main theorem behind the Dolgachev–Lehavi method.

THEOREM 8.1 ([**7**, Theorem 1.1]). *There exists a unique hyperplane $H \subset \mathbb{P}^{2\ell}$ such that*

1. *H contains W, and*
2. *the intersection points of H with the secants $\mathcal{L}_e$ for each nonzero e in S are contained in a subspace N of codimension 3 in H.*

The image of the Weierstrass divisor under the projection $\mathbb{P}^{2\ell} \to \mathbb{P}^3$ with centre N lies on a conic $\mathcal{Q}$ (which may be reducible), and the double cover of $\mathcal{Q}$ ramified over this divisor is a stable curve $\mathcal{X}$ of arithmetic genus 2 such that $\mathcal{J}_{\mathcal{X}} \cong \mathcal{J}_{\mathcal{H}}/S$.

It is crucial to note that Theorem 8.1 is not constructive: it does not in itself yield the hyperplane H, nor the centre N of the projection to $\mathbb{P}^3$. It is noted in [**7**, §3.4] that H is defined by $\phi^*(\Theta_{\mathcal{X}})$, but in our situation we do not yet have an expression for ϕ or $\Theta_{\mathcal{X}}$.

In the case $\ell = 3$, we are saved by a happy coincidence: $2\ell - 1 = 5$, so $H = W$ (we return to this case in §11 below). For $\ell > 3$, we must compute H in some other way; Lemma 8.2, an easy corollary of Lemma 7.1, characterizes the possible hyperplanes.

LEMMA 8.2. *The linear system of all hyperplanes in $\mathbb{P}^{2\ell}$ containing W is generated by the $2\ell - 5$ hyperplanes $V(W_i)$ for $0 \leq i \leq 2\ell - 6$. That is, if $H \supset W$ is a hyperplane in $\mathbb{P}^{2\ell}$, then*

$$H = V(\alpha_0 W_0 + \cdots + \alpha_{2\ell-6} W_{2\ell-6})$$

for some $(\alpha_0 : \cdots : \alpha_{2\ell-6})$ in $\mathbb{P}^{2\ell-6}(\Bbbk)$.

In view of Lemma 8.2, one naïve approach to computing H for $\ell > 3$ would be to take a generic hyperplane $H = V(\sum_{i=0}^{2\ell-6} \alpha_i W_i)$ and compute its intersection with the secants $\mathcal{L}_e$. This yields $(\ell^2 - 1)/2$ points whose coordinates are linear expressions in the α_i. We could then solve for the values of α_i by computing the zero locus of the $(2\ell-2)\times(2\ell-2)$ minors of the matrix formed by the intersections $H\cap\mathcal{L}_e$; but each minor is a degree-$(2\ell-2)$ polynomial in $2\ell-5$ variables, and the number of minors is exponential in ℓ.

Alternatively, we could take a generic set of linear equations determining N inside the generic H; requiring that this centre intersects any one of the $(\ell^2-1)/2$ secants imposes $O(\ell^4)$ quartic polynomial conditions on the $O(\ell)$ unknowns.

In each approach the system is highly overdetermined, and with a clever choice of minors we might hope to get lucky and find solutions for toy examples. However, both approaches already represent a significant undertaking for $\ell=5$, even over finite fields; they are totally impractical for larger ℓ and for infinite fields.

We continue the treatment for general ℓ in §9 and §10, supposing that an equation for H has been found; without such an equation, the `avIsogenies` package [**1**] represents a much more sensible approach for $\ell\geq 5$ (if $\Bbbk$ is finite). For $\ell=3$, the Dolgachev–Lehavi method is as practical as it is interesting; we specialize to this case in §11 and §12.

9. From theory to practice

To compute $\mathcal{X}$ via Theorem 8.1, we must compute the map

$$\Phi := \pi\circ\rho_{2\ell} : \mathcal{H}\to\mathbb{P}^3,$$

where $\pi:\mathbb{P}^{2\ell}\to\mathbb{P}^3$ is the projection with centre N. Suppose that we have an equation

$$H : \sum_i \alpha_i W_i = 0$$

for H. Applying Lemma 6.2, we compute the centre $N=\langle\mathcal{L}_e\cap H : e\in S\setminus\{0\}\rangle$. Since $N\subset H$, we may compute $\nu_{0,0},\ldots,\nu_{0,2\ell},\nu_{1,0},\ldots,\nu_{1,2\ell},\nu_{2,0},\ldots,\nu_{2,2\ell}$ in $\Bbbk$ such that

$$N = V\left(\sum_{i=0}^{2\ell}\nu_{0,i}U_i,\ \sum_{i=0}^{2\ell}\nu_{1,i}U_i,\ \sum_{i=0}^{2\ell}\nu_{2,i}U_i,\ \sum_{i=0}^{2\ell-6}\alpha_iW_i\right).$$

(This amounts to computing the kernel of the matrix whose rows are formed by the coordinates of the $\mathcal{L}_e\cap H$; the choice of $\sum_{i=0}^{6}\alpha_iW_i$ for the fourth defining equation will be convenient later in the procedure.)

Fixing coordinates on $\mathbb{P}^3$, the projection π with centre N is defined by

$$\begin{aligned}\pi : (U_0:\cdots:U_{2\ell}) &\longmapsto (V_0:V_1:V_2:V_3)\\ &= \left(\textstyle\sum_{i=0}^{2\ell}\nu_{0,i}U_i,\ \sum_{i=0}^{2\ell}\nu_{1,i}U_i,\ \sum_{i=0}^{2\ell}\nu_{2,i}U_i,\ \sum_{i=0}^{2\ell-6}\alpha_iW_i\right);\end{aligned}$$

the composed map $\Phi=\pi\circ\rho_{2\ell}$ is then

$$\begin{aligned}\Phi : (X:Y:Z) &\longmapsto (V_0:V_1:V_2:V_3)\\ &= \big(\Phi_0(X,Z):\Phi_1(X,Z):\Phi_2(X,Z):\Phi_3(X,Z)\big),\end{aligned}$$

where

$$\Phi_0 := \sum_{i=0}^{2\ell}\nu_{0,i}X^iZ^{2\ell-i},\quad \Phi_2 := \sum_{i=0}^{2\ell}\nu_{1,i}X^iZ^{2\ell-i},\quad \Phi_2 := \sum_{i=0}^{2\ell}\nu_{2,i}X^iZ^{2\ell-i},$$

and

$$\Phi_3 := \sum_{i=0}^{2\ell-6}\alpha_iX^iZ^{2\ell-6-i}F(X,Z).$$

The image of Φ is a rational curve of degree 2ℓ in $\mathbb{P}^3$. It lies on the Kummer surface $\mathcal{K}_{\mathcal{X}}$ of the unknown codomain Jacobian $\mathcal{J}_{\mathcal{X}}$, and is therefore the intersection of a quadric and a cubic hypersurface in $\mathbb{P}^3$ (see [**9**, Chapter XIII]):

$$\Phi(\mathbb{P}^1) = \widetilde{\mathcal{Q}} \cap \widetilde{\mathcal{C}} \quad \text{where } \widetilde{\mathcal{Q}} = V\left(\widetilde{Q}(V_0, V_1, V_2, V_3)\right) \quad \text{and } \widetilde{\mathcal{C}} = V\left(\widetilde{C}(V_0, V_1, V_2, V_3)\right)$$

for some forms $\widetilde{Q}$ and $\widetilde{C}$ of degree 2 and 3, respectively. The forms $\widetilde{Q}$ and $\widetilde{C}$ generate the elimination ideal

$$(\widetilde{Q}, \widetilde{C}) = (V_0 - \Phi_0, V_1 - \Phi_1, V_2 - \Phi_2, V_3 - \Phi_3) \cap \Bbbk[V_0, V_1, V_2, V_3];$$

note that up to scalar multiples, $\widetilde{Q}$ is uniquely determined while $\widetilde{C}$ is determined modulo $(V_0\widetilde{Q}, V_1\widetilde{Q}, V_2\widetilde{Q}, V_3\widetilde{Q})$.

The Weierstrass points of $\mathcal{H}$ map into the hyperplane $V_3 = 0$, which we identify with $\mathbb{P}^2$. (This simplification motivated our choice of Φ_3.) Theorem 8.1 asserts that a conic $\mathcal{Q}$ passes through the six images, and indeed

$$\mathcal{Q} = V(Q(V_0, V_1, V_2)) \subset \mathbb{P}^2, \quad \text{where} \quad Q(V_0, V_1, V_2) = \widetilde{Q}(V_0, V_1, V_2, 0).$$

The image of the Weierstrass divisor under Φ is therefore $\mathcal{Q} \cap \mathcal{C}$, where

$$\mathcal{C} = V(C(V_0, V_1, V_2)) \subset \mathbb{P}^2 \quad \text{with} \quad C(V_0, V_1, V_2) = \widetilde{C}(V_0, V_1, V_2, 0).$$

We are more interested in the forms Q and C than in $\widetilde{Q}$ and $\widetilde{C}$, and it is a simple matter to interpolate them. To find Q, we compute the six quintic polynomials $\Phi_i\Phi_j(x, 1) \bmod F(x, 1)$ for $0 \le i \le j \le 2$; the unique linear relation between them (and between the $\nu_{i,0}\nu_{j,0}$ if $F_6 = 0$) yields the coefficients of Q. Similarly, to find C we compute the ten quintics $\Phi_i\Phi_j\Phi_k(x, 1) \bmod F(x, 1)$ for $0 \le i \le j \le k \le 2$; any one of the linear relations between them (and the $\nu_{i,0}\nu_{j,0}\nu_{k,0}$ if $F_6 = 0$) gives an equation for a valid cubic C.

10. The codomain curve

The data $(\mathcal{Q}, \mathcal{Q} \cap \mathcal{C})$ specifies a genus 2 curve $\mathcal{X}$ (up to a quadratic twist) as a double cover of $\mathcal{Q}$ ramified over the six points of $\mathcal{Q} \cap \mathcal{C}$. This is the output of the Dolgachev–Lehavi algorithm and of Theorem 8.1, and it is sufficient for computing isomorphism invariants of $\mathcal{X}$ (see, for example, [4] and [**12**]).

In some situations, however, we would like to derive a defining equation for $\mathcal{X}$ itself. When $\mathcal{Q}$ is nonsingular, we recover a hyperelliptic curve; in the degenerate case where $\mathcal{Q}$ is singular, we recover a union of two elliptic curves $\mathcal{X}_+$ and $\mathcal{X}_-$, which are generally defined over a quadratic extension of $\Bbbk$ (in which case they are Galois conjugates). The procedure is essentially standard (cf. [**4**, §2]), but we recall it here for completeness.

Algorithm 10.1. Computes a (possibly reducible) genus 2 curve representing a double cover of a given plane conic, ramified over the intersection of the conic with a plane cubic.

Input: A plane conic $\mathcal{Q} : Q(V_0, V_1, V_2) = 0$ and cubic $\mathcal{C} : C(V_0, V_1, V_2) = 0$.

Output: A genus 2 curve $\mathcal{X}$ forming a double cover of $\mathcal{Q}$ ramified over $\mathcal{Q} \cap \mathcal{C}$. If $\mathcal{Q}$ is singular, then $\mathcal{X}$ will be a one-point union of elliptic curves $\mathcal{X}_+$ and $\mathcal{X}_-$, with $\mathcal{X}_\pm$ ramified over P_0 and $\mathcal{C} \cap \mathcal{L}_\pm$ where $\mathcal{Q} = \mathcal{L}_+ + \mathcal{L}_-$ and $P_0 = \mathcal{L}_+ \cap \mathcal{L}_-$.

1: Let M be the matrix defined by

$$M := \begin{pmatrix} 2q_{0,0} & q_{0,1} & q_{0,2} \\ q_{0,1} & 2q_{1,1} & q_{1,2} \\ q_{0,2} & q_{1,2} & 2q_{2,2} \end{pmatrix}, \quad \text{where} \sum_{0 \le i \le j \le 2} q_{i,j} V_i V_j = Q(V_0, V_1, V_2).$$

2: If $\det(M) = 0$, then $\mathcal{Q}$ is singular.

2a: Compute a diagonal matrix $D = \operatorname{diag}(a, b, 0)$ and an invertible matrix

$$T = \begin{pmatrix} t_{00} & t_{01} & t_{02} \\ t_{10} & t_{11} & t_{12} \\ t_{20} & t_{21} & t_{22} \end{pmatrix}$$

such that $M = TDT^{-1}$.

2b: Set $\delta = \sqrt{-a/b}$, and define cubic forms $C_+(X, Z)$ and $C_-(X, Z)$ by $C_\pm := C\big((t_{00} \pm \delta t_{01})Z + t_{02}X, (t_{10} \pm \delta t_{11})Z + t_{12}X, (t_{20} \pm \delta t_{21})Z + t_{22}X\big)$.

2c: Define elliptic curves $\mathcal{X}_+$ and $\mathcal{X}_-$ over $\Bbbk(\delta)$ in $\mathbb{P}(2, 3, 2)$ by

$$\mathcal{X}_+ : Y^2 = C_+(X, Z) \quad \text{and} \quad \mathcal{X}_- : Y^2 = C_-(X, Z),$$

and return the union of $\mathcal{X}_+$ and $\mathcal{X}_-$ identifying the points at infinity.

3: Otherwise, $\mathcal{Q}$ is nonsingular.

3a: Compute a rational point P in $\mathcal{Q}(\Bbbk)$ (see Remark 10.2), and let $\pi : \mathbb{P}^1 \to \mathcal{Q}$ be the rational parametrization corresponding to P, defined by

$$\pi : (X : Z) \longmapsto (V_0 : V_1 : V_2) = (P_0(X, Z) : P_1(X, Z) : P_2(X, Z))$$

(the P_i are quadratic forms).

3c: Return $\mathcal{X} : Y^2 = C(P_0(X, Z), P_1(X, Z), P_2(X, Z))$.

REMARK 10.2. Step 3a of Algorithm 10.1 requires us to compute a $\Bbbk$-rational point P on the conic $\mathcal{Q}$. If $\mathcal{H}$ has a rational Weierstrass point W_0, then we may take $P = \Psi(W_0)$. Generically, however, $\mathcal{H}$ has no rational Weierstrass points, and then we are obliged to search for a rational point on $\mathcal{Q}$. We are guaranteed that such a rational point exists (cf. [**12**, Lemme 1]). Over a finite field, finding a rational point is straightforward; over the rationals, we can apply (for example) the Cremona–Rusin algorithm [**6**].

11. The algorithm for $\ell = 3$

Consider now the special case $\ell = 3$. The map $\rho_6 : \mathcal{H} \to \mathcal{R}_6 \subset \mathbb{P}^6$ is defined by

$$\rho_6 : (X : Y : Z) \longmapsto (U_0 : U_1 : \cdots : U_5 : U_6) = \big(Z^6 : XZ^5 : \cdots : X^5 Z : X^6\big).$$

The hyperplane H of Theorem 8.1 contains $W = \langle \rho_6(W_\mathcal{H}) \rangle$ by definition; but $\dim H = \dim W = 5$, so $H = W$. Applying Lemma 7.1, we find

$$H = V(W_0) = V\left(\sum_{i=0}^{6} F_i U_i\right) \subset \mathbb{P}^6; \tag{3}$$

this allows us to simplify Lemma 6.2 for the case $\ell = 3$.

PROPOSITION 11.1. *If $e = \langle a, b, d \rangle$ is a nonzero 3-torsion point of $\mathcal{J}_\mathcal{H}$, then*

$$H \cap \mathcal{L}_e = \big(\gamma_0(e) : \cdots : \gamma_6(e)\big),$$

where the γ_i are defined as follows:

(1) *If* $a = 1$, *then* $\gamma_i(e) = 0$ *for* $0 \leq i < 5$, *with* $\gamma_5(e) = F_6$ *and* $\gamma_6(e) = -F_5$.
(2) *If* a *is linear, then* $\gamma_i(e) = 0$ *for* $0 \leq i < 6$, *and* $\gamma_6(e) = 1$.
(3) *If* $a(x) = x^2 + a_1 x + a_0$ *with* $a_1^2 \neq 4a_0$, *then*

$$\gamma_i(e) = \sum_{j=0}^{6} F_j(a_2^j \sigma_{i-j} + a_2^i \sigma_{j-i}) \quad \textit{for} \quad 0 \leq i \leq 6$$

with $\sigma_k = 0$ *for* $k < 1$, $\sigma_1 = 1$, *and* $\sigma_k = -a_1\sigma_{k-1} - a_2\sigma_{k-2}$ *for* $k > 1$.
(4) *If* $a(x) = x^2 + a_1 x + a_0$ *with* $a_1^2 = 4a_0$, *then*

$$\gamma_i(e) = \sum_{j=0}^{6} (i-j)F_j(-a_1/2)^{i+j-1} \quad \textit{for} \quad 0 \leq i \leq 6.$$

PROOF. This follows from Lemma 6.2 with $H = V(\sum_{i=0}^{6} F_i U_i)$, noting that $a(x)$ cannot divide $h(x) = \sum_{i=0}^{6} F_i x^i$ (since otherwise e would have order 2). □

We now present a version of the Dolgachev–Lehavi algorithm for $\ell = 3$ based on the extended Mumford representation. The algorithm requires only elementary matrix algebra and polynomial arithmetic, and should be easily implemented in most computational algebra systems.

ALGORITHM 11.2. A streamlined Dolgachev–Lehavi-style algorithm for $\ell = 3$.

Input: A genus 2 curve $\mathcal{H} : Y^2 = F(X, Z) = \sum_{i=0}^{6} F_i X^i Z^{6-i}$ over $\Bbbk$ and a maximal Weil-isotropic subgroup S of $\mathcal{J}_{\mathcal{H}}[3]$, its elements defined over $\Bbbk$ and presented as in §4.

Output: A genus 2 curve $\mathcal{X}$ over $\Bbbk$ such that there exists an isogeny $\phi : \mathcal{J}_{\mathcal{H}} \to \mathcal{J}_{\mathcal{X}}$ with kernel S (the curve $\mathcal{X}$ is computed up to a quadratic twist, so the isogeny may only be defined over a quadratic extension of $\Bbbk$).

1: Compute a minimal subset $S^{\pm}$ of S such that $S = S^{\pm} \cup \{-e : e \in S^{\pm}\} \cup \{0\}$ (then $\{\mathcal{L}_e : e \in S^{\pm}\} = \{\mathcal{L}_e : e \in S \setminus \{0\}\}$; this avoids redundancy in Steps 2 and 3).

2: For each e in $S^{\pm}$, compute the vector $v_e = (\gamma_0(e), \ldots, \gamma_6(e))$ using the formulæ in Proposition 11.1.

3: Compute vectors $n_i = (\nu_{i,0}, \ldots, \nu_{i,6})$ such that $\{n_0, n_1, n_2, (F_j : 0 \leq j \leq 6)\}$ is a basis for the (left) kernel of the 7×4 matrix $(v_e^t : e \in S^{\pm})$. Set

$$\Phi_i = \sum_{j=0}^{6} \nu_{i,j} X^j Z^{6-j} \quad \text{for } 0 \leq i \leq 2.$$

4: For each $0 \leq i \leq j \leq 2$, compute the vector $r_{i,j}$ of length 6 whose n^{th} entry is the coefficient of x^{n-1} in $(\Phi_i \Phi_j)(x, 1) \bmod F(x, 1)$. If $F_6 = 0$, then take the 6th entry of $r_{i,j}$ to be $\nu_{i,0}\nu_{j,0}$: this allows us to correctly interpolate through the image of the Weierstrass point at infinity.

5: Compute a generator $(q_{i,j} : 0 \leq i \leq j \leq 2)$ for the (left) kernel of the 6×6 matrix whose rows are the $r_{i,j}$ for $0 \leq i \leq j \leq 2$. Set

$$Q(V_0, V_1, V_2) := q_{0,0}V_0^2 + q_{0,1}V_0V_1 + q_{0,2}V_0V_2 + q_{1,1}V_1^2 + q_{1,2}V_1V_2 + q_{2,2}V_2^2.$$

6: For each $0 \leq i \leq j \leq k \leq 2$, compute the vector $s_{i,j,k}$ of length 6 whose n^{th} entry is the coefficient of x^{n-1} in $(\Phi_i \Phi_j \Phi_k)(x, 1) \bmod F(x, 1)$. If $F_6 = 0$, then take the 6th entry of $s_{i,j,k}$ to be $\nu_{i,0}\nu_{j,0}\nu_{k,0}$.

7: Compute any nontrivial element $(c_{i,j,k} : 0 \leq i \leq j \leq k \leq 2)$ of the (left) kernel of the 10×6 matrix whose rows are the $s_{i,j,k}$ for $0 \leq i \leq j \leq k \leq 2$, and set

$$C(V_0, V_1, V_2) := \sum_{0 \leq i \leq j \leq k \leq 2} c_{i,j,k} V_i V_j V_k.$$

8: Return the result $\mathcal{X}$ of Algorithm 10.1 applied to $\mathcal{Q} = V(Q)$ and $\mathcal{C} = V(C)$.

12. The algorithm in practice

We conclude with an example for $\ell = 3$. To avoid a visually overwhelming mass of coefficients, we will work over a small finite field; the curve was chosen at random.

Consider the genus 2 curve over $\mathbb{F}_{997}$ defined by

$$\mathcal{H} : Y^2 = X^6 + 113X^5Z + 99X^4Z^2 + 363X^3Z^3 + 64X^2Z^4 + 503XZ^5 + 630Z^6.$$

Computing the zeta function of $\mathcal{H}$ using Magma, we see that its Weil polynomial is

$$P(T) = T^4 - 31T^3 + 54T^2 - 31 \cdot 997T + 997^2,$$

so $\mathcal{J}_{\mathcal{H}}$ is absolutely simple by the Howe–Zhu criterion [**8**, Theorem 6]. The elements $D_1 = \langle x^2 + 392x + 208, 579x + 603, 2\rangle$ and $D_2 = \langle x^2 + 48x + 527, 918x + 832, 2\rangle$ of $\mathcal{J}_{\mathcal{H}}$ have order 3, and $S = \langle D_1, D_2\rangle$ is a maximal Weil-isotropic subgroup of $\mathcal{J}_{\mathcal{H}}[3]$. Applying Algorithm 11.2, we may take

$$S^{\pm} = \begin{Bmatrix} \langle x^2 + 392x + 208, 579x + 603, 2\rangle, \langle x^2 + 48x + 527, 918x + 832, 2\rangle, \\ \langle x^2 + 428x + 880, 252x + 901, 2\rangle, \langle x^2 + 348x + 292, 596x + 269, 2\rangle \end{Bmatrix}$$

in Step 1. Equation (3) shows that the hyperplane $H \subset \mathbb{P}^6$ is defined by

$$H : 630U_0 + 503U_1 + 64U_2 + 363U_3 + 99U_4 + 113U_5 + U_6 = 0,$$

so the matrix in Step 3 is

$$\begin{pmatrix} 234 & 319 & 906 & 896 \\ 780 & 16 & 29 & 754 \\ 500 & 565 & 703 & 398 \\ 680 & 329 & 823 & 248 \\ 324 & 68 & 779 & 868 \\ 742 & 416 & 468 & 392 \\ 664 & 395 & 698 & 952 \end{pmatrix};$$

computing kernel vectors, we take

$$\begin{aligned} \Phi_0 &= 121X^6 + 742X^5Z + 549X^4Z^2 + XZ^5, \\ \Phi_1 &= 285X^6 + 642X^5Z + 332X^4Z^2 + X^2Z^4, \\ \Phi_2 &= 889X^6 + 701X^5Z + 454X^4Z^2 + X^3Z^3. \end{aligned}$$

The quadratic form of Step 5 is then

$$Q(V_0, V_1, V_2) = V_0^2 + 52V_0V_1 + 361V_1^2 + 548V_0V_2 + 715V_1V_2 + 296V_2^2,$$

and we may take the cubic form in Step 7 to be

$$C(V_0, V_1, V_2) = V_0^3 + 167V_1^3 + 149V_0V_1V_2 + 836V_1^2V_2 + 885V_0V_2^2 + 538V_1V_2^2 + 294V_2^3.$$

We now apply Algorithm 10.1 to $\mathcal{Q} : Q(V_0, V_1, V_2) = 0$ and $\mathcal{C} : C(V_0, V_1, V_2) = 0$. The conic $\mathcal{Q}$ is nonsingular, and $\mathcal{H}$ has a rational Weierstrass point $(-76 : 0 : 1)$

mapping to the point $(-36 : -80 : 1)$ on $\mathcal{Q}$. The associated parametrization $\mathbb{P}^1 \to \mathcal{Q}$ is defined by $(X : Z) \longmapsto (P_0(X,Z) : P_1(X,Z) : P_2(X,Z))$, where

$$\begin{aligned} P_0 &= 36X^2 + 781XZ + 109Z^2, \\ P_1 &= 80X^2 + 865XZ + 17Z^2, \text{ and} \\ P_2 &= 996X^2 + 945XZ + 636Z^2; \end{aligned}$$

substituting the P_i into C, we find that $\mathcal{X}$ has a model

$$\mathcal{X}' : Y^2 = 118X^5Z + 183X^4Z^2 + 613X^3Z^3 + 35X^2Z^4 + 174XZ^5 + 474Z^6.$$

In fact, $\mathcal{X}'$ is the quadratic twist of $\mathcal{X}$: explicit calculation shows that its Weil polynomial is $P(-T)$.

References

[1] G. Bisson, R. Cosset, and D. Robert, *avIsogenies: a library for computing isogenies between abelian varieties.* avisogenies.gforge.inria.fr

[2] W. Bosma, C. Playoust, and J. J. Cannon, *The Magma algebra system. I. The user language.* J. Symbolic Comput. **24** (1997), 235–265 MR1484478

[3] J.-B. Bost and J.-F. Mestre, *Moyenne arithmético-géométrique et périodes des courbes de genre 1 et 2.* Gaz. Math. **38** (1988), 36–64 MR970659 (89k:14072)

[4] G. Cardona and J. Quer, *Field of moduli and field of definition for curves of genus 2.* In *Computational aspects of algebraic curves*, Lecture Notes Ser. Comput. **13**, 71–83. World Sci. Publ., Hackensack, NJ, 2005. MR2181874 (2006h:14036)

[5] J. W. S Cassels and E. V. Flynn, *Prolegomena to a middlebrow arithmetic of curves of genus 2.* London Mathematical Society Lecture Note Series **230**, Cambridge University Press, Cambridge (1996) MR1406090 (97i:11071)

[6] J. E. Cremona and D. Rusin, *Efficient solution of rational conics.* Math. Comp. **72** 473 (2003), 1417–1441 MR1972744 (2004a:11137)

[7] I. Dolgachev and D. Lehavi, *On isogenous principally polarized Abelian surfaces.* In *Curves and abelian varieties*, Contemp. Math. **465** (2008), 51–69 MR2457735 (2009k:14081)

[8] E. W. Howe and H. J. Zhu, *On the existence of absolutely simple abelian varieties of a given dimension over an arbitrary field.* J. Number Theory **92** (2002), 139–163 MR1880590 (2003g:11063)

[9] R. W. H. T. Hudson, *Kummer's quartic surface.* Cambridge University Press, Cambridge (1990) MR1097176 (92e:14033)

[10] D. Lubicz and D. Robert, *Computing isogenies between Abelian Varieties.* To appear in Compos. Math. hal.inria.fr/hal-00446062/en

[11] The Magma computational algebra system. magma.maths.usyd.edu.au

[12] J.-F. Mestre, *Construction de courbes de genre 2 à partir de leurs modules.* In *Effective methods in algebraic geometry (Castiglioncello, 1990)*, Progr. Math. **94**, 313–334. Birkhäuser Boston, Boston, MA, 1991. MR1106431 (92g:14022)

[13] W. A. Stein et al., *Sage mathematics software.* The Sage Development Team, www.sagemath.org

[14] A. Weil, *Zum Beweis des Torellischen Satzes.* Nachr. Akad. Wiss. Göttingen. Math.-Phys. Kl. IIa. **1957** (1957), 33–53 MR0089483 (19:683e)

INRIA Saclay–Île-de-France — and — Laboratoire d'Informatique de l'École polytechnique (LIX), 91128 Palaiseau Cedex, France

E-mail address: smith@lix.polytechnique.fr

URL: http://www.lix.polytechnique.fr/~smith

Contemporary Mathematics
Volume **574**, 2012
http://dx.doi.org/10.1090/conm/574/11432

Hodge classes on certain hyperelliptic prymians

Yuri G. Zarhin

ABSTRACT. Let $n = 2g+2$ be a positive even integer, $f(x)$ a degree n complex polynomial without multiple roots and $C_f : y^2 = f(x)$ the corresponding genus g hyperelliptic curve over the field $\mathbf{C}$ of complex numbers. Let a $(g-1)$-dimensional complex abelian variety P be a Prym variety of C_f that corresponds to a unramified double cover of C_f. Suppose that there exists a subfield K of $\mathbf{C}$ such that $f(x)$ lies in $K[x]$, is irreducible over K and its Galois group is the full symmetric group. Assuming that $g > 2$, we prove that $\mathrm{End}(P)$ is either the ring of integers $\mathbf{Z}$ or the direct sum of two copies of $\mathbf{Z}$; in addition, in both cases the Hodge group of P is "as large as possible". In particular, the Hodge conjecture holds true for all self-products of P.

1. Definitions and statements

Throughout this paper K is a field, $\bar{K}$ its algebraic closure and $\mathrm{Gal}(K) = \mathrm{Aut}(\bar{K}/K)$ the absolute Galois group of K. (If $K_s \subset \bar{K}$ is the separable closure of K then the natural homomorphism $\mathrm{Aut}(\bar{K}/K) \to \mathrm{Aut}(K_s/K) = \mathrm{Gal}(K_s/K)$ is an isomorphism.)

If X is an abelian variety over $\bar{K}$ then we write $\mathrm{End}(X)$ for the ring of all its $\bar{K}$-endomorphisms and $\mathrm{End}^0(X)$ for the corresponding $\mathbf{Q}$-algebra $\mathrm{End}(X) \otimes \mathbf{Q}$; the notation 1_X stands for the identity automorphism of X. If Y is an abelian variety over $\bar{K}$ then we write $\mathrm{Hom}(X, Y)$ for the corresponding group of all $\bar{K}$-homomorphisms.

Let $f(x) \in K[x]$ be a polynomial of degree $n \geq 2$ with coefficients in K and without multiple roots, $\mathfrak{R}_f \subset \bar{K}_a$ the (n-element) set of roots of f and $K(\mathfrak{R}_f) \subset \bar{K}_a$ the splitting field of f. We write $\mathrm{Gal}(f) = \mathrm{Gal}(f/K)$ for the Galois group $\mathrm{Gal}(K(\mathfrak{R}_f)/K)$ and call it the Galois group of $f(x)$ over K; it permutes roots of f and may be viewed as a certain permutation group of $\mathfrak{R}_f$, i.e., as a subgroup of the group $\mathrm{Perm}(\mathfrak{R}_f) \cong \mathbf{S}_n$ of permutation of $\mathfrak{R}_f$. (It is well known that $\mathrm{Gal}(f)$ is transitive if and only if f is irreducible.) Let us put

$$g = \left[\frac{n-1}{2}\right].$$

Clearly, g is a nonnegative integer and either $n = 2g+1$ or $n = 2g+2$.

Let us assume that $\mathrm{char}(K) \neq 2$. We write C_f for the genus g hyperelliptic K-curve $y^2 = f(x)$ and $J(C_f)$ for its jacobian. Clearly, $J(C_f)$ is a g-dimensional

---

2010 *Mathematics Subject Classification.* 14H40, 14C30, 11G10.

abelian variety that is defined over K. In particular, $J(C_f) = \{0\}$ if and only if $n = 2$. The abelian variety $J(C_f)$ is an elliptic curve if and only if $n = 3$ or 4.

Let us assume that K is a subfield of the field $\mathbf{C}$ of complex numbers (and $\bar{K}$ is the algebraic closure of K in $\mathbf{C}$). Then one may view $J(C_f)$ as a complex abelian variety and consider its first rational homology group $\mathrm{H}_1(J(C_f), \mathbf{Q})$ and the Hodge group $\mathrm{Hdg}(J(C_f))$ of $J(C_f)$, which is a certain connected reductive algebraic $\mathbf{Q}$-subgroup of the general linear group $\mathrm{GL}(\mathrm{H}_1(J(C_f), \mathbf{Q}))$ [**10, 11, 3, 15, 19, 9**]. The canonical principal polarization on $J(C_f)$ gives rise to the nondegenerate alternating bilinear form

$$\mathrm{H}_1(J(C_f), \mathbf{Q}) \times \mathrm{H}_1(J(C_f), \mathbf{Q}) \to \mathbf{Q}$$

and the corresponding symplectic group $\mathrm{Sp}(\mathrm{H}_1(J(C_f), \mathbf{Q}))$ contains $\mathrm{Hdg}(J(C_f))$ as a (closed) algebraic $\mathbf{Q}$-subgroup. In addition, $\mathrm{End}(J(C_f))$ coincides with the endomorphism ring of the complex abelian variety $J(C_f)$ and $\mathrm{End}^0(J(C_f))$ coincides with the centralizer of $\mathrm{Hdg}(J(C_f))$ in $\mathrm{End}_{\mathbf{Q}}(\mathrm{H}_1(J(C_f), \mathbf{Q}))$ [**11, 15, 9**]. Recall [**10, 11**] that the Mumford-Tate group $\mathrm{MT}(J(C_f))$ of $J(C_f)$ is the reductive algebraic $\mathbf{Q}$-subgroup of $\mathrm{GL}(\mathrm{H}_1(J(C_f), \mathbf{Q}))$ that is generated by $\mathrm{Hdg}(J(C_f))$ and the group of homotheties $\mathbf{G}_m$; the natural homomorphism of $\mathbf{Q}$-algebraic (sub)groups

$$\mathbf{G}_m \times \mathrm{Hdg}(J(C_f)) \to \mathrm{MT}(J(C_f)), \ (a, u) \mapsto au$$

is an *isogeny*, i.e., a surjection with finite kernel. In particular, the $\mathbf{Q}$-Lie algebra mt of $\mathrm{MT}(J(C_f))$ coincides with a direct sum $\mathbf{Q}\mathrm{Id} \oplus \mathrm{hdg}$ of the scalars $\mathbf{Q}\mathrm{Id}$ and the $\mathbf{Q}$-Lie algebra hdg of the Hodge group.

The following result was obtained by the author in [**20**, Th. 2.1], [**21**, Sect. 10]. (See also [**25**], [**23, 24**].)

THEOREM 1.1. *Suppose that* $K \subset \mathbf{C}$, $n \geq 5$ *(i.e.,* $g \geq 2$*) and* $\mathrm{Gal}(f) = \mathbf{S}_n$ *or the alternating group* $\mathbf{A}_n$. *Then* $\mathrm{End}(J(C_f)) = \mathbf{Z}$ *and* $\mathrm{Hdg}(J(C_f)) = \mathrm{Sp}(\mathrm{H}_1(J(C_f), \mathbf{Q}))$. *Every Hodge class on each self-product of* $J(C_f)$ *can be presented as a linear combination of products of divisor classes. In particular, the Hodge conjecture is valid for each self-product of* $J(C_f)$.

REMARK 1.2. The assertion that $\mathrm{Hdg}(J(C_f)) = \mathrm{Sp}(\mathrm{H}_1(J(C_f), \mathbf{Q}))$ was not stated explicitly in [**21**]. However, it follows immediately from the description of the Lie algebra mt of the corresponding Mumford-Tate group [**21**, p. 429] as the direct sum of the scalars $\mathbf{Q}\mathrm{Id}$ and the Lie algebra of the symplectic group, because the Lie algebra of the Hodge group coincides with the intersection of Lie algebras of the Mumford-Tate group and the symplectic group. (The same arguments prove the equality $\mathrm{Hdg}(J(C_f)) = \mathrm{Sp}(\mathrm{H}_1(J(C_f), \mathbf{Q}))$ for all $f(x)$ that satisfy the conditions of Theorem 10.1 of [**21**].)

Our next result that was obtained in [**22**, Th. 1.2 and Theorem 2.5] deals with homomorphisms of hyperelliptic jacobians.

THEOREM 1.3. *Suppose that* $\mathrm{char}(K) \neq 2$, $n \geq 3$ *and* $m \geq 3$ *are integers and let* $f(x)$ *and* $h(x)$ *be irreducible separable polynomials over* K *of degree* n *and* m *respectively. Suppose that*

$$\mathrm{Gal}(f) = \mathbf{S}_n, \ \mathrm{Gal}(h) = \mathbf{S}_m$$

and the corresponding splitting fields $K(\mathfrak{R}_f)$ *and* $K(\mathfrak{R}_h)$ *are linearly disjoint over* K. *Then either*

$$\mathrm{Hom}(J(C_f), J(C_h)) = \{0\}, \ \mathrm{Hom}(J(C_h), J(C_f)) = \{0\}$$

or $\mathrm{char}(K) > 0$ *and both* $J(C_f)$ *and* $J(C_h)$ *are supersingular abelian varieties.*

REMARK 1.4. If $K \subset \mathbf{C}$ then Theorem 1.3 implies that (under its assumptions) there are no nonzero homomorphisms between complex abelian varieties $J(C_f)$ and $J(C_h)$.

The main results of the present paper are the following statements.

THEOREM 1.5. *Suppose that* $n = 2g + 2 = \deg(f) \geq 8$. *Let* $\tilde{C}_f \to C_f$ *be an unramified double cover of complex smooth projective irreducible curves and let* P *be the corresponding Prym variety, which is a* $(g-1)$*-dimensional (principally polarized) complex abelian variety.*

If $\mathrm{Gal}(f) = \mathbf{S}_n$ *then:*

- $\mathrm{End}(P) = \mathbf{Z}$ *or* $\mathbf{Z} \oplus \mathbf{Z}$.
- *Every Hodge class on each self-product of* P *can be presented as a linear combination of products of divisor classes. In particular, the Hodge conjecture holds true for each self-product of* P.

THEOREM 1.6. *Suppose that* $n = 2g + 2 \geq 10$, $K \subset \mathbf{C}$ *and* $f(x) = (x-a)h(x)$ *where* $a \in K$ *and* $h(x) \in K[x]$ *is an irreducible degree* $n-1$ *polynomial with* $\mathrm{Gal}(h) = \mathbf{S}_{n-1}$. *Let* $\tilde{C}_f \to C_f$ *be an unramified double cover of complex smooth projective irreducible curves and let* P *be the corresponding Prym variety, which is a* $(g-1)$*-dimensional (principally polarized) complex abelian variety.*

Then:

- $\mathrm{End}(P) = \mathbf{Z}$ *or* $\mathbf{Z} \oplus \mathbf{Z}$.
- *every Hodge class on each self-product of* P *can be presented as a linear combination of products of divisor classes. In particular, the Hodge conjecture holds true for each self-product of* P.

Our proof is based on the explicit description of Prym varieties of hyperelliptic curves [**12, 2**] and our results about Hodge groups of hyperelliptic jacobians mentioned above.

REMARK 1.7. If $n = 2g + 2 \leq 10$ then $\dim(P) = g - 1 \leq 3$. Notice that if A is a complex abelian varietiy of dimension ≤ 3 then it is well known that every Hodge class on each self-product of A can be presented as a linear combination of products of divisor classes [**9**, Th. 0.1(iv)]. In particular, the Hodge conjecture holds true for each self-product of A.

REMARK 1.8. Notice that the method of this paper (and previous author's publications on the same topic during the last dozen of years) allows us to construct easily a plenty of explicit examples of abelian varieties over $\bar{\mathbf{Q}}$ (and even $\mathbf{Q}$) that are in "general position" with respect to a certain property (e.g., "trivial/small" endomorphism rings or "trivial" Hodge rings).

The paper is organized as follows. In Section 2 we discuss an elementary construction from Galois theory and apply it in Section 3 to homomorphisms of hyperelliptic jacobians. Section 4 deals with Hodge groups of hyperelliptic jacobians. In Section 5 we discuss hyperelliptic prymians and prove the main results.

2. Galois theory

Throughout this Section, K is an arbitrary field and $n \geq 3$ is an integer, $f(x) \in K[x]$ is a degree n irreducible polynomial without multiple roots, whose

Galois group

$$\mathrm{Gal}(f) = \mathrm{Gal}(K(\mathfrak{R}_f)/K)$$

is the full symmetric group $\mathrm{Perm}(\mathfrak{R}_f) = \mathbf{S}_n$. If $T \subset \mathfrak{R}_f$ is a non-empty subset then we put

$$f_T(x) = \prod_{\alpha \in T} (x - \alpha) \in K(\mathfrak{R}_f)[x].$$

By definition,

$$\deg(f_T) = \#(T),\ \mathfrak{R}_{f_T} = T.$$

We view $\mathrm{Perm}(T)$ as a subgroup of $\mathrm{Perm}(\mathfrak{R}_f) = \mathrm{Gal}(f)$ that consists of all permutations that leave invariant every element outside T.

REMARK 2.1. Let us consider the subfield $E_0 = K(\mathfrak{R}_f)^{\mathrm{Perm}(T)}$ of $\mathrm{Perm}(T)$-invariants. Since $\mathrm{Perm}(T)$ leaves invariant $T = \mathfrak{R}_{f_T}$,

$$f_T(x) \in E_0[x].$$

Clearly,

$$\mathrm{Gal}(K(\mathfrak{R}_f)/E_0) = \mathrm{Perm}(T) \subset \mathrm{Perm}(\mathfrak{R}_f) = \mathrm{Gal}(\mathfrak{R}_f/K).$$

Let us prove that the splitting field $E_0(\mathfrak{R}_{f_T}) = E_0(T)$ of $f_T(x)$ over E_0 coincides with $K(\mathfrak{R}_f)$. Indeed, $\mathrm{Gal}(K(\mathfrak{R}_f))/E_0(T))$ consists of all elements of $\mathrm{Perm}(T) = \mathrm{Gal}(K(\mathfrak{R}_f)/E_0)$ that leave invariant every element of T. Since every element of $\mathrm{Perm}(T)$ leaves invariant every element of $\mathfrak{R}_f \setminus T$, $\mathrm{Gal}(K(\mathfrak{R}_f))/E_0(T)) = \{1\}$, i.e., $K(\mathfrak{R}_f) = E_0(T) = E_0(\mathfrak{R}_{f_T})$. This implies that the Galois group

$$\mathrm{Gal}(E_0(\mathfrak{R}_{f_T})/E_0) = \mathrm{Gal}(K(\mathfrak{R}_f)/E_0) = \mathrm{Perm}(T).$$

LEMMA 2.2. *Let T and S be two nonempty disjoint subsets of $\mathfrak{R}_f$. Then there exists a field subextension $E/K \subset K(\mathfrak{R}_f)/K$ that enjoys the following properties.*

(i) *The Galois group $\mathrm{Gal}(K(\mathfrak{R}_f)/E)$ coincides with the subgroup*

$$\mathrm{Gal}(T) \times \mathrm{Gal}(S) \subset \mathrm{Perm}(\mathfrak{R}_f) = \mathrm{Gal}(K(\mathfrak{R}_f)/K),$$

which consists of all permutations that leave invariant T, S and every element outside $T \sqcup S$.

(ii) *Both $f_T(x)$ and $f_S(x)$ lie in $E[x]$, i.e., all their coefficients belong to E.*

(iii) *Let $E(\mathfrak{R}_T) = E(T)$ and $E(\mathfrak{R}_S) = E(S)$ be the splitting fields over E of $f_T(x)$ and $f_S(x)$ respectively. Then the natural injective homomorphisms*

$$\mathrm{Gal}(E(T)/E) \hookrightarrow \mathrm{Perm}(T),\ \mathrm{Gal}(E(S)/E) \hookrightarrow \mathrm{Perm}(S)$$

are group isomorphisms, i.e.,

$$\mathrm{Gal}(E(T)/E) = \mathrm{Perm}(T),\ \mathrm{Gal}(E(S)/E) = \mathrm{Perm}(S).$$

(iv) *$E(T)$ and $E(S)$ are linearly disjoint over E.*

(v) *The compositum $E(T)E(S)$ coincides with $K(\mathfrak{R}_f)$.*

PROOF. Recall that

$$\mathfrak{R}_{f_T} = T,\ \mathfrak{R}_{f_S} = S.$$

We define E as the subfield $K(\mathfrak{R}_f)^{\mathrm{Perm}(T) \times \mathrm{Perm}(S)}$ of $\mathrm{Perm}(T) \times \mathrm{Perm}(S)$-invariants. Now Galois theory gives us (i). The subgroup $\mathrm{Perm}(T) \times \mathrm{Perm}(S)$ leaves invariant both sets $T = \mathfrak{R}_T$ and $S = \mathfrak{R}_S$. This implies that all the coefficients of $f_T(x)$ and $f_S(x)$ are $[\mathrm{Perm}(T) \times \mathrm{Perm}(S)]$-invariant, i.e., lie in E. This proves (ii). Clearly,

$$[K(\mathfrak{R}_f) : E] = \#(\mathrm{Perm}(T) \times \mathrm{Perm}(S)).$$

Clearly, the subgroup of $\mathrm{Perm}(T) \times \mathrm{Perm}(S)$ that consists of all permutations that act identically on S coincides with $\mathrm{Perm}(T)$. Similarly, the subgroup of $\mathrm{Perm}(T) \times \mathrm{Perm}(S)$ that consists of all permutations that act identically on T coincides with $\mathrm{Perm}(S)$. This implies that

$$\mathrm{Gal}(E(T)/E) = [\mathrm{Perm}(T) \times \mathrm{Perm}(S)]/\mathrm{Perm}(S) = \mathrm{Perm}(T),$$

$$\mathrm{Gal}(E(S)/E) = [\mathrm{Perm}(T) \times \mathrm{Perm}(S)]/\mathrm{Perm}(T) = \mathrm{Perm}(S),$$

which proves (iii). This implies that

$$[E(T) : E] = \#(\mathrm{Perm}(T)), \ [E(S) : E] = \#(\mathrm{Perm}(S)).$$

Let L be the compositum $E(T)E(S)$. Clearly, L contains T, S and E. Therefore $\mathrm{Gal}(K(\mathfrak{R}_f)/L)$ consists of elements of $\mathrm{Perm}(T) \times \mathrm{Perm}(S) = \mathrm{Gal}(K(\mathfrak{R}_f)/E)$ that act identically on T and T. Since all elements of $\mathrm{Perm}(T) \times \mathrm{Perm}(S)$ act identically on the complement to $T \sqcup S$, we conclude that $\mathrm{Gal}(K(\mathfrak{R}_f)/L) = \{1\}$, i.e.,

$$K(\mathfrak{R}_f) = L = E(T)E(S).$$

This proves (v). We also obtain that

$$[E(T)E(S) : E] = [K(\mathfrak{R}_f) : E] = \#(\mathrm{Perm}(T) \times \mathrm{Perm}(S)) = [E(T) : E][E(S) : E],$$

i.e.

$$[E(T)E(S) : E] = [E(T) : E][E(S) : E],$$

which means that $E(T)/E$ and $E(S)/E$ are linearly disjoint. This proves (iv). $\square$

3. Homomorphisms of hyperelliptic jacobians

We keep the notation and assumptions of Section 2. Also we assume that $\mathrm{char}(K) \neq 2$.

THEOREM 3.1. *Let T and S be disjoint nonempty subsets of $\mathfrak{R}_f$ and consider the hyperelliptic curves*

$$C_{f_T} : y^2 = f_T(x), \ C_{f_S} : y^2 = f_S(x)$$

and their jacobians $J(C_{f_T})$ and $J(C_{f_S})$. Then either

$$\mathrm{Hom}(J(C_{f_T}), J(C_{f_S})) = \{0\}, \ \mathrm{Hom}(J(C_{f_S}), J(C_{f_T})) = \{0\}$$

or $\mathrm{char}(K) > 0$ *and both $J(C_{f_T})$ and $J(C_{f_S})$ are supersingular abelian varieties.*

PROOF. If $\#(T) < 3$ (resp. $\#(S) < 3$) then C_{f_T} (resp. C_{f_S}) has genus zero and $J(C_{f_T}) = 0$ (resp. $J(C_{f_S}) = 0$), which implies that there are no nonzero homomorphisms between $J(C_{f_T})$ and $J(C_{f_S})$. So, further we assume that

$$n_1 := \#(T) \geq 3, \ n_2 := \#(S) \geq 3.$$

By Lemma 2.2, there exists a field E such that both $f_T(x)$ and $f_S(x)$ lie in $E[x]$, their Galois groups are $\mathrm{Perm}(T) \cong \mathbf{S}_{n_1}$ and $\mathrm{Perm}(S) \cong \mathbf{S}_{n_2}$ respectively. In addition, their splitting fields are linearly disjoint over E. Now the result follows from Theorem 1.3 applied to $E, f_T(x), f_S(x)$ instead of $K, f(x), h(x)$. $\square$

REMARK 3.2. Suppose that $m := \#(S) = 2r+1$ is odd and let b be an arbitrary element of K. Let us consider the hyperelliptic curve $C_{f_S}^b : y^2 = (x-b)f_S(x)$. By Remark 2.1, there exists a field $E_0 \subset K(\mathfrak{R}_f)$ such that $f_S(x)$ lies in $E_0[x]$ and $\mathrm{Gal}(E_0(\mathfrak{R}_{f_S})/E_0) = \mathrm{Perm}(S) = \mathbf{S}_m$. Then the standard substitution [**25**, p. 25]

$$x_1 = \frac{1}{x-b}, \; y_1 := \frac{y}{(x-b)^{r+1}}$$

gives us a degree m irreducible polynomial $h(x_1) \in E[x_1]$ such that

$$E_0(\mathfrak{R}_h) = E_0(\mathfrak{R}_{f_S}), \; \mathrm{Gal}(E_0(\mathfrak{R}_h)/E) = \mathrm{Gal}(E_0(\mathfrak{R}_{f_S})/E) = \mathbf{S}_m$$

and $C_{f_S}^b$ is E_0-birationally isomorphic to the hyperelliptic curve $C_h : y_1^2 = h(x_1)$. (It is assumed in [**25**, p. 25] that $m \ge 5$ but the substitution works for any positive odd m.)

COROLLARY 3.3. *Let T and S be disjoint nonempty subsets of $\mathfrak{R}_f$ and assume that $\#(S)$ is odd. Let b be an arbitrary element of K. Let us consider the hyperelliptic curves*

$$C_{f_T} : y^2 = f_T(x), \; C_{f_S}^b : y^2 = (x-b)f_S(x)$$

and their jacobians $J(C_{f_T})$ and $J(C_{f_S}^b)$. Then either

$$\mathrm{Hom}(J(C_{f_T}), J(C_{f_S}^b)) = \{0\}, \; \mathrm{Hom}(J(C_{f_S}^b), J(C_{f_T})) = \{0\}$$

or $\mathrm{char}(K) > 0$ *and both $J(C_{f_T})$ and $J(C_{f_S}^b)$ are supersingular abelian varieties.*

PROOF. We may assume that both $m_1 = \#(T)$ and $m_2 = \#(S)$ are, at least, 3. Let E be as in Lemma 2.2. In particular, the splitting fields $E(T)$ and $E(S)$ are linearly disjoint over E and

$$\mathrm{Gal}(E(T)/E) \cong \mathbf{S}_{m_1}, \; \mathrm{Gal}(E(S)/E) \cong \mathbf{S}_{m_2}.$$

Using Remark 3.2 over E (instead of E_0), we obtain that there is a degree m irreducible polynomial $h(x) \in E[x]$ such that

$$E(\mathfrak{R}_h) = E(\mathfrak{R}_{f_S}), \; \mathrm{Gal}(E(\mathfrak{R}_h)/E) = \mathrm{Gal}(E(\mathfrak{R}_{f_S})/E) = \mathbf{S}_{m_2}$$

and $C_{f_S}^b$ is birationally E-isomorphic to the hyperelliptic curve $C_h : y^2 = h(x)$. Clearly, the jacobians $J(C_{f_S}^b)$ and $J(C_h)$ are isomorphic. Applying Theorem 1.3 to $E, f_T(x), h(x)$, we conclude that either

$$\mathrm{Hom}(J(C_{f_T}), J(C_h)) = \{0\}, \; \mathrm{Hom}(J(C_h)), J(C_{f_T})) = \{0\}$$

or $\mathrm{char}(K) > 0$ and both $J(C_{f_T})$ and $J(C_{f_S})$ are supersingular abelian varieties. Since $J(C_{f_S}^b)$ and $J(C_h)$ are isomorphic, we are done. □

THEOREM 3.4. *Let T and S be disjoint nonempty subsets of $\mathfrak{R}_f$ and assume that both $\#(T)$ and $\#(S)$ are odd. Let a and b be arbitrary (not necessarily distinct) elements of K. Let us consider the hyperelliptic curves*

$$C_{f_T}^a : y^2 = (x-a)f_T(x), \; C_{f_S}^b : y^2 = (x-b)f_S(x)$$

and their jacobians $J(C_{f_T}^a)$ and $J(C_{f_S}^b)$. Then either

$$\mathrm{Hom}(J(C_{f_T}^a), J(C_{f_S}^b)) = \{0\}, \; \mathrm{Hom}(J(C_{f_S}^b), J(C_{f_T}^a)) = \{0\}$$

or $\mathrm{char}(K) > 0$ *and both $J(C_{f_T}^b)$ and $J(C_{f_S}^a)$ are supersingular abelian varieties.*

PROOF. We may assume that both $m_1 := \#(T)$ and $m_2 := \#(S)$ are, at least, 3. Again, let E be as in Lemma 2.2. Applying Remark 3.2 two times over E (instead of E_0) to the polynomials $(x-a)f_T(x)$ and $(x-b)f_S(x)$, we conclude that there are degree m irreducible polynomials $h_1(x) \in E[x]$ and $h_2(x) \in E[x]$ such that

$$E(\mathfrak{R}_{h_1}) = E(T), \ E(\mathfrak{R}_{h_2}) = E(S),$$
$$\mathrm{Gal}(E(\mathfrak{R}_{h_1})/E) = \mathbf{S}_{m_1}, \ \mathrm{Gal}(E(\mathfrak{R}_{h_2})/E) = \mathbf{S}_{m_2},$$

$C^a_{f_T}$ is E-birationally isomorphic to C_{h_1} and $C^b_{f_T}$ is E-birationally isomorphic to C_{h_2}. Clearly, $J(C^a_{f_T}) \cong J(C_{h_1})$ and $J(C^b_{f_S}) \cong J(C_{h_2})$. Applying Theorem 1.3 to $E, h_1(x), h_2(x)$, we conclude that either

$$\mathrm{Hom}(J(C_{h_1}), J(C_{h_2})) = \{0\}, \ \mathrm{Hom}(J(C_{h_2})), J(C_{h_1})) = \{0\}$$

or $\mathrm{char}(K) > 0$ and both $J(C_{h_1})$ and $J(C_{h_2})$ are supersingular abelian varieties. The rest is clear.

□

REMARK 3.5. Let K_2/K be the only quadratic subextension of $K(\mathfrak{R}_f)/K$. Clearly, $K_2(\mathfrak{R}_f) = K(\mathfrak{R}_f)$ and the Galois group $\mathrm{Gal}(K_2(\mathfrak{R}_f)/K_2)$ coincides with the alternating group $\mathbf{A}_n$.

THEOREM 3.6. *Suppose that* $\mathrm{char}(K) \neq 2$. *Let* $T \subset \mathfrak{R}_f$ *be a* 4*-element subset. Let us consider the corresponding elliptic curve*

$$C_{f_T} : y^2 = f_T(x)$$

and its jacobian $J(C_{f_T})$. *If* $n \geq 8$ *then one of the following conditions holds:*

- $\mathrm{End}(J(C_{f_T})) = \mathbf{Z}$ *for all* T.
- $\mathrm{char}(K) > 0$ *and all* $J(C_{f_T})$*'s are supersingular elliptic curves mutually isomorphic over* $\bar{K}$.

PROOF. Let j_T be the j-invariant of the elliptic curve $J(C_{f_T})$ ([**18**], [**7**, Ch. III, Sect. 2]). Clearly,

$$j_T \in K(T) \subset K(\mathfrak{R}_f)$$

and

$$j_{\sigma T} = \sigma j_T \ \forall \sigma \in \mathrm{Gal}(K(\mathfrak{R}_f)/K) = \mathrm{Gal}(f).$$

Suppose that $J(C_{f_T})$ admits complex multiplication. Then one of the following two conditions holds.

(i) $p = \mathrm{char}(K) > 0$. Then a classical result of M. Deuring asserts that j_T is *algebraic*, i.e., lies in a finite field $\mathbf{F}_q$ where q is a power of the prime p. (See [**4**], [**14**, Sect. 3.2], [**8**, Ch. 13, Sect. 5].) In particular, $K(j_T)/K$ is an abelian field extension.

(ii) $\mathrm{char}(K) = 0$. Then there exists an imaginary quadratic field k such that $\mathrm{End}^0(J(C_{f_T})) = k$. In addition, a classical result of the theory of complex multiplication asserts that j_T is an algebraic number such that the field extension $k(j_T)/k$ is abelian. (See [**17**, Sect. 5.4], [**8**, Ch. 10, Sect. 3].)

Let us consider the overfield K' of K that is defined as follows. If $\mathrm{char}(K) > 0$ then $K' = K_2$. If $\mathrm{char}(K) = 0$ then K' is the compositum K_2k of K_2 and the imaginary quadratic field k; in particular, K' contains k.

Since $\mathbf{A}_n = \mathrm{Gal}(K'(\mathfrak{R}_f)/K')$ is simple nonabelian, the field extension $K'(\mathfrak{R}_f)/K'$ does not contain nontrivial abelian subextensions. However, $j_T \in$

$K'(\mathfrak{R}_f)$ and the field (sub)extension $K'(j_T)/K'$ is abelian. This implies that this subextension is trivial, i.e., $j_T \in K'$. This means that for all $\sigma \in \mathrm{Gal}(K'(\mathfrak{R}_f)/K') = \mathbf{A}_n$

$$j_T = \sigma j_T = j_{\sigma T}.$$

Since $n \geq 8$, the permutatation group $\mathbf{A}_n$ is 4-transitive and therefore the jacobians $J(C_{f_T})$'s are mutually isomorphic over $\bar{K}$ for all 4-element subsets $T \subset \mathfrak{R}_f$.

Let T_1 and T_2 be two *disjoint* 4-element subsets of $\mathfrak{R}_f$. (Since $n \geq 8$, such T_1 and T_2 do exist.) Applying Theorem 3.1 to T_1 and T_2 (instead of T and S) and taking into account that $J(C_{f_{T_1}})$ and $J(C_{f_{T_2}})$ are isomorphic over $\bar{K}$ (i.e., $\mathrm{Hom}(J(C_{f_{T_1}}), J(C_{f_{T_2}})) \neq \{0\}$), we conclude that $\mathrm{char}(K) > 0$ and both $J(C_{f_{T_1}})$ and $J(C_{f_{T_2}})$) are supersingular elliptic curves. □

THEOREM 3.7. *Suppose that* $\mathrm{char}(K) \neq 2$. *Let a be an arbitrary element of K. Let $T \subset \mathfrak{R}_f$ be a 3-element subset. Let us consider the corresponding elliptic curve*

$$C^a_{f_T} : y^2 = (x-a) f_T(x)$$

and its jacobian $J(C^a_{f_T})$. If $n \geq 6$ then (for given a) one of the following conditions holds:

- $\mathrm{End}(J(C^a_{f_T})) = \mathbf{Z}$ *for all T.*
- $\mathrm{char}(K) > 0$ *and all $J(C^a_{f_T})$'s are supersingular elliptic curves mutually isomorphic over $\bar{K}$.*

PROOF. Let $j_{T,a}$ be the j-invariant of the elliptic curve $J(C^a_{f_T})$. Clearly,

$$j_{T,a} \in K(T) \subset K(\mathfrak{R}_f)$$

and

$$j_{\sigma T,a} = \sigma j_{T,a} \; \forall \sigma \in \mathrm{Gal}(K(\mathfrak{R}_f)/K) = \mathrm{Gal}(f).$$

Suppose that $J(C^a_{f_T})$ admits complex multiplication. Then, as in the proof of Theorem 3.6, there exists an overfield $K' \supset K_2$ such that either $K' = K_2$ or K' is a quadratic extension of K_2 and in both cases $K'(j_{T,a}) \subset K'(\mathfrak{R}_f)$ and the field (sub)extension $K'(j_{T,a})/K'$ is abelian. Again, $\mathbf{A}_n = \mathrm{Gal}(K'(\mathfrak{R}_f)/K')$ is simple nonabelian and therefore there are no nontrivial abelian subextensions of $K'(\mathfrak{R}_f)/K'$. This implies that $j_{T,a} \in K'$, i.e., for all $\sigma \in \mathrm{Gal}(K'(\mathfrak{R}_f)/K') = \mathbf{A}_n$

$$j_{T,a} = \sigma j_{T,a} = j_{\sigma T,a}.$$

Since $n \geq 6$, the permutatation group $\mathbf{A}_n$ is 3-transitive and therefore the jacobians $J(C^a_{f_T})$'s are mutually isomorphic over $\bar{K}$ for all 3-element subsets $T \subset \mathfrak{R}_f$. Let T_1 and T_2 be two *disjoint* 3-element subsets of $\mathfrak{R}_f$. (Since $n \geq 6$, such T_1 and T_2 do exist.) Applying Theorem 3.4 to T_1, a and T_2, a (instead of T, a and S, b) and taking into account that $J(C^a_{f_{T_1}})$ and $J(C^a_{f_{T_2}})$ are isomorphic over $\bar{K}$ (i.e., $\mathrm{Hom}(J(C_{f_{T_1}}), J(C_{f_{T_2}})) \neq \{0\}$), we conclude that $\mathrm{char}(K) > 0$ and both $J(C^a_{f_{T_1}})$ and $J(C^a_{f_{T_2}})$) are supersingular elliptic curves. The rest is clear. □

4. Hodge groups of hyperelliptic jacobians

We keep the notation and assumptions of Sections 2 and 3. Also we assume that $K \subset \mathbf{C}$.

DEFINITION 4.1. We say (as in [**9**, Sect. 1.8]) that a complex abelian variety X satisfies property (D) if every Hodge class on each self-product X^r of X can be presented as a linear combination of products of divisor classes. If this condition is satisfied then the Hodge conjecture is true for all X^r.

REMARK 4.2. Abelian varieties that satisfy (D) are also called *stably nondegenerate* [**5**]; see also [**13**].

EXAMPLE 4.3. If Y is an elliptic curve over $\mathbf{C}$ with $\mathrm{End}(Y) = \mathbf{Z}$ then it is well known [**9**, Th. 0.1(iv)] that Y satisfies (D) and $\mathrm{Hdg}(Y) = \mathrm{Sp}(\mathrm{H}_1(Y, \mathbf{Q}))$.

THEOREM 4.4. *Let X_1 and X_2 be complex abelian varieties of positive dimension and $X = X_1 \times X_2$. Suppose that*

$$\mathrm{End}(X_1) = \mathbf{Z},\ \mathrm{End}(X_2) = \mathbf{Z},\ \mathrm{Hom}(X_1, X_2) = \{0\}.$$

Then:

(1) $\mathrm{End}(X_1 \times X_2) = \mathbf{Z} \oplus \mathbf{Z}$.
(2) *If both X_1 and X_2 satisfy (D) then* $\mathrm{Hdg}(X) = \mathrm{Hdg}(X_1) \times \mathrm{Hdg}(X_2)$ *and X satisfies (D).*

PROOF. (i) is obvious. (ii) follows from [**6**, Th. 0.1 and Prop. 1.8] (see also Theorem 3.2(i) of [**9**]). □

REMARK 4.5. Since $\mathrm{End}(X_i) = \mathbf{Z}$ and X_i satisfies (D),

$$\mathrm{Hdg}(X_i) = \mathrm{Sp}(\mathrm{H}_1(X_i, \mathbf{Q}))$$

[**5, 13**]. (See also [**9**, Sect. 1.8].)

LEMMA 4.6. *Suppose that T is a subset of $\mathfrak{R}_f$ with $\#(T) \geq 5$. Let us consider the hyperelliptic curve $C_{f_T} : y^2 = f_T(x)$ and its jacobian $J(C_{f_T})$. Then* $\mathrm{End}(J(C_{f_T})) = \mathbf{Z}$ *and* $\mathrm{Hdg}(J(C_{f_T})) = \mathrm{Sp}(\mathrm{H}_1(J(C_{f_T}), \mathbf{Q}))$. *In addition, $J(C_{f_T})$ satisfies (D).*

PROOF. Let us put $m = \#(T)$. We have $m \geq 5$ and $\deg(f_T) = m \geq 5$.
By Remark 2.1, there exists a (sub)field

$$E_0 \subset K(\mathfrak{R}_f) \subset \bar{K} \subset \mathbf{C}$$

such that $f_T(x) \in E_0(T)$ and the Galois group of $f_T(x)$ over E_0 is $\mathrm{Perm}(T) \cong \mathbf{S}_m$. Now the result follows from Theorem 1.1 applied to $m, E_0, f_T(x)$ instead of $n, K, f(x)$. □

LEMMA 4.7. *Suppose that T is a subset of $\mathfrak{R}_f$ with $m := \#(T) \geq 5$. Suppose that m is odd and let a be an arbitrary element of K. Let us consider the hyperelliptic curve $C_{f_T}^a : y^2 = (x-a)f_T(x)$ and its jacobian $J(C_{f_T}^a)$. Then* $\mathrm{End}(J(C_{f_T}^a)) = \mathbf{Z}$ *and* $\mathrm{Hdg}(J(C_{f_T}^a)) = \mathrm{Sp}(\mathrm{H}_1(J(C_{f_T}^a), \mathbf{Q}))$. *In addition, $J(C_{f_T}^a)$ satisfies (D).*

PROOF. By Remark 3.2, there exists a field

$$E_0 \subset K(\mathfrak{R}_f) \subset \bar{K} \subset \mathbf{C}$$

and a degree m irreducible polynomial $h(x) \in E_0[x]$ such that

$$E(\mathfrak{R}_h) = E(\mathfrak{R}_{f_T}),\ \mathrm{Gal}(E(\mathfrak{R}_h)/E) = \mathrm{Gal}(E(\mathfrak{R}_{f_T})/E) = \mathbf{S}_m$$

and $C_{f_S}^a$ is birationally E-isomorphic to to the hyperelliptic curve $C_h : y^2 = h(x)$. Clearly, the jacobians $J(C_{f_S}^a)$ and $J(C_h)$ are isomorphic. It follows from Lemma 4.6

applied to $m, E_0, h(x)$ (instead of $n, K, f(x)$) that $\mathrm{End}(J(C_h)) = \mathbf{Z}$, $\mathrm{Hdg}(J(C_h) = \mathrm{Sp}(\mathrm{H}_1(J(C_h), \mathbf{Q}))$ and $J(C_h)$ satisfies (D). Since $J(C^a_{f_S})$ and $J(C_h)$ are isomorphic, we are done. □

5. Prym varieties

Following [**12, 2**], let us give an explicit description of hyperelliptic prymians P, assuming that $\mathrm{char}(K) \neq 2$. Suppose that $n = 2g + 2 \geq 6$ and

$$f(x) \in K[x] \subset \bar{K}[x]$$

is a degree n polynomial without multiple roots. Let us split the n-element set $\mathfrak{R}_f$ of roots of $f(x)$ into a *disjoint* union

$$\mathfrak{R}_f = \mathfrak{R}_1 \sqcup \mathfrak{R}_2$$

of *non-empty* sets $\mathfrak{R}_1$ and $\mathfrak{R}_2$ of *even* cardinalities n_1 and n_2 respectively. Further we assume that

$$n_1 \geq n_2 \geq 2.$$

and put

$$f_1(x) = \prod_{\alpha \in \mathfrak{R}_1} (x - \alpha), \ f_2(x) = \prod_{\alpha \in \mathfrak{R}_2} (x - \alpha).$$

We have $n_1 + n_2 = n$ and define nonnegative integers g_1 and g_2 by

$$n_1 = 2g_1 + 2, \ n_2 = 2g_2 + 2.$$

Clearly,

$$g_1 + g_2 = g - 1.$$

Let us consider the hyperelliptic curves $C_{f_1} : y^2 = f_1(x)$ and $C_{f_2} : y^2 = f_2(x)$ of genus g_1 and g_2 respectively and corresponding hyperellptic jacobians $J(C_{f_1})$ and $J(C_{f_2})$ of dimension g_1 and g_2 respectively. Then the prymians P of $C_f : y^2 = f(x)$ are just the products $J(C_{f_1}) \times J(C_{f_2})$ for all the partitions $\mathfrak{R}_f = \mathfrak{R}_1 \sqcup \mathfrak{R}_2$.

Now Theorem 1.5 becomes an immediate corollary of the following statement.

Theorem 5.1. *Suppose that* $n = 2g + 2 \geq 8$, $K \subset \mathbf{C}$ *and* $\mathrm{Gal}(f) = \mathbf{S}_n$. *Let us put* $P = J(C_{f_1}) \times J(C_{f_2})$. *Then:*

(i) $\mathrm{Hom}(J(C_{f_1}), J(C_{f_2})) = \{0\}, \ \mathrm{Hom}(J(C_{f_2}), J(C_{f_1})) = \{0\}$.

(ii) *Suppose that* $g_i \geq 1$, *i.e.*, $n_i \geq 4$. *Then*

$$\mathrm{End}(J(C_{f_i})) = \mathbf{Z}, \ \mathrm{Hdg}(J(C_{f_i})) = \mathrm{Sp}(\mathrm{H}_1(J(C_{f_i}), \mathbf{Q}))$$

and $J(C_{f_i})$ *satisfies (D).*

(iii) *If* $n_2 = 2$ *then* $P = J(C_{f_1})$. *In particular,* $\mathrm{End}(P) = \mathbf{Z}$, $\mathrm{Hdg}(P) = \mathrm{Sp}(\mathrm{H}_1(P, \mathbf{Q}))$ *and* P *satisfies (D).*

(iv) *If* $n_2 \geq 4$ *then* $\mathrm{End}(P) = \mathbf{Z} \oplus \mathbf{Z}$,

$$\mathrm{Hdg}(P) = \mathrm{Hdg}(J(C_{f_1})) \times \mathrm{Hdg}(J(C_{f_2})) = \mathrm{Sp}(\mathrm{H}_1(J(C_{f_1}), \mathbf{Q})) \times \mathrm{Sp}(\mathrm{H}_1(J(C_{f_2}), \mathbf{Q}))$$

and P *satisfies (D).*

Proof of Theorem 5.1. The assertion (i) follows from Theorem 3.1.

If $n_i \geq 6$ then (ii) follows from Lemma 4.6. Suppose that $n_i = 4$. Then $J(C_{f_i})$ is an elliptic curve. It follows from Theorem 3.6 that $\mathrm{End}(J(C_{f_i})) = \mathbf{Z}$. Now the assertion about its Hodge group and property (D) follows from Example 4.3. This completes the proof of (ii).

Let us prove (iii). If $n_2 = 2$ then $J(C_{f_2}) = 0$ and therefore $P = J(C_{f_1})$. Now the assertion follows from (ii).

Let us prove (iv). We assume that

$$n_1 \geq n_2 \geq 4.$$

By already proven (i) and (ii),

$$\mathrm{End}(J(C_{f_1})) = \mathbf{Z}, \ \mathrm{End}(J(C_{f_2})) = \mathbf{Z}, \ \mathrm{Hom}(J(C_{f_1}), J(C_{f_2})) = \{0\}.$$

Now (iv) follows from Theorem 4.4 applied to $X_1 = J(C_{f_1})$, $X_2 = J(C_{f_2})$ and $X = P$. □

Theorem 1.6 is an immediate corollary of the following statement.

THEOREM 5.2. *Suppose that $n = 2g+2 \geq 10$, $K \subset \mathbf{C}$ and $f(x) = (x-a)h(x)$ where $a \in K$ and $h(x) \in K[x]$ is an irreducible degree $(n-1)$ polynomial with* $\mathrm{Gal}(h) = \mathbf{S}_{n-1}$. *Let us put $P = J(C_{f_1}) \times J(C_{f_2})$. Then:*

(i) $\mathrm{Hom}(J(C_{f_1}), J(C_{f_2})) = \{0\}, \ \mathrm{Hom}(J(C_{f_2}), J(C_{f_1})) = \{0\}$.

(ii) *Suppose that $g_i \geq 1$, i.e., $n_i \geq 4$. Then*

$$\mathrm{End}(J(C_{f_i})) = \mathbf{Z}, \ \mathrm{Hdg}(J(C_{f_i})) = \mathrm{Sp}(\mathrm{H}_1(J(C_{f_i}), \mathbf{Q}))$$

and $J(C_{f_i})$ satisfies (D).

(iii) *If $n_2 = 2$ then $P = J(C_{f_1})$. In particular,* $\mathrm{End}(P) = \mathbf{Z}$, $\mathrm{Hdg}(P) = \mathrm{Sp}(\mathrm{H}_1(P, \mathbf{Q}))$ *and P satisfies (D).*

(iv) *If $n_2 \geq 4$ then* $\mathrm{End}(P) = \mathbf{Z} \oplus \mathbf{Z}$,

$$\mathrm{Hdg}(P) = \mathrm{Hdg}(J(C_{f_1})) \times \mathrm{Hdg}(J(C_{f_2})) = \mathrm{Sp}(\mathrm{H}_1(J(C_{f_1}), \mathbf{Q})) \times \mathrm{Sp}(\mathrm{H}_1(J(C_{f_2}), \mathbf{Q}))$$

and P satisfies (D).

PROOF. Clearly, $\mathfrak{R}_f = \mathfrak{R}_h \sqcup \{a\}$; in particular, a belongs to precisely one of $\mathfrak{R}_1$ and $\mathfrak{R}_2$. Suppose that a lies in $\mathfrak{R}_j$ and does *not* belong to $\mathfrak{R}_k$ and put

$$T = \mathfrak{R}_k \subset \mathfrak{R}_h, \ S = \mathfrak{R}_j \setminus \{a\} \subset \mathfrak{R}_h.$$

Now the assertion (i) follows from Corollary 3.3 applied to $h(x)$ (instead of $f(x)$).

If $n_k \geq 6$ then the assertion (ii) for $J(C_{f_k})$ follows from Lemma 4.6. Suppose that $n_k = 4$, i.e., $J(C_{f_k})$ is an elliptic curve. Then it follows from Theorem 3.6 applied to $m = n-1 \geq 9$ and $h(x)$ (instead of $f(x)$) that $\mathrm{End}(J(C_{f_k})) = \mathbf{Z}$. Now the assertion about its Hodge group and property (D) follows from Example 4.3.

If $n_j \geq 6$ then the assertion (ii) for $J(C_{f_j})$ follows from Lemma 4.7. Suppose that $n_j = 4$, i.e., $J(C_{f_j})$ is an elliptic curve. Then it follows from Theorem 3.7 applied to $m = n-1$ and $h(x)$ (instead of $f(x)$) that $\mathrm{End}(J(C_{f_j})) = \mathbf{Z}$. Now the assertion about its Hodge group and property (D) follows from Example 4.3. This ends the proof of (ii).

The proof of the remaining assertions (iii) and (iv) goes literally as the proof of the corresponding assertions of Theorem 5.1. □

EXAMPLES 5.3. Let us take $K = \mathbf{Q}$ and $f_n(x) = x^n - x - 1$. It is known [**16**, p. 42] that $\mathrm{Gal}(f_n) = \mathbf{S}_n$. Let a be a rational number. Suppose that $n = 2g+2$ and let us consider the hyperelliptic genus g curves $C_{f_n} : y^2 = f_n(x)$ and $C^a_{f_{n-1}} : y^2 = (x-a)f_{n-1}(x)$. Then:

(i) If $n = 2g+2 \geq 8$ then all $(2^{2g}-1)$ Prym varieties P of C_{f_n} satisfy (D). Among them there are exactly $n(n-1)/2$ complex abelian varieties with $\mathrm{End}(P) = \mathbf{Z}$; for all others $\mathrm{End}(P) = \mathbf{Z} \oplus \mathbf{Z}$.

(ii) If $n = 2g+2 \geq 10$ then all $(2^{2g}-1)$ Prym varieties P of $C^a_{f_{n-1}}$ satisfy (D). Among them there are exactly $n(n-1)/2$ complex abelian varieties with $\mathrm{End}(P) = \mathbf{Z}$; for all others $\mathrm{End}(P) = \mathbf{Z} \oplus \mathbf{Z}$.

EXAMPLES 5.4. Let $z_1, \ldots, z_n$ be algebraically independent (transcendental) complex numbers and $L = \bar{\mathbf{Q}}(z_1, \ldots, z_n) \subset \mathbf{C}$ the corresponding subfield of $\mathbf{C}$, which is isomorphic to the field of rational functions in n variables over $\bar{\mathbf{Q}}$. Let $K \subset L$ be the (sub)field of symmetric rational functions. Then

$$h_n(x) = \prod_{i=1}^{n}(x - z_i) \in K[x], \ \mathfrak{R}_f = \{z_1, \ldots, z_n\}, \ \mathrm{Gal}(f) = \mathbf{S}_n.$$

Let $a \in \bar{\mathbf{Q}}$ be an algebraic number Suppose that $n = 2g+2$ and let us consider the hyperelliptic genus g curves $C_{h_n} : y^2 = h_n(x)$ and and $C^a_{h_{n-1}} : y^2 = (x-a)h_{n-1}(x)$. Then:

(i) If $g \geq 3$ (i.e., $n \geq 8$) then all $(2^{2g}-1)$ Prym varieties P of C_{h_n} satisfy (D). Among them there are exactly $n(n-1)/2$ complex abelian varieties with $\mathrm{End}(P) = \mathbf{Z}$; for all others $\mathrm{End}(P) = \mathbf{Z} \oplus \mathbf{Z}$.

If $n = 6$ (i.e., $g = 2$) then all fifteen Prym varieties P are elliptic curves $y^2 = \prod_{z \in T}(x - z)$ where T is a 4-element subset of $\{z_1, \ldots, z_6\}$. The algebraic independence of $z_1, \ldots, z_6$ implies that the j-invariants of these elliptic curves are transcendental numbers and therefore all P have no complex multiplication, i.e., $\mathrm{End}(P) = \mathbf{Z}$.

(ii) If $n = 2g+2 \geq 10$ then all $(2^{2g}-1)$ Prym varieties P of $C^a_{h_{n-1}}$ satisfy (D). Among them there are exactly $n(n-1)/2$ complex abelian varieties with $\mathrm{End}(P) = \mathbf{Z}$; for all others $\mathrm{End}(P) = \mathbf{Z} \oplus \mathbf{Z}$.

REMARK 5.5. The property (D) and equality $\mathrm{End}(P) = \mathbf{Z}$ for *general* (not necessarily unramified) Prym varieties P of arbitrary smooth projective curves were proven in [**1**].

References

[1] I. Biswas, K. H. Paranjape, *The Hodge conjecture for general Prym varieties.* J. Algebraic Geometry **11** (2002), 33–39. MR1865912 (2003j:14008)

[2] S. G. Dalaljan, *The Prym variety of an unramified double covering of a hyperelliptic curve.* (Russian) Uspehi Mat. Nauk 29 (1974), no. 6(180), 165–166. MR0404270 (53:8073)

[3] P. Deligne, *Hodge cycles on abelian varieties* (notes by J.S. Milne). Lecture Notes in Math., vol. **900** (Springer-Verlag, 1982), pp. 9–100.

[4] M. Deuring, *Die Typen der Multiplikatorenringe elliptischer Funktionenkörper.* Abh. Math. Sem. Hansischen Univ. **14** (1941), 197–272. MR0005125 (3,104f)

[5] F. Hazama, *Algebraic cycles on certain abelian varieties and powers of special surfaces.* J. Fac. Sci. Univ. Tokyo Sect. IA Math. **31** (1985), no. 3, 487–520. MR776690 (86k:14004)

[6] F. Hazama, *Algebraic cycles on nonsimple abelian varieties.* Duke Math. J. **58** (1989), 31–37. MR1016412 (91g:14005)

[7] A. Knapp, Elliptic curves. Princeton University Press, Princeton, 1992. MR1193029 (93j:11032)

[8] S. Lang, Elliptic functions, Second edition. Springer Verlag, New York, 1987. MR890960 (88c:11028)

[9] B. Moonen, Yu. G. Zarhin, *Hodge classed on abelian varieties of low dimension.* Math. Ann. **315** (1999), 711–733. MR1731466 (2000j:14018)

[10] D. Mumford, *Families of abelian varieties.* In: Algebraic Groups and Discontinuous Subgroups. Amer. Math. Soc. Proc. Sympos. Pure Math. **9** (1966), 347–351. MR0206003 (34:5828)

[11] D. Mumford, *A note of Shimura's paper "Discontinuous groups and abelian varieties"*. Math. Ann. **181** (1969), 345–351. MR0248146 (40:1400)

[12] D. Mumford, *Prym varieties* I. In: Contributions to Analysis, pp. 325–350, Academic Press, 1974; Selected Papers, vol. I, pp. 545–570, Springer Verlag, New York, 2004. MR0379510 (52:415)

[13] V. Kumar Murty, *Exceptional Hodge classes on certain abelian varieties.* Math. Ann. 268 (1984), no. 2, 197–206. MR744607 (85m:14063)

[14] F. Oort, *The isogeny class of a CM-type abelian variety is defined over a finite extension of the prime field.* J. Pure Appl. Algebra **3** (1973), 399–408. MR0330175 (48:8513)

[15] K. Ribet, *Hodge classes on certain abelian varieties.* Amer. J. Math. **105** (1983), 523–538. MR701568 (85a:14030)

[16] J.-P. Serre, Topics in Galois Theory. Jones and Bartlett Publishers, Boston-London, 1992. MR1162313 (94d:12006)

[17] G. Shimura, Introduction to the arithmetic theory of automorphic functions. Publ. Math. Soc. Japan **11**, Iwanami Shoten and Princeton University Press, Princeton, 1971. MR0314766 (47:3318)

[18] J. Tate, *Algebraic formulas in arbitrary characteristic.* Appendix 1 to [**8**]. pp. 299–306.

[19] Yu.G. Zarhin, *Weights of simple Lie algebras in the cohomology of algebraic varieties.* Izv. Akad. Nauk SSSR Ser. Mat. **48** (1984), 264–304; Math. USSR Izv. **24** (1985), 245 - 281. MR740792 (85k:14012)

[20] Yu.G. Zarhin, *Hyperelliptic jacobians without complex multiplication.* Math. Res. Letters **7** (2000), 123–132. MR1748293 (2001a:11097)

[21] Yu.G. Zarhin, *Very simple 2-adic representations and hyperelliptic jacobians.* Moscow Math. J. **2** (2002), issue 2, 403-431. MR1944511 (2003k:11098)

[22] Yu.G. Zarhin, *Homomorphisms of hyperelliptic Jacobians.* In: Number Theory, Algebra and Algebraic Geometry (Shafarevich Festschrift). Trudy Mat. Inst. Steklov **241** (2003), 90–104; Proc. Steklov Inst. Math. **241** (2003), 79–92. MR2024045 (2004m:14056)

[23] Yu.G. Zarhin, *Non-supersingular hyperelliptic Jacobians.* Bull. Soc. Math. France **132** (2004), no. 4, 617–634. MR2131907 (2006c:14051)

[24] Yu.G. Zarhin, *Homomorphisms of abelian varieties.* In: Y. Aubry, G. Lachaud (ed.), Arithmetic, Geometry and Coding Theory (AGCT 2003), Séminaires et Congrés **11** (2005), 189–215. MR2182844 (2006k:14077)

[25] Yu.G. Zarhin, *Families of absolutely simple hyperelliptic Jacobians.* Proc. Lond. Math. Soc. (3) **100** (2010), no. 1, 24–54, MR2578467 (2011d:14055)

DEPARTMENT OF MATHEMATICS, PENNSYLVANIA STATE UNIVERSITY, UNIVERSITY PARK, PENNSYLVANIA 16802

E-mail address: zarhin@math.psu.edu